I0036850

Titanium Dioxide Photocatalysis

Titanium Dioxide Photocatalysis

Special Issue Editors

Vladimiro Dal Santo
Alberto Naldoni

MDPI • Basel • Beijing • Wuhan • Barcelona • Belgrade

MDPI

Special Issue Editors

Vladimiro Dal Santo
CNR-Istituto di Scienze e Tecnologie Molecolari
Italy

Alberto Naldoni
CNR-Istituto di Scienze e Tecnologie Molecolari
Italy

Editorial Office
MDPI
St. Alban-Anlage 66
4052 Basel, Switzerland

This is a reprint of articles from the Special Issue published online in the open access journal *Catalysts* (ISSN 2073-4344) from 2017 to 2018 (available at: https://www.mdpi.com/journal/catalysts/special_issues/titanium_dioxide)

For citation purposes, cite each article independently as indicated on the article page online and as indicated below:

LastName, A.A.; LastName, B.B.; LastName, C.C. Article Title. *Journal Name* **Year**, *Article Number, Page Range.*

ISBN 978-3-03897-694-3 (Pbk)
ISBN 978-3-03897-695-0 (PDF)

© 2019 by the authors. Articles in this book are Open Access and distributed under the Creative Commons Attribution (CC BY) license, which allows users to download, copy and build upon published articles, as long as the author and publisher are properly credited, which ensures maximum dissemination and a wider impact of our publications.

The book as a whole is distributed by MDPI under the terms and conditions of the Creative Commons license CC BY-NC-ND.

Contents

About the Special Issue Editors

Vladimiro Dal Santo earned a diploma in chemistry from the University of Milano in 1997 and a PhD in Chemical Sciences from the same university in 2002. In 2001 he joined the Italian National Research Council, working as researcher from 2001 to 2018, at the CSMTBO Center (2001–2002) and the Institute of Molecular Science and Technology (2002–2018). From 2018, he has been employed as a Senior Researcher. From 2016 to 2019 he taught Chemistry at the University of Milano, Bicocca as an adjunct professor. Besides his more recent research activities in photo(electro)catalysis, his main research interests since 2001 have been heterogenous catalysis applied to hydrogenations and hydrogen production by the Steam Reformation of methane, alcohols, polyols, and organic acids. In his career, V.D.S. has published more than 90 peer-reviewed publications. His works have been cited more than 2800 times, corresponding to a H-index of 27 (Source: Web of Science, 12 February 2019). He has co-supervised five PhD students in addition to numerous Post-Doctorate, Master, and Bachelor students in their projects. V.D.S. is Vice-President and has been a member of the Governing Board of the Lombardy Energy and Cleantech Cluster (LE2C) since 2018.

Alberto Naldoni is co-leader of the Nano-Photoelectrochemistry Group at the Regional Center of Advanced Technologies and Materials of Palacký University Olomouc (Czech Republic). He graduated in Photochemistry and Chemistry of Materials (with honors) from the University of Bologna in 2007. Afterwards, A.N. earned his PhD (2010) in Chemical Sciences from the University of Milan before moving to the Italian National Research Council to study photocatalysis and photoelectrochemical water splitting. He spent three years as a visiting faculty member in the Nanophotonics Group at the Birck Nanotechnology Center of Purdue University (United States). A.N.'s research interests include nanomaterials for energy and environment with a special emphasis on photocatalysis, electrocatalysis, plasmon-enhanced chemical transformations, defects, and doping in metal oxides, as well as charge transfer at solid-solid and solid-liquid interfaces. A.N. has published 60 papers in reputed journals including *Science, JACS, Advanced Materials*, and *Angewandte Chemie*. His works have been cited more than 2000 times, corresponding to a H-index of 21 (Google Scholar).

Preface to "Titanium Dioxide Photocatalysis"

This book contains the contributions of the Special Issue of *Catalysts* on "Titanium Dioxide Photocatalysis", with the aim of presenting the current state-of-the-art in the use of titanium dioxide (TiO_2) as a photocatalyst, with a special emphasis on new TiO_2 nanomaterials for photocatalytic hydrogen production, photoelectrochemical water splitting, and environmental remediation. For this Special Issue, we invited contributions from leading groups in the field with the aim of giving a balanced view of the current state-of-the-art in this discipline.

Dating from the seminal work of Fujishima et al. issued in 1971, TiO_2 is at the center of intense research devoted to the development of efficient photocatalysts. Among the many candidates for photocatalytic applications, TiO_2 is almost the only material suitable for industrial use. This is because TiO_2 has a good trade-off between efficient photoactivity, high stability, and low cost. The rational design elements of interests for efficient TiO_2 catalysts are optical properties, nanocrystal shape, and organization in superstructures. The main drawback of TiO_2 photocatalysts remains their inability to achieve visible light absorption and photoconversion, and most recent research activities have been devoted to the improvement of the optical absorption properties of TiO_2 nanomaterials. Here we present some examples of strategies to enhance the final efficiency of TiO_2-based materials. These approaches include doping, metal co-catalyst deposition, and the realization of composites with plasmonic materials, other semiconductors, and graphene. On the other hand, the precise crystal shape (and homogeneous size) and the organization in superstructures from ultrathin films to hierarchical nanostructures have been demonstrated to be critical for obtaining photocatalysts with high efficiency and selectivity, as showcased in the presented articles. Finally, the theoretical modeling of TiO_2 nanoparticles in real experimental conditions and the reactivity of photoelectrons are discussed in two contributions from Kohtani et al. and di Valentin et al.

The review by Monai and the papers by Nunes, Liu, and Zelny address the synthesis of novel nanostructures. Monai et al. provide a comprehensive review on brookite, describing the most advanced synthetic methodologies to produce pure brookite and brookite-containing composites, together with some guidelines for characterization. Finally, structure/activity relations are summarized and a perspective on the future development of brookite nanostructured materials is given. Nunes et al. describe the synthesis of TiO_2 nanorods, spheres, powders, and arrays by microwave irradiation. The synthesis of large-scale pinecone nanostructured TiO_2 films, active under solar irradiation, in the photo-oxidation of organic pollutants and in hydrogen production by a fast anodizing method is reported by Liu et al. Furthermore, Zelny et al. report a detailed investigation of mechanical and adhesion properties of Ti films sputtered at different temperatures, showing that the most active sample in photoelectrochemical water splitting was obtained at 150 °C.

Other important strategies to increase TiO_2 photocatalytic efficiency include non-metal doping, metal co-catalyst deposition, the formation of composites with carbon-based nanomaterials, and the preparation of plasmonic nanoparticles. Cravanzola et al. report the synthesis of S-doped TiO_2 photocatalysts, which were then tested for methylene blue photodegradation. An extensive FTIR investigation shines light on the structure–activity relationship of the prepared materials. In contrast, Bernareggi et al. report a strategy based on flame spray pyrolysis to produce Cu- and Cu–Pt-modified TiO_2 for photocatalytic hydrogen production. An optimal loading of 0.05% Cu was found for the most active photocatalyst. Interestingly, copper-modified TiO_2 nanomaterials were also the focus of the review by Janczarek and Kowalska. In particular, they describe the performance enhancement by copper species for oxidative reactions, and identify two key factors: plasmonic properties of zero-valent copper and heterojunctions between titania and copper oxides. In another report, Zabihi et al. described composites made by a semiconductor and graphene as promising materials to enhance photogenerated charge separation due to the high electrical conductivity of graphene-based nanomaterials. A new route to couple graphene to TiO_2 is reported, showing the possibility of using ultrasonication to increase the processability and scalability of composite materials for enhanced photocurrent generation and photocatalytic dye degradation. Finally, the review by Kohtani et al. summarizes the recent progress in the research on electron transfer in photoexcited TiO_2 and highlights the use of highly uniform TiO_2 nanocrystals with specific exposure of the reactive facets. In particular, the authors point out the key role of the precise control of the structural properties, that is, the maximization of surface shallow traps and the minimization of density of deep traps as well as inner (bulk) traps.

To conclude, we would like to express my sincerest gratitude to all authors for the valuable contributions, without which this book would not have been possible.

Vladimiro Dal Santo, Alberto Naldoni
Special Issue Editors

catalysts

MDPI

Editorial

Titanium Dioxide Photocatalysis

Vladimiro Dal Santo [1,*] and Alberto Naldoni [2,*]

[1] CNR-Istituto di Scienze e Tecnologie Molecolari, Via C. Golgi 19, 20133 Milano, Italy
[2] Regional Centre of Advanced Technologies and Materials, Šlechtitelů 27, 78371 Olomouc, Czech Republic
* Correspondence: vladimiro.dalsanto@istm.cnr.it (V.D.S.); alberto.naldoni@upol.cz (A.N.)

Received: 22 November 2018; Accepted: 28 November 2018; Published: 29 November 2018

1. Definitions, Historical Aspects, and Perspectives

Dating from the seminal work of Fujishima et al. issued in 1971 [1], titanium dioxide (TiO_2) is at the center of intense research devoted to the development of efficient photocatalysts. Among the many candidates for photocatalytic applications, TiO_2 is almost the only material suitable for industrial use. This is because TiO_2 shows a good trade-off between efficient photoactivity, high stability, and low cost [1,2]. The principal applications deal with the use of TiO_2 as a photocatalyst for environmental remediation both in polluted air and waste water treatment [3] and as a material in solar cells [3,4].

The main drawback of TiO_2 photocatalysts still remains their inability for visible light absorption and photoconversion, and most recent research activities have been devoted to the improvement of the optical absorption properties of TiO_2 nanomaterials. Strategies including doping; self-doping; and the realization of composites with plasmonic materials, 2D materials, other semiconductors, and quantum dots are of particular interest [1,2]. Black-TiO_2 visible light active photocatalysts [5], antimicrobial materials [6], photoelectrochemical devices for water splitting, and CO_2 photoreduction [7] are among the hot topics. The rational design elements of interests for efficient TiO_2 catalysts are optical properties, nanocrystal shape, and organization in superstructures. On the other hand, precise crystal shape (and homogeneous size) and organization in superstructure from ultrathin films to hierarchical nanostructures have been demonstrated to be critical for obtaining photocatalyst with high efficiency and selectivity.

The present Special Issue of Catalysts is aimed at presenting the current state of the art in the use of TiO_2 as a photocatalyst, with a special emphasis on new TiO_2 nanomaterials (both powdered catalysts and photoelectrodes) for photocatalytic water splitting, CO_2 reduction, and environmental remediation. In the present Special Issue, we have invited contributions from leading groups in the field with the aim of giving a balanced view of the current state of the art in this discipline.

2. This Special Issue

Dr. Alberto Naldoni and I were honored to accept the kind invitation by Assistant Editor Shelly Liu to act as editors of this Special Issue. We tried to acquire possible authors able to contribute with high-level papers and reviews and we hope we succeeded in this task. This is particularly due to the wonderful and uncomplicated cooperation of Assistant Editor Shelly Liu and her competent team. Moreover, I owe particular thanks to all the authors who contributed their excellent papers to this Special Issue that is comprised of 11 articles, among them 3 reviews, covering key aspects of this topic together with a variety of innovative approaches.

Three comprehensive reviews cover most recent advances in key areas, such as electron transfer dynamics, brookite-based photocatalysts, and copper-modified titania.

The review by Kohtani et al. [8] summarizes the recent progress in the research on electron transfer in photoexcited TiO_2. In particular, the authors point out the key role of the precise control of the structural properties, that is, the maximization of surface shallow traps and minimization of density of deep traps as well as inner (bulk) traps in the development of highly active photocatalysts. The authors also highlight, as a promising strategy, the use of highly uniform TiO$_2$ nanocrystals with specific exposure of the reactive facets.

Monai et al. [9] provides a comprehensive review of the advancement in the applications of brookite as a photocatalyst. First, the most advanced synthetic methodologies to produce pure brookite and well-defined brookite-containing composites are presented, together with some guidelines for thorough characterization of the materials. Finally, structure/activity relations are summarized and a perspective on the future development of brookite nanostructured materials is given.

The review by Janczarek and Kowalska [10] focuses on the performance enhancement by copper species for oxidative reactions due to their importance in environmental remediation. Two key factors are identified and discussed: plasmonic properties of zero-valent copper and heterojunctions between semiconductors (titania and copper oxides) including novel systems of cascade heterojunctions. The role of particle morphology (faceted particles, core-shell structures) is also described. Finally, future trends of research on copper-modified titania are discussed.

Synthesis of novel nanostructures by different preparation routes is addressed in the papers by Nunes, Liu, and Zelny [11]. Microwave irradiation proved to be an effective synthesis route to produce TiO$_2$ nanorod sphere powders and arrays at low process temperatures using water as a solvent. The remarkable photocatalytic activity under UV and solar irradiation was ascribed to the presence of brookite but also depends on the nanorod, sphere, and aggregate sizes.

A fast anodizing method [12] was employed to synthesize large-scale (e.g., 300 × 360 mm) pinecone nanostructured TiO$_2$ films. The pinecone TiO$_2$ possesses strong solar absorption and exhibits high photocatalytic activities in photo-oxidizing organic pollutants in wastewater, producing hydrogen from water under natural sunlight. This work shows a promising future for the practical utilization of anodized TiO$_2$ films in renewable energy and clean environment applications.

A promising approach to fabricate nanostructured TiO$_2$ films on transparent substrates is self-ordering by the anodizing of thin metal films on fluorine-doped tin oxide (FTO) coupling pulsed direct current (DC) magnetron sputtering for the deposition of titanium thin films on conductive glass substrates and anodization and annealing for the TiO$_2$ nanotube array [13]. Zelny et al. reported a detailed investigation of mechanical and adhesion properties of Ti films sputtered at different temperatures, showing that a more active TiO$_2$ nanotube sample towards photoelectrochemical water splitting was obtained from a Ti substrate sputtered at 150 °C, showing the lowest crystallite size, best degree of self-organization, and enhanced charge transfer at the semiconductor/liquid interface.

The use of plasmonic nanomaterials in photocatalysis [14] has gained great attention due to their ability to enhance the reaction yield of semiconductor photocatalysts. In this contribution, Bao et al. coupled plasmonic Ag nanoparticles to high-surface-area TiO$_2$ nanofibers to achieve a very active photocatalyst toward dye molecule degradation, showing enhanced performance when using the plasmonic Ag/TiO$_2$ material.

Composites made by semiconductor and graphene [15] are particularly promising to enhance photogenerated charge separation due to the high electrical conductivity of graphene-based nanomaterials. In this article, a new route to couple graphene to TiO$_2$ was reported, showing the possibility of using ultrasonication to increase the processability and scalability of composite materials for enhanced photocurrent generation and photocatalytic dye degradation as well.

Bernareggi et al. [16] report a strategy based on flame spray pyrolysis to produce Cu- and Cu–Pt-modified TiO$_2$ for photocatalytic hydrogen production. An optimal loading of 0.05% Cu was found for the most active photocatalyst, which only contained Cu.

Nonmetal doping [17] is a very common approach to increase the light absorption and therefore the photocatalytic efficiency of TiO$_2$. In this report, S-doped TiO$_2$ photocatalysts were synthesized and tested for methylene blue photodegradation. An extensive FTIR investigation shined light on the structure–activity relationship of the prepared materials.

The article by Selli et al. [18] provides a new approach for the computational modeling of large titanium dioxide nanoparticles with diameters from 1.5 nm (~300 atoms) to 4.4 nm (~4000 atoms), usually too demanding for theoretical calculation. The authors investigated photoexcitation and photoemission processes involving electron/hole pair formation, separation, trapping, and recombination and provided a description of the titania/water multilayer interface—a relevant case study for photocatalytic systems.

In conclusion, the special issue "Titanium Dioxide Photocatalysis" should be of great interest for all of those involved in the various aspects of this topic, which are discussed in the contributions and review articles. They introduce new synthetic procedures, modeling of structures and reactivity, novel nanostructures, and plasmonic composites, thereby meeting the state of the art of both scientific and technical standards.

References

1. Schneider, J.; Matsuoka, M.; Takeuchi, M.; Zhang, J.; Horiuchi, Y.U.; Anpo, M.; Bahnemann, D.W. Understanding TiO$_2$ Photocatalysis: Mechanisms and Materials. *Chem. Rev.* **2014**, *114*, 9919–9986. [CrossRef] [PubMed]
2. Ge, M.; Cao, C.; Huang, J.; Li, S.; Chen, Z.; Zhang, K.-Q.; Al-Dey, S.S.; Lai, Y. A review of one-dimensional TiO$_2$ nanostructured materials for environmental and energy applications. *J. Mater. Chem. A* **2016**, *4*, 6772–6801. [CrossRef]
3. Shahrezaei, M.; Babaluo, A.A.; Habibzadeh, S.; Haghighi, M. Photocatalytic Properties of 1D TiO$_2$ Nanostructures Prepared from Polyacrylamide Gel–TiO$_2$ Nanopowders by Hydrothermal Synthesis. *Eur. J. Inorg. Chem.* **2017**, *3*, 694–703. [CrossRef]
4. Kment, S.; Riboni, F.; Pausova, S.; Wang, L.; Wang, L.; Han, H.; Hubicka, Z.; Krysa, J.; Schmuki, P.; Zboril, R. Photoanodes based on TiO$_2$ and α-Fe$_2$O$_3$ for solar water splitting—Superior role of 1D nanoarchitectures and of combined heterostructures. *Chem. Soc. Rev.* **2017**, *46*, 3716–3769. [CrossRef] [PubMed]
5. Yan, X.; Li, Y.; Xia, T. Black Titanium Dioxide Nanomaterials in Photocatalysis. *Int. J. Photoenergy* **2017**, *2017*, 8529851. [CrossRef]
6. Fu, G.; Vary, P.S.; Lin, C.-T. Anatase TiO$_2$ Nanocomposites for Antimicrobial Coatings. *J. Phys. Chem. B* **2005**, *109*, 8889–8898. [CrossRef] [PubMed]
7. Zhang, L.; Can, M.; Ragsdale, S.W.; Armstrong, F.A. Fast and Selective Photoreduction of CO$_2$ to CO Catalyzed by a Complex of Carbon Monoxide Dehydrogenase, TiO$_2$, and Ag Nanoclusters. *ACS Catal.* **2018**, *8*, 2789–2795. [CrossRef]
8. Kohtani, S.; Kawashima, A.; Miyabe, H. Reactivity of Trapped and Accumulated Electrons in Titanium Dioxide Photocatalysis. *Catalysts* **2017**, *7*, 303. [CrossRef]
9. Monai, M.; Montini, T.; Fornasiero, P. Brookite: Nothing New under the Sun? *Catalysts* **2017**, *7*, 304. [CrossRef]
10. Janczarek, M.; Kowalska, E. On the Origin of Enhanced Photocatalytic Activity of Copper-Modified Titania in the Oxidative Reaction Systems. *Catalysts* **2017**, *7*, 317. [CrossRef]
11. Nunes, D.; Pimentel, A.; Santos, L.; Barquinha, P.; Fortunato, E.; Martins, R. Photocatalytic TiO$_2$ Nanorod Spheres and Arrays Compatible with Flexible Applications. *Catalysts* **2017**, *7*, 60. [CrossRef]
12. Liu, Y.; Zhang, Y.; Wang, L.; Yang, G.; Shen, F.; Deng, S.; Zhang, X.; He, Y.; Hu, Y.; Chen, X. Fast and Large-Scale Anodizing Synthesis of Pine-Cone TiO$_2$ for Solar-Driven Photocatalysis. *Catalysts* **2017**, *7*, 229. [CrossRef]
13. Zelny, M.; Kment, S.; Ctvrtlik, R.; Pausova, S.; Kmentova, H.; Tomastik, J.; Hubicka, Z.; Rambabu, Y.; Krysa, J.; Naldoni, A.; et al. TiO$_2$ Nanotubes on Transparent Substrates: Control of Film Microstructure and Photoelectrochemical Water Splitting Performance. *Catalysts* **2018**, *8*, 25. [CrossRef]
14. Bao, N.; Miao, X.; Hu, X.; Zhang, Q.; Jie, X.; Zheng, X. Novel Synthesis of Plasmonic Ag/AgCl@TiO$_2$ Continues Fibers with Enhanced Broadband Photocatalytic Performance. *Catalysts* **2017**, *7*, 117. [CrossRef]
15. Zabihi, F.; Ahmadian-Yazdi, M.; Eslamian, M. Photocatalytic Graphene-TiO$_2$ Thin Films Fabricated by Low-Temperature Ultrasonic Vibration-Assisted Spin and Spray Coating in a Sol-Gel Process. *Catalysts* **2017**, *7*, 136. [CrossRef]
16. Bernareggi, M.; Dozzi, M.; Bettini, L.; Ferretti, A.; Chiarello, G.; Selli, E. Flame-Made Cu/TiO$_2$ and Cu-Pt/TiO$_2$ Photocatalysts for Hydrogen Production. *Catalysts* **2017**, *7*, 301. [CrossRef]

17. Cravanzola, S.; Cesano, F.; Gaziano, F.; Scarano, D. Sulfur-Doped TiO2: Structure and Surface Properties. *Catalysts* **2017**, *7*, 214. [CrossRef]

18. Selli, D.; Fazio, G.; Di Valentin, C. Using Density Functional Theory to Model Realistic TiO$_2$ Nanoparticles, Their Photoactivation and Interaction with Water. *Catalysts* **2017**, *7*, 357. [CrossRef]

© 2018 by the authors. Licensee MDPI, Basel, Switzerland. This article is an open access article distributed under the terms and conditions of the Creative Commons Attribution (CC BY) license (http://creativecommons.org/licenses/by/4.0/).

MDPI

Review

Reactivity of Trapped and Accumulated Electrons in Titanium Dioxide Photocatalysis

Shigeru Kohtani *, Akira Kawashima and Hideto Miyabe

Department of Pharmacy, School of Pharmacy, Hyogo University of Health Sciences, 1-3-6 Minatojima, Chuo-ku, Kobe 650-8530, Japan; ak-kawashima@huhs.ac.jp (A.K.); miyabe@huhs.ac.jp (H.M.)
* Correspondence: kohtani@huhs.ac.jp; Tel.: +81-78-304-3158

Received: 20 September 2017; Accepted: 8 October 2017; Published: 13 October 2017

Abstract: Electrons, photogenerated in conduction bands (CB) and trapped in electron trap defects (Ti_{ds}) in titanium dioxide (TiO_2), play crucial roles in characteristic reductive reactions. This review summarizes the recent progress in the research on electron transfer in photo-excited TiO_2. Particularly, the reactivity of electrons accumulated in CB and trapped at Ti_{ds} on TiO_2 is highlighted in the reduction of molecular oxygen and molecular nitrogen, and the hydrogenation and dehalogenation of organic substrates. Finally, the prospects for developing highly active TiO_2 photocatalysts are discussed.

Keywords: titanium dioxide; photocatalysis; surface defects; bulk defects; trapped electrons; accumulated electrons

1. Introduction

Since Fujishima and Honda discovered photoelectrochemical water splitting on titanium dioxide (TiO_2) photoelectrodes in the early 1970s [1], TiO_2 photocatalysis has been applied in various fields, such as the storage of solar energy [2–5], environmental purification [6], organic synthesis [7–11], anti-bacterial applications [12], and anti-fogging treatments [12,13]. These characteristic photo-functionalities are induced by incident light, in which the behavior of photogenerated electrons and holes, as well as the roles of defects formed on surface and in lattice, are of particular importance. The defect sites are the recombination centers for the photogenerated electrons and holes, because photocatalytic activities decrease with increasing the amount of defects created [2,6,8]. However, Amano et al. reported that the introduction of defect states in TiO_2 with H_2 reduction treatment greatly enhanced the photocatalytic activity for the water oxidation reaction in aqueous solution [14,15]. Moreover, Kong and coworkers claimed that tuning the relative concentration ratio of bulk defects/surface defects in TiO_2 nanocrystal improves the separation efficiencies of photogenerated electrons and holes, thereby enhancing the photocatalytic activity [16]. Thus, further understanding of the defects in TiO_2 necessitates the development of highly active photocatalysts.

The properties of defects—such as energy levels, structures, and interactions with adsorbates—have been reviewed by Diebold [17], Henderson [18], and Nowotny [19,20] in detail, but many unanswered questions remain. Recent studies in this field have made the considerable progress during the last decade. This review summarizes the recent progress in the research on the defects in TiO_2. Herein, we focus on the properties of electron trap defects formed within the bandgap of TiO_2 associated with Ti defects, specifically the intra-bandgap Ti states (Ti_{ds}). Firstly, the fate of photogenerated electron and holes in TiO_2 are described with respect to Ti_{ds} and hole trap sites in Section 2. Next, the origin of Ti_{ds} and their energy distribution in TiO_2 are considered in Section 3. In Section 4, the reactivity of electrons trapped at Ti_{ds} and accumulated in the conduction band (CB) on the representative reductive reactions are highlighted. Finally, the prospect for developing a highly active TiO_2 catalyst is discussed.

2. Fate of Photogenerated Electrons and Holes in TiO$_2$

Although several models exist for the charge transport, trapping, and the reaction of photogenerated electrons and holes on photoexcited TiO$_2$, we adopted a schematic model for the anatase TiO$_2$ based on the recent selected reviews and reports as illustrated in Figure 1 [11,21–25].

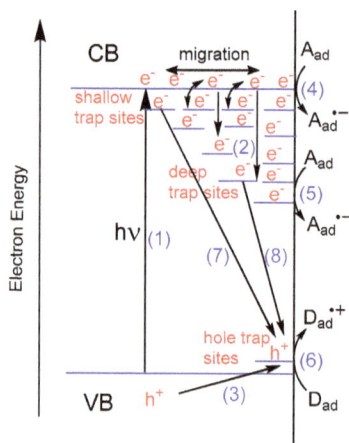

Figure 1. Schematic model of the earlier stage of photocatalysis in the anatase titanium dioxide (TiO$_2$). CB: conduction band; VB: valence band; A$_{ad}$: adsorbed electron acceptor; D$_{ad}$: adsorbed electron donor.

This model consists of several steps:

Step 1. Electron–hole pair generation

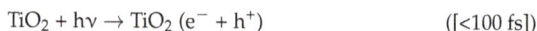

$$TiO_2 + h\nu \rightarrow TiO_2\ (e^- + h^+) \qquad\qquad ([<100\ fs])$$

Step 2. Trapping CB electrons (e_{cb}^-) at defect Ti^{4+} sites

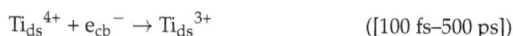

$$Ti_{ds}^{4+} + e_{cb}^- \rightarrow Ti_{ds}^{3+} \qquad\qquad ([100\ fs–500\ ps])$$

Step 3. Trapping valence band holes (h_{vb}^+) at terminal Ti–OH or surface Ti–O–Ti sites

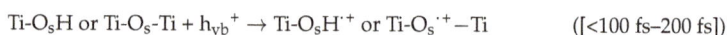

$$Ti\text{-}O_sH\ or\ Ti\text{-}O_s\text{-}Ti + h_{vb}^+ \rightarrow Ti\text{-}O_sH^{\cdot+}\ or\ Ti\text{-}O_s^{\cdot+}\text{-}Ti \qquad\qquad ([<100\ fs–200\ fs])$$

Step 4. Reduction of adsorbed electron acceptor (A$_{ad}$) with e_{cb}^- at reduction sites

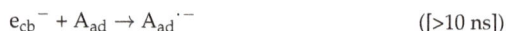

$$e_{cb}^- + A_{ad} \rightarrow A_{ad}^{\cdot-} \qquad\qquad ([>10\ ns])$$

Step 5. Reduction of A$_{ad}$ with electrons trapped at defect sites (Ti$_{ds}^{3+}$)

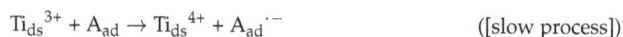

$$Ti_{ds}^{3+} + A_{ad} \rightarrow Ti_{ds}^{4+} + A_{ad}^{\cdot-} \qquad\qquad ([slow\ process])$$

Step 6. Oxidation of adsorbed electron donor (D$_{ad}$) by trapped holes at oxidation sites

$$Ti\text{-}O_sH^{\cdot+}\ or\ Ti\text{-}O_s^{\cdot+}\text{-}Ti + D_{ad} \rightarrow Ti\text{-}O_sH\ or\ Ti\text{-}O_s\text{-}Ti + D_{ad}^{\cdot+} \qquad\qquad ([100\ ps–10\ ns])$$

Step 7. Recombination of e_{cb}^- with trapped holes

$$e_{cb}^- + Ti\text{-}O_sH^{\cdot +} \text{ or } Ti\text{-}O_s^{\cdot +}\text{-}Ti \rightarrow Ti\text{-}O_sH \text{ or } Ti\text{-}O_s\text{-}Ti \qquad ([1\text{--}10 \text{ ps}])$$

Step 8. Recombination of Ti_{ds}^{3+} with trapped holes

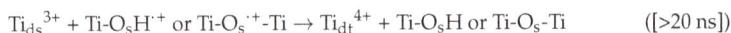

$$Ti_{ds}^{3+} + Ti\text{-}O_sH^{\cdot +} \text{ or } Ti\text{-}O_s^{\cdot +}\text{-}Ti \rightarrow Ti_{dt}^{4+} + Ti\text{-}O_sH \text{ or } Ti\text{-}O_s\text{-}Ti \qquad ([>20 \text{ ns}])$$

where time scales for each step are described in brackets [21–24]. The time scales depend on the crystalline phases, crystallinity, specific surface area, and the presence of bulk and surface defect states in TiO$_2$. The following assumptions were applied to this model: (a) CB electrons (e_{cb}^-) contain both electrons in CB and electrons trapped at shallow sites, located just below the CB edge of TiO$_2$ within 0–0.05 eV. These electrons were assumed to be in thermal equilibrium in the bulk CB and at the shallow trap sites; (b) Valence band holes (h_{vb}^+) are rapidly transported to the surface hole trap sites (Ti–O$_s$H or Ti–O$_s$–Ti) (Step 3); (c) trapped holes (Ti–O$_s$H$^{\cdot +}$ or Ti–O$_s^{\cdot +}$–Ti) are the main oxidants for the adsorbed electron donor (D_{ad}) (Step 6); and (d) charge carrier recombination occurs between e_{cb}^- and holes trapped at the surface trap sites (Step 7), as well as between electrons trapped at Ti defect states (Ti_{ds}^{3+}) and holes trapped at the surface trap sites (Step 8), whereas the interband electron–hole carrier recombination ($e^- + h^+ \rightarrow h\nu$ or heat) is negligible.

These assumptions can be justified as follows. Tamaki and coworkers described the charge carrier dynamics under weak excitation conditions for nano-crystalline anatase TiO$_2$ samples in femtosecond to microsecond time scales [22,23], which should be compatible with the actual photocatalytic reactions under the usual UV irradiation conditions. They observed the e_{cb}^- and h_{vb}^+ pair generation within 100 fs, and the e_{cb}^- migration between CB and shallow trap sites in equilibrium. These electrons then relaxed to deep trap sites (Ti$_{ds}$) with an approximate 500 ps time constant. Meanwhile, h_{vb}^+ was rapidly trapped to the surface terminal Ti–O$_s$H sites within 100 fs to create Ti–O$_s$H$^{\cdot +}$ [22,23]. If the photoinduced event occurred in alcohols, the lifetime of the Ti–O$_s$H$^{\cdot +}$ generated on the TiO$_2$ surface would be in the nanosecond or sub-nanosecond time scale (approximately 0.1–3 ns in alcohols) due to the fast reaction of Ti–O$_s$H$^{\cdot +}$ with the abundant alcohol adsorbed on the TiO$_2$ surface [24]. Therefore, the free h_{vb}^+ rarely presents in the bulk or on the surface of TiO$_2$, so that e_{cb}^- may recombine only with the trapped holes.

3. Origin and Energy Distribution of Electron Trap Defects (Ti$_{ds}$)

The bulk and surface Ti$_{ds}$ are formed in reduced or doped TiO$_2$ in both rutile and anatase phases [17–20]. As depicted in Diebold's review [17], the bulk Ti$_{ds}$ are easily created in the rutile single crystal by thermal annealing in a vacuum, resulting in the formation of blue color centers, indicating high conductivity. Therefore, TiO$_2$ is classified as an n-type semiconductor. The H$_2$ reduction of TiO$_2$ creates both oxygen vacancies and Ti^{3+} ions, which is an electron trapped in a Ti^{4+} lattice site, as described in Reaction (1) using Kröger–Vink notation [14,15,19,20]

$$O_O^x + 2Ti_{Ti}^x + H_2 \rightarrow V_O^{\bullet\bullet} + 2Ti'_{Ti} + H_2O \qquad (1)$$

where O_O^x is an O^{2-} ion in the oxygen lattice site, $V_O^{\bullet\bullet}$ is an oxygen vacancy with a double positive charge, and Ti'$_{Ti}$ is a Ti^{3+} ion in the titanium lattice site. The two Ti'$_{Ti}$ that are created per $V_O^{\bullet\bullet}$ have two excess electrons, which are responsible for the n-type conductivity, the blue-black colorization, and the enhancement of photocatalytic activity on TiO$_2$. The H$_2$ reduction on TiO$_2$ can also induce a disordered structure in the surface layer of TiO$_2$ nanocrystals, indicated by black TiO$_2$ [26–28]. Black TiO$_2$ exhibits high photocatalytic performance in decomposing organic pollutants and in generating hydrogen gas in an aqueous methanol solution under solar light irradiation. The other titanium oxides that have Ti$_{ds}$ are the F-doped or Nb-doped TiO$_2$, in which oxygen atoms are substituted with fluorine atoms or Ti atoms are replaced with Nb atoms, respectively [29]. Another type of Ti defect in TiO$_2$

is titanium interstitials ($Ti_i^{\bullet\bullet\bullet}$) possessing excess Ti atoms or ions in the lattice or in the near-surface region on TiO_2 surface [17–20,30].

Facile laser ablation and processing techniques have been developed to introduce the defects into TiO_2 nanocrystals and colloids in liquid [31–34]. In a typical procedure, TiO_2 suspensions are irradiated by a high-intensity pulsed laser with frequent repetition rates to produce the characteristic blue-black TiO_2. The obtained TiO_2 nanoparticles enhanced the photocatalytic activities in decomposing an organic dye [31] and in a water splitting reaction [32].

The electronic energy of Ti_{ds} is located just below the CB edge in the band gap of TiO_2 in a broad range of 0–1.8 eV [35]. Di Valentin and co-workers theoretically calculated the energy levels of point defects in bulk anatase TiO_2, which are located at 0.3, 0.4, 0.7, and 0.8 eV below the CB edge, for six-fold-coordinated Ti introduced by F- or Nb-doping, Ti–OH species associated with hydrogen doping, five-coordinated Ti'_{Ti} associated with the oxygen vacancy site, and titanium interstitials $Ti_i^{\bullet\bullet\bullet}$, respectively [29]. Deskins et al. calculated the relative energies of Ti_{ds} formed in the {110}-terminated rutile TiO_2 surface by means of the density functional theory (DFT), known as the DFT + U method [36,37]. They modeled the formation of Ti_{ds} at various Ti sites, such as the five-coordinated Ti and oxygen vacancies [37]. The calculation for the five-coordinated Ti in the presence of surface hydroxyls indicated that deep Ti_{ds} sites may exist in the second Ti layer from the surface or under the five-coordinated Ti rows [36].

The presence of these Ti_{ds} species, such as Ti'_{Ti} and $Ti_i^{\bullet\bullet\bullet}$, can be experimentally confirmed by means of electron spin resonance (ESR) [38–42], infrared radiation (IR) [40,42–48], ultraviolet-visible absorption (UV–vis) [14,42,49–51], photoluminescence (PL) [52], photoacoustic [53–56], and photoelectron spectroscopies [17,18,30]. Scanning tunneling microscopy (STM) and atomic force microscopy (AFM) are powerful tools for the direct observation of surface Ti_{ds} [17,18,30,57–59]. The oxygen vacancy (Ti'_{Ti} accompanied by $V_O^{\bullet\bullet}$) [57–59] and interstitial Ti ($Ti_i^{\bullet\bullet\bullet}$) [30] sites were directly observed on the TiO_2 surface.

The Ti_{ds} described above are not the only type of point defects. There are many types of lattice defects including step edges, line defects, grain boundaries, and impurities [17]. The Ti_{ds} energies can be strongly affected by site heterogeneity due to the local structures. Furthermore, under actual photocatalytic conditions, the TiO_2 surface is always covered by adsorbates, especially solvent molecules; thus, the energy of Ti_{ds} should depend on the adsorbed species [19,20]. Therefore, the information from the theoretical calculations and the photoelectron spectroscopy, applied to the clean catalyst surfaces under ultra-high vacuum conditions, may be limited for actual photocatalytic systems. To address this issue, Ohtani et al. developed a powerful tool for measuring the energy-resolved distribution of electron trap states for many types of TiO_2 powders, composed of anatase, rutile, and brookite in methanol-containing gas phase, by means of reversed double-beam photoacoustic spectroscopy (RDB-PAS) [56]. They showed that the electron energy of the trap states is distributed around the CB edge of TiO_2 and the distribution range of anatase TiO_2 is relatively broader than that of rutile TiO_2 (Figure 2). The energy distribution patterns for both anatase and rutile TiO_2 powders are similar to those obtained by the photochemical method, which uses the surface reaction of the trapped electrons with methyl viologen to release its cation radical in de-aerated aqueous solution containing methanol as a sacrificial reagent [60]. They also revealed that the total density of the traps is well-correlated to the specific surface area of TiO_2 powders, suggesting that the electron trap sites are predominantly located on the surface of TiO_2, and they do not depend on the type of crystallites in anatase, rutile, or their mixtures.

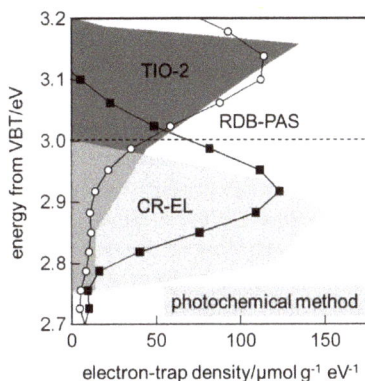

Figure 2. Comparison of energy-resolved distribution of electron traps (ERDT) patterns measured using the photochemical method shown by the grey patterns [60], and reversed double-beam photoacoustic spectroscopy (RDB-PAS), shown by the plots [56], for representative anatase (TIO^{-2}; 18 m^{2}g^{-1}) and rutile (7 m^{2}g^{-1}) samples. The top line, at 3.2 eV, and the dashed line show the conduction bands (CB) edge positions of anatase and rutile estimated by the reported bandgaps at 3.2 and 3.0 eV, respectively. Reprinted from reference [56], an open access article under conditions of the Creative Commons Attribution (CC BY) license.

The energy distribution of Ti$_{ds}$ can also be obtained by an electrochemical method [61,62]. In this method, the potential variation in accumulated charge at the TiO$_2$ electrode can be measured in aqueous solution. By calculating the derivative of the accumulated charge (Q) versus the applied potential (U), the energy density of Ti$_{ds}$ is directly proportional to dQ/dU and the plot of dQ/dU vs. U reflects the energy distribution of Ti$_{ds}$. The maximum density of Ti$_{ds}$ is located around 0.25–0.4 eV below the CB edge for nanostructured anatase TiO$_2$ and P25 TiO$_2$ samples with a ratio of anatase to rutile of approximately 80:20. Thus, the distribution of Ti$_{ds}$ within 0–0.4 eV is predominant, so that electrons trapped in these trap states may participate in the reductive reaction on TiO$_2$.

4. Reactivity of Trapped and Accumulated Electrons

This section highlights the reactivity of electrons trapped at Ti$_{ds}$ and accumulated in CB in the reductions of molecular oxygen O$_2$ and molecular nitrogen N$_2$, and the hydrogenation and dehalogenation of selected organic substrates occurring on TiO$_2$. The reductive reactions associated with Ti$_{ds}$ have been extensively investigated under various experimental conditions and through theoretical calculation methods. Although many studies have been performed for clean surfaces on TiO$_2$ under high-vacuum conditions [17,18,30,57–59], here we focus on the reactions occurring on powder or colloidal TiO$_2$ under conventional gas or liquid phase conditions.

Here we define the terms 'accumulated electrons' (e$_{cb}{}^-$) and 'trapped electrons' (Ti$_{ds}{}^{3+}$) to distinguish them. Accumulated electrons contain both electrons in CB and electrons trapped at shallow Ti states, located just below the CB edge of TiO$_2$, within 0–0.05 eV. These electrons can be in thermal equilibrium between the bulk CB and at the shallow trap sites, and easily migrate through these states. The accumulation of electrons in these states should occur after saturation of the intra-bandgap Ti states (Ti$_{ds}$) during UV irradiation. These electrons are highly reactive at the TiO$_2$ interface (Step 4 in Figure 1).The trapped electrons Ti$_{ds}{}^{3+}$ mean the electrons trapped at Ti$_{ds}$ are located in relatively deep energy from the CB edge. Therefore, the trapped electrons Ti$_{ds}{}^{3+}$ cannot be excited thermally to the CB or the shallow states, exhibiting either low or no reactivity at the TiO$_2$ interface (Step 5 in Figure 1).

The trapped electrons Ti$_{ds}{}^{3+}$ and the accumulated electrons e$_{cb}{}^-$ are quite stable in the presence of a good sacrificial hole scavenger, such as alcohols or amines, and in the absence of electron acceptors such as O_2, the lifetime may exceed several hours [50,51,63]. This extremely long lifetime of Ti_{ds}^{3+} and e_{cb}^- is attributable to the excellent hole scavenging ability of the sacrificial reagents on the TiO_2 surface [24], which prevents the recombination of Ti_{ds}^{3+} and e_{cb}^- with the surface trapped holes. In other words, the hole scavengers inhibit Steps 7 and 8 in Figure 1. The excess charges caused by electrons Ti_{ds}^{3+} and e_{cb}^- on the irradiated TiO_2 are balanced by the insertion (intercalation) of protons into the TiO_2 lattice [47,62,64,65]. The electrons Ti_{ds}^{3+} and e_{cb}^- show the unique blue-black coloration from the visible to the IR region [14,40,42–51], which enables the tracing of the lifetimes of the species generated on the irradiated TiO_2. Interestingly, the electrons Ti_{ds}^{3+} and e_{cb}^- are distinguishable by measuring IR spectra. The free electrons e_{cb}^- exhibit the typical exponential frequency-dependent spectrum that is attributed to the intra-CB transition [43,44,46,48], whereas the trapped electrons Ti_{ds}^{3+} are characterized by a broad absorption in the mid-IR region that is ascribed to a direct optical transition [46,48].

4.1. Reduction of Molecular Oxygen and Hydrogen Peroxide

The interfacial electron transfer between molecular oxygen (O_2) and electrons Ti_{ds}^{3+} or e_{cb}^- on nanocrystalline TiO_2 films was examined using a transient UV–vis absorption spectroscopy in gas phase [66]. In the presence of ethanol, as a sacrificial hole scavenger under ethanol-saturated conditions (5.8%), the half-life ($t_{50\%}$) of the electron-species Ti_{ds}^{3+} and e_{cb}^- was approximately 0.5 s in the absence of O_2. The $t_{50\%}$ value drastically decreased with increasing O_2 concentration, to approximately 12 µs in an oxygen concentration of 21% (air saturated conditions). Thus, the efficient electron transfer of molecular oxygen in a gaseous phase occurred on TiO_2 films by using ethanol as the hole scavenger. The dynamics of the electron transfer between O_2 and the nanosized TiO_2 particles in a liquid phase were also investigated by employing a simple and facile stopped flow technique [50,51]. With methanol as the hole scavenger, the electrons on the TiO_2 particles, in an argon-purged and de-aerated aqueous solution, were accumulated by pre-UV irradiation, causing blue colorization characterized by a broad absorption band in the visible light region of 400–800 nm. In the stopped flow experiment, the TiO_2 solution containing the electron-species Ti_{ds}^{3+} and e_{cb}^- was mixed with the aqueous solution containing O_2 at pH2.3, and the change in absorbance of the electrons was recorded at 600 nm. The absorbance signal decreased slowly with a rate constant of 8.9×10^{-7} mol L^{-1} s^{-1} in the absence of O_2, whereas the signal rapidly disappeared within a few seconds under O_2-saturated conditions. Thus, efficient electron transfer from the accumulated TiO_2 to O_2 proceeded even in the aqueous solution. The reduction of hydrogen peroxide (H_2O_2) was also confirmed by the same stopped flow experiment [50,51]. The decay rate of the electron-species Ti_{ds}^{3+} and e_{cb}^- in the H_2O_2 reduction was slower than in the O_2 reduction under similar conditions.

The reduction of O_2 on TiO_2 results in the formation of reactive oxygen species (ROS), such as superoxide anion radicals ($O_2^{\cdot-}$), hydroperoxy radicals ($^{\cdot}OOH$), hydrogen peroxide (H_2O_2), and hydroxyl radicals ($^{\cdot}OH$), under both aqueous and aerated conditions [6,21]. These ROS play a crucial role in the photocatalysis on TiO_2 for water purification, air cleaning, self-cleaning, self-sterilization, etc. The sequential ROS generation on photo-irradiated TiO_2 under acidic conditions can be depicted as follows [6,21,50,51]:

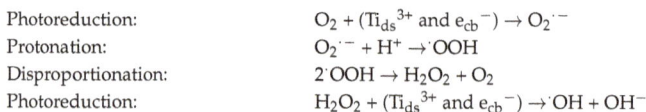

Photoreduction:	$O_2 + (Ti_{ds}^{3+} \text{ and } e_{cb}^-) \rightarrow O_2^{\cdot-}$
Protonation:	$O_2^{\cdot-} + H^+ \rightarrow {}^{\cdot}OOH$
Disproportionation:	$2\,{}^{\cdot}OOH \rightarrow H_2O_2 + O_2$
Photoreduction:	$H_2O_2 + (Ti_{ds}^{3+} \text{ and } e_{cb}^-) \rightarrow {}^{\cdot}OH + OH^-$

Though the CB edge of anatase TiO_2 is located at -0.27 V vs. standard hydrogen electrode (SHE) at pH2.3 [67], anatase TiO_2 produced the superoxide anion radial by photoexcitation; $O_2/O_2^{\cdot-}$ was approximately 0.33 V vs. SHE [68]. This could be due to the strong adsorption of O_2 at the Ti_{ds} sites, such as $V_O^{\bullet\bullet}$, which may lead to a positive shift of the redox potential ($O_2/O_2^{\cdot-}$). In this electron transfer step, an electron seems to be transferred from the Ti 3d orbital to the π^* orbital of the adsorbed O_2.

4.2. Reduction of Molecular Nitrogen, Nitrate, and Nitrite Ions

Molecular nitrogen (N_2) is chemically stable, so photocatalytic reduction of N_2 to ammonia (NH_3) under ambient temperature and pressure is challenging. Hirakawa and coworkers recently reported that the photocatalytic conversion of N_2 to ammonia with water occurred on the bare TiO_2 powders under ambient conditions [69]. They stated that the active sites for N_2 reduction are the Ti_{ds} with oxygen vacancies mainly formed on the rutile {110} surface. They investigated the photocatalytic reductions of nitrate (NO_3^-) and nitrite (NO_2^-) ions to ammonia and N_2 on bare TiO_2 under ambient conditions [70]. They proposed that the Ti_{ds} sites selectively promoted the eight-electron reduction of NO_3^- to NH_3 ($NO_3^- + 9H^+ + 8e^- \rightarrow NH_3 + 3H_2O$), while the Lewis acid site promoted nonselective reduction, resulting in N_2 and NH_3 formation. Thus, TiO_2 with many Ti_{ds} and a small number of Lewis acid sites produced ammonia with very high selectivity. The use of artificial fertilizers in agriculture has caused a great deal of concern for water pollution caused by the production of NO_3^- and NO_2^- ions from fertilizers [71]. Therefore, a chemical process for the reduction of NO_3^- and NO_2^- ions on TiO_2 may be useful for an environmental recycling process.

4.3. Hydrogenation of Carbonyl Compounds

Kohtani et al. examined whether electrons Ti_{ds}^{3+} and e_{cb}^- transfer to acetophenone (AP) derivatives adsorbed on TiO_2 [63,72]. The photoreductive hydrogenation of several aromatic carbonyl compounds was confirmed to occur on UV-irradiated P25 TiO_2 as illustrated in Scheme 1 [72,73]. They evaluated the number of transferred electrons in the injection experiment using a pre-irradiated TiO_2 suspension. When the P25 TiO_2 powder was dispersed in de-aerated ethanol as a hole scavenger and irradiated with UV light for 2 h, the white color of TiO_2 powder changed to blue-gray. After the blue-gray color change was confirmed, a large amount of AP derivatives was injected into this TiO_2 suspension in the dark. In this experiment, ethanol acted not only as a solvent but also as a hole scavenger.

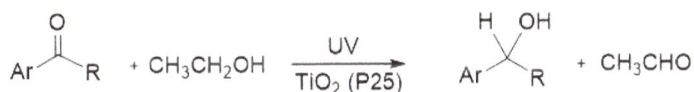

Scheme 1. Photocatalytic hydrogenation of aromatic carbonyl compounds: Ar = aromatic ring, R = H, Me, Et, *i*-Pr, or CF_3. Reprinted with permission from reference [72]. Copyright (2014) The Royal Society of Chemistry.

Figure 3a shows that the blue-gray color of the pre-irradiated P25 TiO_2 suspension remained for a few days in the absence of AP derivatives [63], meaning that the electrons accumulated on the P25 TiO_2 surface are quite stable in the de-aerated ethanol. Figure 3b,c show the color change induced by the addition of aromatic carbonyl compounds. The blue-gray color of TiO_2 rapidly changed after the injection of 2,2,2-trifluoacetophenone (TFAP) where the aromatic ring (Ar) was C_6H_5 and the R was CF_3 (Scheme 1). The change from blue-gray to white was completed within 3 h in the TFAP solution as shown in Figure 3c. This result indicates that all the trapped and accumulated electrons on TiO_2 were consumed in the reduction of TFAP within 3 h. On the other hand, with AP, a part of the blue-gray species on TiO_2 was remarkably stable even after 50 h, as shown in Figure 3b, which may be due to the remaining electrons trapped at the deep states of TiO_2.

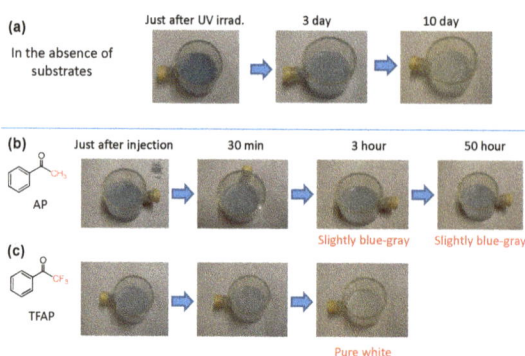

Figure 3. Color changes of the pre-irradiated P25 TiO$_2$ powder dispersed in de-aerated ethanol (**a**) in the absence of substrates, and after injection of (**b**) 300 µmol acetophenone (AP), or (**c**) 2,2,2-trifluoacetophenone (TFAP). Adapted with permission from reference [63]. Copyright 2012 American Chemical Society. UV irrad.: UV irraditaion.

Figure 4 depicts the time evolution of the secondary alcohols 1-phenylethanol (AP-OH) or 1-phenyl-2,2,2-trifluoroethanol (TFAP-OH), as products after the injection of substrates AP or TFAP, into the sufficient pre-irradiated TiO$_2$ suspension, respectively [63]. The amount of each product quickly grew within 0.5 h, which agrees with the observation of color change in the TiO$_2$ suspension (Figure 3b,c). The amount of TFAP-OH product obtained from the reactive substrate TFAP rapidly increased, and reached 5.4 µmoL within 1 h. Assuming all Ti$_{ds}$$^{3+}$ and e$_{cb}$$^{-}$ electrons were consumed in the reductive hydrogenation of TFAP, the total amount of Ti$_{ds}$$^{3+}$ and e$_{cb}$$^{-}$ on the TiO$_2$ powder was estimated to be about 100 µmol·g^{-1}. The time evolution of AP-OH from the less reactive substrate AP consisted of a fast component within 0.5 h, and a slow component after 0.5 h, which increased gradually to reach 4.1 µmol (Figure 4). The slow component represents the slow electron transfer event from middle Ti$_{ds}$, between shallow and deep states, to AP adsorbed on TiO$_2$. The total amount of AP-OH production was about 25% smaller than that of TFAP-OH production. In the reduction of the less reactive substrate AP, the deep Ti$_{ds}$$^{3+}$ species remained on the TiO$_2$ surface for a long time (>20 h) after the injection of AP. The amount of deep Ti$_{ds}$$^{3+}$ that remained on TiO$_2$ was roughly estimated to be 25% given the difference between the amounts of TFAP-OH (5.4 µmoL) and AP-OH (4.1 µmoL). Thus, the residual 25% electrons could not react with AP and remained at the deep trap sites.

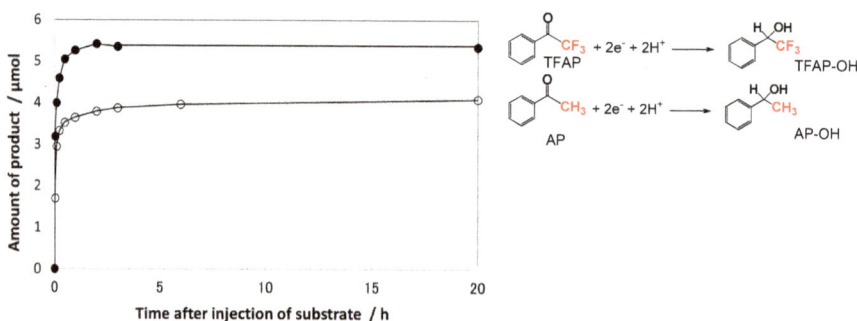

Figure 4. Time evolutions of the amount of AP-OH (○) and TFAP-OH (●) after the 300 µmol injection of substrate (AP or TFAP) into the 2 h pre-irradiated TiO$_2$ suspension at 32 °C Adapted with permission from reference [63]. Copyright (2012) American Chemical Society.

Assuming that all Ti_{ds}^{3+} and e_{cb}^{-} electrons react with TFAP, the percentages of reacted electrons were estimated for other AP derivatives, as summarized in Table 1 [72]. The number of reacted electrons showed a tendency to decrease with decreasing the reduction potential (E_{red}) of the substrates according to the dependence on the actual reaction rates as listed in Table 1. This implies that the rates of photocatalytic hydrogenation of AP derivatives are governed by the electron transfer efficiency from the Ti_{ds} sites to the adsorbed AP sites. Notably, the relative position between the Ti_{ds} energy distributed within the bandgap and the acceptor level of the AP derivatives (the solid Gaussian curve) should be appropriate (Figure 5) [72]. The energy distribution of Ti_{ds} in TiO_2 powders and colloids can be obtained by photochemical [60], electrochemical [61,62], and RDB-PAS [56] methods. The acceptor levels of AP derivatives can be estimated by the Marcus theory [72,74–76].

Table 1. Reduction potentials, amount of reacted electrons, percentages of reacted electrons, and reaction rates at maximum concentration of substrates [72]. Adapted with permission from reference [72]. Copyright (2014) The Royal Society of Chemistry.

Substrate	E_{red}/V [a]	Amount of Reacted Electrons [b]/μmoL	Percentage [c]/%	Reaction Rate [d]/mMh^{-1}
1 (TFAP)	−1.35	10.2	100	–
2	−1.59	8.22	81	3.4 ± 0.2
3	−1.62	6.32	62	2.2 ± 0.2
4	−1.80	6.09	60	2.0 ± 0.1
5 (AP)	−1.89	7.38	72	1.9 ± 0.1
6	−0.92	5.70	56	1.2 ± 0.1
7	−1.94	4.76	47	0.75 ± 0.05

[a] Reduction potentials vs. SHE (standard hydrogen electrode). [b] Molar number of reacted electrons estimated by the injection experiment using a pre-irradiated TiO_2 suspension. [c] Percentage of the reacted electrons per the total amount of Ti_{st}^{3+} and e_{cb}^{-} (10.2 μmol) generated on 0.10 g TiO_2. [d] Reaction rates at the maximum concentration of substrates.

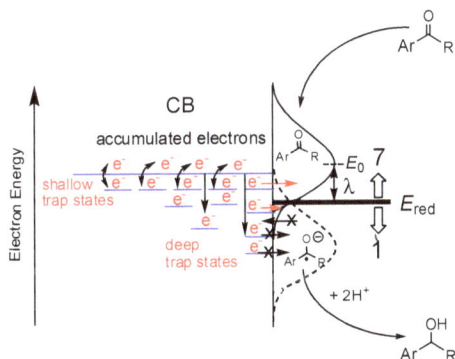

Figure 5. Schematic illustration of the electron transfer reaction from the Ti_{ds} states to the adsorbed AP derivatives, where E_{red} is the reduction potential of AP derivatives, λ is the reorganization energy (approximately 0.7 eV for AP), and E_0 is the energy at the top of curve for the acceptor level, calculated by $qE_{red} - \lambda$ and shown by the solid line. The dotted line indicates the donor energy level for anionic species. Reprinted with permission from reference [72]. Copyright (2014) The Royal Society of Chemistry.

Kohtani et al. also reported the photohydrogenation of AP derivatives on P25 TiO_2, modified with metal-free organic dyes such as rhodamine B, fluorescein, and coumarin derivatives [77,78]. The use of these organic dyes successfully extended the UV response of TiO_2 to the visible light region, though these reaction rates were much slower than the hydrogenation rate using UV excitation of non-modified TiO_2. In this dye-sensitized system, the electron injection from dye into TiO_2 can take place in two different ways: (1) injection via a lowest unoccupied molecular orbital (LUMO) level of the excited dye to the CB of TiO_2, and (2) direct injection of TiO_2 to CB on the excitation of the charge transfer complex ($TiO_2^{\delta-}$ dye$^{\delta+}$) [77]. The injected electrons should then be distributed to the Ti_{ds} sites on the P25 TiO_2 surface. The accumulated electrons were observed with the blue-gray color for all dye-TiO_2 powders during visible light irradiation.

4.4. Defluorination of Fluorinated AP Derivatives

Compounds containing fluorine atoms are often used as pharmaceutical and agrochemical reagents. Since the C–F bond is one of the strongest bonds, C–F bond activation and cleavage is a field of current interest in organic chemistry [79], although less is known about the catalytic activation and cleavage methods. Photocatalytic reaction is one of the promising methods to promote the activation and cleavage of the C–F bond of fluorinated compounds under mild conditions. Therefore, an attempt was made to use trapped and accumulated electrons on TiO_2 for the sequential multi-step electron transfer in the reduction of fluorinated AP derivatives (Figure 6) [80]. The reaction of fluorinated AP derivatives on TiO_2 showed the two photocatalytic reductive transformations, i.e., the defluorination and reduction of the carbonyl group. The reduction of 2-fluoromethylacetophenone (MFAP) only provided the ketone AP because of the C–F bond cleavage, whereas the reaction of TFAP only provided the alcohol TFAP-OH as a result of the reduction of the carbonyl group. Interestingly, the reduction of 2,2-difluoromethyacetophenone (DFAP), possessing characteristics between those of MFAP and TFAP, gave the defluorinated ketones, MFAP and AP, as well as hydrogenated alcohol 1-phenyl-2,2-difluoroethanol (DFAP-OH), as shown in Figure 6. The defluorination reactions became unfavorable with increasing the number of fluorine atoms on the substrates. This mainly arises because of the increase in the bond dissociation energy of the C–F bond and the positive shift of the reduction potential of fluorinated AP derivatives with the increasing number of fluorine atoms [80].

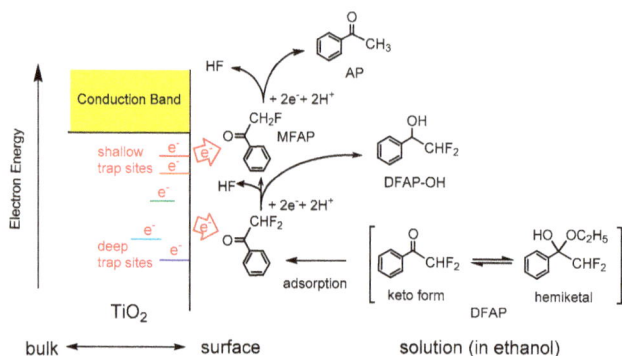

Figure 6. Schematic illustration on the photocatalytic reduction of 2,2-difluoromethyacetophenone (DFAP) on the UV irradiated TiO_2. Reprinted with permission from reference [80]. Copyright (2016) Elsevier B.V.

4.5. Hydrogenation of Nitroaromatic Compounds

Several organic nitroaromatic compounds can be easily hydrogenated to create the corresponding amino compounds on the UV-irradiated TiO_2 in the presence of 2-propanol as a sacrificial hole scavenger (Scheme 2).

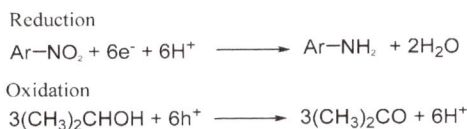

Reduction

$$Ar-NO_2 + 6e^- + 6H^+ \longrightarrow Ar-NH_2 + 2H_2O$$

Oxidation

$$3(CH_3)_2CHOH + 6h^+ \longrightarrow 3(CH_3)_2CO + 6H^+$$

Scheme 2. Photocatalytic reduction of nitroaromatic compounds and oxidation of 2-propanol as a sacrificial hole scavenger.

Shiraishi et al. reported that some kinds of rutile TiO_2 particles promote the highly efficient photocatalytic hydrogenation of nitroaromatic compounds [81,82]. They claimed that the oxygen vacancy sites on the rutile {110} behave as the adsorption sites for the nitroaromatic compounds and the electron trap sites, resulting in the formation of aniline derivatives with significantly high yields of greater than 25% at 370 nm [81]. They also found that the activity of rutile particles depends on the number of defects on the particles [82]. The inner (bulk) defects behave as the deactivation sites for the recombination of electrons and holes, whereas the surface defects behave as the active reaction sites as well as the deactivation sites. As a result, the reaction rate is proportional to the ratio of the amount of surface defects to that of total defects ($N_{surface}/N_{total}$) in the rutile TiO_2 particles, which aligns with the report of Kong and coworkers [16].

Molinari and coworkers examined the selective hydrogenation of $NO_2-C_6H_4-CHO$, bearing the two reducible functional groups, $-NO_2$ and $-CHO$ [62]. They found that the nitro group was easily reduced by the trapped electrons at Ti_{ds} within the bandgap, whereas the aldehyde group was reduced by electrons accumulated on CB. Therefore, the chemoselective reduction of functional groups can be controlled by the energy distribution patterns of Ti_{ds}, which may depend on the type of TiO_2 powders. The selective reduction of functional groups is difficult to achieve through conventional thermal catalysis. Therefore, this topic is an interesting issue for chemoselective photocatalysis.

5. Summary and Potential for the Development of Efficient Photocatalysis

According to the reports of Kong et al. [16] and Shiraishi et al. [82], the photocatalytic efficiencies increased with increasing the ratio of the amount of surface to bulk defects ($N_{surface}/N_{bulk}$) or the amount of surface to total defects ($N_{surface}/N_{total}$). For example, molecular oxygen (O_2) in a gaseous phase would be easily adsorbed on the surface Ti_{ds} and reduced by electrons Ti_{ds}^{3+} and e_{cb}^-, resulting in the efficient formation of ROS, which oxidize benzene efficiently [16]. Further, the surface Ti_{ds} behaves as the active reaction site in the efficient reduction of nitrobenzene [81]. Thus, the surface defects favorably act as adsorption sites as well as reaction sites.

In addition, the relative position between the energy distributions of Ti_{ds} and the acceptor level (reduction potential) of substrates should be appropriate as indicated in Figure 5 [72]. The electrons accumulated in CB and trapped at shallow Ti_{ds} can easily participate in the reaction, whereas those trapped at deep Ti_{ds} cannot [63]. These unreacted electrons remain at the deep Ti_{ds} sites and exhibit an extremely long lifetime in alcohols. Thus, the shallow traps enhance photocatalytic activity, while the deep traps cause a reduction. Furthermore, Amano et al. proposed that the creation of shallow Ti_{ds} greatly enhances electrical conductivity, thereby facilitating the charge transport and separation caused by the formation of band bending in the space charge layer at the TiO_2-liquid interface [14,15].

In conclusion, the development of highly active photocatalysts necessitates precise control of the structural properties; the density of surface shallow traps should be maximized and the density of deep traps as well as inner (bulk) traps minimized, as proposed by Ohtani [83]. One of the promising strategies for meeting these requirements is the use of highly uniform TiO_2 nanocrystals with specific exposure of the reactive facets [42,84–87]. In particular, the anatase {101} and {001} facets have been reported to be favorable for the reductive and oxidative reactions in TiO_2 photocatalysis, respectively [87–91]. If the reductive and oxidative facets could be selectively covered with a large amount of the active shallow Ti_{ds} and the terminal $Ti–O_sH$ hole trapping sites, respectively, the photocatalytic activity on the TiO_2 nanocrystals would be greatly enhanced by the effective electron-hole charge separation, followed by the subsequent charge transfer reactions at the specific reductive and oxidative sites. Therefore, special attention should be directed toward the development of TiO_2 nanocrystals with precisely controlled facets.

Acknowledgments: The works in references [63,72,73,77,78,80] were supported by JSPS KAKENHI Grant Numbers 21590052 and 24590067.

Conflicts of Interest: The author declares no conflict of interest.

References

1. Fujishima, A.; Honda, K. Electrochemical photolysis of water at a semiconductor electrode. *Nature* **1972**, *238*, 37–38. [CrossRef] [PubMed]
2. Kudo, A.; Miseki, Y. Heterogeneous photocatalyst materials for water splitting. *Chem. Soc. Rev.* **2009**, *38*, 253–278. [CrossRef] [PubMed]
3. Fujishima, A.; Zhang, X.; Tryk, D. Heterogeneous photocatalysis: From water photolysis to applications in environmental cleanup. *Int. J. Hydrog. Energy* **2007**, *32*, 2664–2672. [CrossRef]
4. Grätzel, M. Recent advances in sensitized mesoscopic solar cells. *Acc. Chem. Res.* **2009**, *42*, 1788–1798. [CrossRef] [PubMed]
5. Hagfeldt, A.; Boschloo, G.; Sun, L.; Kloo, L.; Pettersson, H. Dye-sensitized solar cells. *Chem. Rev.* **2010**, *110*, 6595–6663. [CrossRef] [PubMed]
6. Hoffmann, M.; Martin, S.; Choi, W.; Bahnemann, D. Environmental applications of semiconductor photocatalysis. *Chem. Rev.* **1995**, *95*, 69–96. [CrossRef]
7. Sakata, T. Photocatalysis of irradiated semiconductor surfaces: Its application to water splitting and some organic reactions. *J. Photochem.* **1985**, *29*, 205–215. [CrossRef]
8. Kisch, H. Semiconductor photocatalysis for organic synthesis advances in photochemistry. *Adv. Photochem.* **2001**, *26*, 93–143.
9. Palmisano, G.; Augugliaro, V.; Pagliaro, M.; Palmisano, L. Photocatalysis: A promising route for 21st century organic chemistry. *Chem. Commun.* **2007**, 3425–3437. [CrossRef] [PubMed]
10. Kohtani, S.; Yoshioka, E.; Miyabe, H. Photocatalytic hydrogenation on semiconductor particles. *Hydrogenation* **2012**, 291–308. [CrossRef]
11. Kisch, H. Semiconductor photocatalysis-mechanistic and synthetic aspects. *Angew. Chem. Int. Ed.* **2013**, *52*, 812–847. [CrossRef] [PubMed]
12. Fujishima, A.; Rao, T.; Tryk, D. Titanium dioxide photocatalysis. *J. Photochem. Photobiol. C* **2000**, *1*, 1–21. [CrossRef]
13. Thompson, T.; Yates, J. Surface science studies of the photoactivation of TiO_2 new photochemical processes. *Chem. Rev.* **2006**, *106*, 4428–4453. [CrossRef] [PubMed]
14. Amano, F.; Nakata, M.; Yamamoto, A.; Tanaka, T. Effect of Ti^{3+} ions and conduction band electrons on photocatalytic and photoelectrochemical activity of rutile titania for water oxidation. *J. Phys. Chem. C* **2016**, *120*, 6467–6474. [CrossRef]
15. Amano, F.; Nakata, M.; Yamamoto, A.; Tanaka, T. Rutile titanium dioxide prepared by hydrogen reduction of degussa P25 for highly efficient photocatalytic hydrogen evolution. *Catal. Sci. Technol.* **2016**, *6*, 5693–5699. [CrossRef]
16. Kong, M.; Li, Y.; Chen, X.; Tian, T.; Fang, P.; Zheng, F.; Zhao, X. Tuning the relative concentration ratio of bulk defects to surface defects in TiO_2 nanocrystals leads to high photocatalytic efficiency. *J. Am. Chem. Soc.* **2011**, *133*, 16414–16417. [CrossRef] [PubMed]
17. Diebold, U. The surface science of titanium dioxide. *Surf. Sci. Rep.* **2003**, *48*, 53–229. [CrossRef]

18. Henderson, M. A Surface science perspective on TiO$_2$ photocatalysis. *Surf. Sci. Rep.* **2011**, *66*, 185–297. [CrossRef]

19. Nowotny, M.K.; Sheppard, L.R.; Bak, T.; Nowotny, J. Defect chemistry of titanium dioxide. Application of defect engineering in processing of TiO$_2$-based photocatalysts. *J. Phys. Chem. C* **2008**, *112*, 5275–5300. [CrossRef]

20. Nowotny, J.; Alim, M.; Bak, T.; Idris, M.; Ionescu, M.; Prince, K.; Sahdan, M.; Sopian, K.; Mat Teridi, M.; Sigmund, W. Defect chemistry and defect engineering of TiO$_2$-based semiconductors for solar energy conversion. *Chem. Soc. Rev.* **2015**, *44*, 8424–8442. [CrossRef] [PubMed]

21. Schneider, J.; Matsuoka, M.; Takeuchi, M.; Zhang, J.; Horiuchi, Y.; Anpo, M.; Bahnemann, D. Understanding TiO$_2$ photocatalysis: Mechanisms and materials. *Chem. Rev.* **2014**, *114*, 9919–9986. [CrossRef] [PubMed]

22. Tamaki, Y.; Furube, A.; Murai, M.; Hara, K.; Katoh, R.; Tachiya, M. Dynamics of efficient electron-hole separation in TiO$_2$ nanoparticles revealed by femtosecond transient absorption spectroscopy under the weak-excitation condition. *Phys. Chem. Chem. Phys.* **2007**, *9*, 1453–1460. [CrossRef] [PubMed]

23. Tamaki, Y.; Hara, K.; Katoh, R.; Tachiya, M.; Furube, A. Femtosecond visible-to-IR spectroscopy of TiO$_2$ nanocrystalline films: Elucidation of the electron mobility before deep trapping. *J. Phys. Chem. C* **2009**, *113*, 11741–11746. [CrossRef]

24. Tamaki, Y.; Furube, A.; Murai, M.; Hara, K.; Katoh, R.; Tachiya, M. Direct observation of reactive trapped holes in TiO$_2$ undergoing photocatalytic oxidation of adsorbed alcohols: Evaluation of the reaction rates and yields. *J. Am. Chem. Soc.* **2006**, *128*, 416–417. [CrossRef] [PubMed]

25. Ma, Y.; Wang, X.; Jia, Y.; Chen, X.; Han, H.; Li, C. Titanium dioxide-based nanomaterials for photocatalytic fuel generations. *Chem. Rev.* **2014**, *114*, 9987–10043. [CrossRef] [PubMed]

26. Chen, X.; Liu, L.; Yu, P.Y.; Mao, S.S. Increasing solar absorption for photocatalysis with black hydrogenated titanium dioxide nanocrystals. *Science* **2011**, *331*, 746–750. [CrossRef] [PubMed]

27. Chen, X.; Liu, L.; Huang, F. Black titanium dioxide (TiO$_2$) nanomaterials. *Chem. Soc. Rev.* **2015**, *44*, 1861–1885. [CrossRef] [PubMed]

28. Zhou, W.; Li, W.; Wang, J.-Q.; Qu, Y.; Yang, Y.; Xie, Y.; Zhang, K.; Wang, L.; Fu, H.; Zhao, D. Ordered mesoporous black TiO$_2$ as highly efficient hydrogen evolution photocatalyst. *J. Am. Chem. Soc.* **2014**, *136*, 9280–9283. [CrossRef] [PubMed]

29. Di Valentin, C.; Pacchioni, G.; Selloni, A. Reduced and n-type doped TiO$_2$: Nature of Ti^{3+} species. *J. Phys. Chem. C* **2009**, *113*, 20543–20552. [CrossRef]

30. Wendt, S.; Sprunger, P.T.; Lira, E.; Madsen, G.K.H.; Li, Z.; Hansen, J.O.; Matthiesen, J.; Blekinge Rasmussen, A.; Laegsgaard, E.; Hammer, B.; et al. The role of interstitial sites in the Ti*3d* defect state in the band gap of titania. *Science* **2008**, *320*, 1755–1759. [CrossRef] [PubMed]

31. Chen, X.; Zhao, D.; Liu, K.; Wang, C.; Liu, L.; Li, B.; Zhang, Z.; Shen, D. Laser-modified black titanium oxide nanospheres and their photocatalytic activities under visible light. *ACS Appl. Mater. Interfaces* **2015**, *7*, 16070–16077. [CrossRef] [PubMed]

32. Filice, S.; Compagnini, G.; Fiorenza, R.; Scirè, S.; D'Urso, L.; Fragalà, M.E.; Russo, P.; Fazio, E.; Scalese, S. Laser processing of TiO$_2$ colloids for an enhanced photocatalytic water splitting activity. *J. Colloid Interface Sci.* **2017**, *489*, 131–137. [CrossRef] [PubMed]

33. Russo, P.; Liang, R.; He, R.X.; Zhou, Y.N. Phase transformation of TiO$_2$ nanoparticles by femtosecond laser ablation in aqueous solutions and deposition on conductive substrates. *Nanoscale* **2017**, *9*, 6167–6177. [CrossRef] [PubMed]

34. Zhang, D.; Liu, J.; Li, P.; Tian, Z.; Liang, C. Recent advances in surfactant-free, surface-charged, and defect-rich catalysts developed by laser ablation and processing in liquids. *ChemNanoMat* **2017**, *3*, 512–533. [CrossRef]

35. Weiler, B.; Gagliardi, A.; Lugli, P. Kinetic monte carlo simulations of defects in anatase titanium dioxide. *J. Phys. Chem. C* **2016**, *120*, 10062–10077. [CrossRef]

36. Deskins, N.A.; Rousseau, R.; Dupuis, M. Localized electronic states from surface hydroxyls and polarons in TiO$_2$ (110). *J. Phys. Chem. C* **2009**, *113*, 14583–14586. [CrossRef]

37. Deskins, N.A.; Rousseau, R.; Dupuis, M. Distribution of Ti^{3+} surface sites in reduced TiO$_2$. *J. Phys. Chem. C* **2011**, *115*, 7562–7572. [CrossRef]

38. Howe, R.; Grätzel, M. EPR observation of trapped electrons in colloidal titanium dioxide. *J. Phys. Chem.* **1985**, *89*, 4495–4499. [CrossRef]

39. Hurum, D.; Agrios, A.; Gray, K.; Rajh, T.; Thurnauer, M. Explaining the enhanced photocatalytic activity of degussa P25 mixed-phase TiO$_2$ using EPR. *J. Phys. Chem. B* **2003**, *107*, 4545–4549. [CrossRef]
40. Berger, T.; Sterrer, M.; Diwald, O.; Knözinger, E.; Panayotov, D.; Thompson, T.L.; Yates, J.T. Light-induced charge separation in anatase TiO$_2$ particles. *J. Phys. Chem. B* **2005**, *109*, 6061–6068. [CrossRef] [PubMed]
41. Li, G.; Dimitrijevic, N.; Chen, L.; Nichols, J.; Rajh, T.; Gray, K. The important role of tetrahedral Ti^{4+} sites in the phase transformation and photocatalytic activity of TiO$_2$ nanocomposites. *J. Am. Chem. Soc.* **2008**, *130*, 5402–5403. [CrossRef] [PubMed]
42. Gordon, T.; Cargnello, M.; Paik, T.; Mangolini, F.; Weber, R.; Fornasiero, P.; Murray, C. Nonaqueous synthesis of TiO$_2$ nanocrystals using TiF$_4$ to engineer morphology, oxygen vacancy concentration, and photocatalytic activity. *J. Am. Chem. Soc.* **2012**, *134*, 6751–6761. [CrossRef] [PubMed]
43. Szczepankiewicz, S.; Colussi, A.J.; Hoffmann, M. Infrared spectra of photoinduced species on hydroxylated titania surfaces. *J. Phys. Chem. B* **2000**, *104*, 9842–9850. [CrossRef]
44. Szczepankiewicz, S.; Moss, J.; Hoffmann, M. Slow surface charge trapping kinetics on irradiated TiO$_2$. *J. Phys. Chem. B* **2002**, *106*, 2922–2927. [CrossRef]
45. Takeuchi, M.; Martra, G.; Coluccia, S.; Anpo, M. Verification of the photoadsorption of H$_2$O molecules on TiO$_2$ semiconductor surfaces by vibrational absorption spectroscopy. *J. Phys. Chem. C* **2007**, *111*, 9811–9817. [CrossRef]
46. Panayotov, D.; Burrows, S.; Morris, J. Infrared spectroscopic studies of conduction band and trapped electrons in UV-photoexcited, H-atom n-doped, and thermally reduced TiO$_2$. *J. Phys. Chem. C* **2012**, *116*, 4535–4544. [CrossRef]
47. Savory, D.; McQuillan, A.J. Influence of formate adsorption and protons on shallow trap infrared absorption (STIRA) of anatase TiO$_2$ during photocatalysis. *J. Phys. Chem. C* **2013**, *117*, 23645–23656. [CrossRef]
48. Litke, A.; Hensen, E.J.M.; Hofmann, J. Role of dissociatively adsorbed water on the formation of shallow trapped electrons in TiO$_2$ photocatalysts. *J. Phys. Chem. C* **2017**, *121*, 10153–10162. [CrossRef] [PubMed]
49. Takeuchi, M.; Deguchi, J.; Sakai, S.; Anpo, M. Effect of H$_2$O vapor addition on the photocatalytic oxidation of ethanol, acetaldehyde and acetic acid in the gas phase on TiO$_2$ semiconductor powders. *Appl. Catal. B Environ.* **2010**, *96*, 218–223. [CrossRef]
50. Mohamed, H.; Dillert, R.; Bahnemann, D. Reaction dynamics of the transfer of stored electrons on TiO$_2$ nanoparticles: A stopped flow study. *J. Photochem. Photobiol. A Chem.* **2011**, *217*, 271–274. [CrossRef]
51. Mohamed, H.; Mendive, C.; Dillert, R.; Bahnemann, D. Kinetic and mechanistic investigations of multielectron transfer reactions induced by stored electrons in TiO$_2$ nanoparticles: A stopped flow study. *J. Phys. Chem. A* **2011**, *115*, 2139–2147. [CrossRef] [PubMed]
52. Knorr, F.; Mercado, C.; McHale, J. Trap-state distributions and carrier transport in pure and mixed-phase TiO$_2$: Influence of contacting solvent and interphasial electron transfer. *J. Phys. Chem. C* **2008**, *112*, 12786–12794. [CrossRef]
53. Leytner, S.; Hupp, J. Evaluation of the energetics of electron trap states at the nanocrystalline titanium dioxide/aqueous solution interface via time-resolved photoacoustic spectroscopy. *Chem. Phys. Lett.* **2000**, *330*, 231–236. [CrossRef]
54. Murakami, N.; Prieto Mahaney, O.; Torimoto, T.; Ohtani, B. Photoacoustic spectroscopic analysis of photoinduced change in absorption of titanium(IV) oxide photocatalyst powders: A novel feasible technique for measurement of defect density. *Chem. Phys. Lett.* **2006**, *426*, 204–208. [CrossRef]
55. Murakami, N.; Prieto Mahaney, O.; Abe, R.; Torimoto, T.; Ohtani, B. Double-beam photoacoustic spectroscopic studies on transient absorption of titanium(IV) oxide photocatalyst powders. *J. Phys. Chem. C* **2007**, *111*, 11927–11935. [CrossRef]
56. Nitta, A.; Takase, M.; Takashima, M.; Murakami, N.; Ohtani, B. A fingerprint of metal-oxide powders: Energy-resolved distribution of electron traps. *Chem. Commun.* **2016**, *52*, 12096–12099. [CrossRef] [PubMed]
57. Pang, C.; Lun Pang, C.; Lindsay, R.; Thornton, G. Chemical reactions on rutile TiO$_2$ (110). *Chem. Soc. Rev.* **2008**, *37*, 2328–2353. [CrossRef] [PubMed]
58. Papageorigiou, A.; Papageorgiou, A.C.; Beglitis, N.S.; Pang, C.L.; Teobaldi, G.; Cabailh, G.; Chen, Q.; Fisher, A.J.; Hofer, W.A.; Thornton, G. Electron traps and their effect on the surface chemistry of TiO$_2$ (110). *Proc. Natl. Acad. Sci. USA* **2010**, *107*, 2391–2396. [CrossRef] [PubMed]
59. Setvin, M.; Aschauer, U.; Scheiber, P.; Li, Y.F.; Hou, W.; Schmid, M.; Selloni, A.; Diebold, U. Reaction of O$_2$ with subsurface oxygen vacancies on TiO$_2$ anatase (101). *Science* **2013**, *341*, 988–991. [CrossRef] [PubMed]

60. Ikeda, S.; Sugiyama, N.; Murakami, S.; Kominami, H.; Kera, Y.; Noguchi, H.; Uosaki, K.; Torimoto, T.; Ohtani, B. Quantitative analysis of defective sites in titanium(IV) oxide photocatalyst powders. *Phys. Chem. Chem. Phys.* **2003**, *5*, 778–783. [CrossRef]

61. Wang, H.; He, J.; Boschloo, G.; Lindström, H.; Hagfeldt, A.; Lindquist, S.-E. Electrochemical investigation of traps in a nanostructured TiO$_2$ film. *J. Phys. Chem. B* **2001**, *105*, 2529–2533. [CrossRef]

62. Molinari, A.; Maldotti, A.; Amadelli, R. Probing the role of surface energetics of electrons and their accumulation in photoreduction processes on TiO$_2$. *Chem. Eur. J.* **2014**, *20*, 7759–7765. [CrossRef] [PubMed]

63. Kohtani, S.; Yoshioka, E.; Saito, K.; Kudo, A.; Miyabe, H. Adsorptive and kinetic properties on photocatalytic hydrogenation of aromatic ketones upon UV irradiated polycrystalline titanium dioxide: Differences between acetophenone and its trifluoromethylated derivative. *J. Phys. Chem. C* **2012**, *116*, 17705–17713. [CrossRef]

64. Lemon, B.; Hupp, J. Photochemical quartz crystal microbalance study of the nanocrystalline titanium dioxide semiconductor electrode/water interface: Simultaneous photoaccumulation of electrons and protons. *J. Phys. Chem.* **1996**, *100*, 14578–14580. [CrossRef]

65. Jimenez, J.; Jiménez, J.; Bourret, G.; Berger, T.; McKenna, K. Modification of charge trapping at particle/particle interfaces by electrochemical hydrogen doping of nanocrystalline TiO$_2$. *J. Am. Chem. Soc.* **2016**, *138*, 15956–15964. [CrossRef] [PubMed]

66. Peiró, A.; Colombo, C.; Doyle, G.; Nelson, J.; Mills, A.; Durrant, J. Photochemical reduction of oxygen adsorbed to nanocrystalline TiO$_2$ films: A transient absorption and oxygen scavenging study of different TiO$_2$ preparations. *J. Phys. Chem. B* **2006**, *110*, 23255–23263. [CrossRef] [PubMed]

67. Dung, D.; Ramsden, J.; Grätzel, M. Dynamics of interfacial electron-transfer processes in colloidal semiconductor systems. *J. Am. Chem. Soc.* **1982**, *104*, 2977–2985. [CrossRef]

68. Ilan, Y.; Meisel, D.; Czapski, G. The redox potential of the O$_2$/O$_2^-$ system in aqueous media. *Isr. J. Chem.* **1974**, *12*, 891–895. [CrossRef]

69. Hirakawa, H.; Hashimoto, M.; Shiraishi, Y.; Hirai, T. Photocatalytic conversion of nitrogen to ammonia with water on surface oxygen vacancies of titanium dioxide. *J. Am. Chem. Soc.* **2017**, *139*, 10929–10936. [CrossRef] [PubMed]

70. Hirakawa, H.; Hashimoto, M.; Shiraishi, Y.; Hirai, T. Selective nitrate-to-ammonia transformation on surface defects of titanium dioxide photocatalysts. *ACS Catal.* **2017**, *7*, 3713–3720. [CrossRef]

71. Burt, T.P.; Howden, N.J.K.; Worrall, F.; Whelan, M.J. Long-term monitoring of river water nitrate: How much data do we need? *J. Environ. Monit.* **2010**, *12*, 71–79. [CrossRef] [PubMed]

72. Kohtani, S.; Kamoi, Y.; Yoshioka, E.; Miyabe, H. Kinetic study on photocatalytic hydrogenation of acetophenone derivatives on titanium dioxide. *Catal. Sci. Technol.* **2014**, *4*, 1084–1091. [CrossRef]

73. Kohtani, S.; Yoshioka, E.; Saito, K.; Kudo, A.; Miyabe, H. Photocatalytic hydrogenation of acetophenone derivatives and diaryl ketones on polycrystalline titanium dioxide. *Catal. Commun.* **2010**, *11*, 1049–1053. [CrossRef]

74. Lewis, N. Progress in understanding electron-transfer reactions at semiconductor/liquid interfaces. *J. Phys. Chem. B* **1998**, *102*, 4843–4855. [CrossRef]

75. Hamann, T.; Gstrein, F.; Brunschwig, B.; Lewis, N. Measurement of the free-energy dependence of interfacial charge-transfer rate constants using ZnO/H$_2$O semiconductor/liquid contacts. *J. Am. Chem. Soc.* **2005**, *127*, 7815–7824. [CrossRef] [PubMed]

76. Ondersma, J.; Hamann, T. Measurements and modeling of recombination from nanoparticle TiO$_2$ electrodes. *J. Am. Chem. Soc.* **2011**, *133*, 8264–8271. [CrossRef] [PubMed]

77. Kohtani, S.; Nishioka, S.; Yoshioka, E.; Miyabe, H. Dye-sensitized photo-hydrogenation of aromatic ketones on titanium dioxide under visible light irradiation. *Catal. Commun.* **2014**, *43*, 61–65. [CrossRef]

78. Kohtani, S.; Mori, M.; Yoshioka, E.; Miyabe, H. Photohydrogenation of acetophenone using coumarin dye-sensitized titanium dioxide under visible light irradiation. *Catalysts* **2015**, *5*, 1417–1424. [CrossRef]

79. Beier, P.; Alexandrova, A.; Zibinsky, M.; Surya Prakash, G.K. Nucleophilic difluoromethylation and difluoromethylenation of aldehydes and ketones using diethyl difluoromethylphosphonate. *Tetrahedron* **2008**, *64*, 10977–10985. [CrossRef] [PubMed]

80. Kohtani, S.; Kurokawa, T.; Yoshioka, E.; Miyabe, H. Photoreductive transformation of fluorinated acetophenone derivatives on titanium dioxide: Defluorination vs. reduction of carbonyl group. *Appl. Catal. A Genel.* **2016**, *521*, 68–74. [CrossRef]

81. Shiraishi, Y.; Togawa, Y.; Tsukamoto, D.; Tanaka, S.; Hirai, T. Highly efficient and selective hydrogenation of nitroaromatics on photoactivated rutile titanium dioxide. *ACS Catal.* **2012**, *2*, 2475–2481. [CrossRef]
82. Shiraishi, Y.; Hirakawa, H.; Togawa, Y.; Sugano, Y.; Ichikawa, S.; Hirai, T. Rutile crystallites isolated from Degussa (Evonik) P25 TiO₂: Highly efficient photocatalyst for chemoselective hydrogenation of nitroaromatics. *ACS Catal.* **2013**, *3*, 2318–2326. [CrossRef]
83. Ohtani, B. Titania photocatalysis beyond recombination: A critical review. *Catalysts* **2013**, *3*, 942–953. [CrossRef]
84. Yang, H.; Sun, C.; Qiao, S.; Zou, J.; Liu, G.; Smith, S.; Cheng, H.; Lu, G. Anatase TiO₂ single crystals with a large percentage of reactive facets. *Nature* **2008**, *453*, 638–641. [CrossRef] [PubMed]
85. Amano, F.; Yasumoto, T.; Prieto Mahaney, O.-O.; Uchida, S.; Shibayama, T.; Ohtani, B. Photocatalytic activity of octahedral single-crystalline mesoparticles of anatase titanium(IV) oxide. *Chem. Commun.* **2009**, 2311–2313. [CrossRef] [PubMed]
86. Liu, G.; Yang, H.; Pan, J.; Yang, Y.; Lu, G.Q.; Cheng, H.-M. Titanium dioxide crystals with tailored facets. *Chem. Rev.* **2014**, *114*, 9559–9612. [CrossRef] [PubMed]
87. Liu, J.; Olds, D.; Peng, R.; Yu, L.; Foo, G.; Qian, S.; Keum, J.; Guiton, B.; Wu, Z.; Page, K. Quantitative analysis of the morphology of {101} and {001} faceted anatase TiO₂ nanocrystals and its implication on photocatalytic activity. *Chem. Mater.* **2017**, *29*, 5591–5604. [CrossRef]
88. Roy, N.; Sohn, Y.; Pradhan, D. Synergy of low-energy {101} and high-energy {001} TiO₂ crystal facets for enhanced photocatalysis. *ACS Nano* **2013**, *7*, 2532–2540. [CrossRef] [PubMed]
89. Li, C.; Koenigsmann, C.; Ding, W.; Rudshteyn, B.; Yang, K.; Regan, K.; Konezny, S.; Batista, V.; Brudvig, G.; Schmuttenmaer, C.; Kim, J.-H. Facet-dependent photoelectrochemical performance of TiO₂ nanostructures: An experimental and computational study. *J. Am. Chem. Soc.* **2015**, *137*, 1520–1529. [CrossRef] [PubMed]
90. Zhou, P.; Zhang, H.; Ji, H.; Ma, W.; Chen, C.; Zhao, J. Modulating the photocatalytic redox preferences between anatase TiO₂ {001} and {101} surfaces. *Chem. Commun.* **2017**, *53*, 787–790. [CrossRef] [PubMed]
91. Chamtouri, M.; Kenens, B.; Aubert, R.; Lu, G.; Inose, T.; Fujita, Y.; Masuhara, A.; Hofkens, J.; Uji-i, H. Facet-dependent diol-induced density of states of anatase TiO₂ crystal surface. *ACS Omega* **2017**, *2*, 4032–4038. [CrossRef] [PubMed]

© 2017 by the authors. Licensee MDPI, Basel, Switzerland. This article is an open access article distributed under the terms and conditions of the Creative Commons Attribution (CC BY) license (http://creativecommons.org/licenses/by/4.0/).

catalysts

MDPI

Review

Brookite: Nothing New under the Sun?

Matteo Monai *, Tiziano Montini and Paolo Fornasiero

Department of Chemical and Pharmaceutical Sciences ICCOM-CNR Trieste URT, Consortium INSTM Trieste Research Unit, University of Trieste, via L. Giorgieri 1, 34127 Trieste, Italy; tmontini@units.it (T.M.); pfornasiero@units.it (P.F.)
* Correspondence: matteo.monai@phd.units.it; Tel.: +39-040-5583973

Received: 2 October 2017; Accepted: 11 October 2017; Published: 13 October 2017

Abstract: Advances in the synthesis of pure brookite and brookite-based TiO_2 materials have opened the way to fundamental and applicative studies of the once least known TiO_2 polymorph. Brookite is now recognized as an active phase, in some cases showing enhanced performance with respect to anatase, rutile or their mixture. The peculiar structure of brookite determines its distinct electronic properties, such as band gap, charge–carrier lifetime and mobility, trapping sites, surface energetics, surface atom arrangements and adsorption sites. Understanding the relationship between these properties and the photocatalytic performances of brookite compared to other TiO_2 polymorphs is still a formidable challenge, because of the interplay of many factors contributing to the observed efficiency of a given photocatalyst. Here, the most recent advances in brookite TiO_2 material synthesis and applications are summarized, focusing on structure/activity relation studies of phase and morphology-controlled materials. Many questions remain unanswered regarding brookite, but one answer is clear: Is it still worth studying such a hard-to-synthesize, elusive TiO_2 polymorph? Yes.

Keywords: TiO_2; brookite; polymorph; polymorphism; photocatalysis

---

1. Introduction

Polymorphism—the ability of a solid material to exist in more than one crystal structure—is a common property of metal oxides [1]. Different polymorphs have distinct chemico-physical properties, such as electronic and optical properties, magnetism [2], ion conductivity [3], photo/electro-chromism [4], and surface energy and atom arrangement, which in turn affect the material performance in various applications, such as sensing [5] and photocatalysis [6,7]. Rationalizing the effect of crystal structure on the performance of a photocatalytic material is a pivotal research goal, since both Fe_2O_3 and TiO_2, the most investigated and widely used photocatalysts, may exist in different crystal structures [7,8]. Nonetheless, studying such a relation is not a trivial task, because of the many interconnected factors which influence the rate of photochemical reactions (e.g., particle size, shape, crystallinity, number/type of defects, electron–hole recombination, surface area, reactant and intermediate species adsorption on the catalyst surface) and the difficulty of synthesizing materials with a single crystal structure. The latest advances in nanotechnology are shedding new light on the field of photocatalysis, helping to understand the role of crystal structure and morphology in photocatalytic processes.

There are four commonly known polymorphs of TiO_2: anatase (tetragonal), rutile (tetragonal, the most thermodynamically stable phase), brookite (orthorhombic) and TiO_2 (B) (monoclinic) [9]. The structural parameters of these polymorphs are listed in Table 1. Such different lattice structures cause different mass densities and electronic band structures in TiO_2 polymorphs. The exact position of the conduction and valance band edges of TiO_2 polymorphs is still debated, mostly because it is dependent on the purity of the material, its crystallinity and its particle dimensions. While a consensus has been reached for an approximate band gap value, E_g, of anatase (3.2 eV) and rutile (3.02 eV), the reported experimental E_g values of brookite range from 3.1 to 3.4 eV: both smaller and larger than that of anatase [10]. Even more uncertain is the value of E_g of monoclinic TiO$_2$ (B), reported in the range of 3.1–4.1 [11–13]. Theoretical approaches (e.g., hybrid functionals, GW methods, the Bethe-Salpeter equation (BSE) method) are also not able to accurately estimate both the fundamental and optical band gap of these materials [10,13,14]. Nonetheless, determining the band gap, the flatband potential and the energy level of trap states, defects and impurities is crucial for the design of materials with improved photocatalytic properties, so future theoretical and experimental detailed studies are needed in order to solve these issues [15].

Table 1. Structural parameters of the main TiO$_2$ polymorphs. Date reproduce from reference [9], with permission of the American Chemical Society, 2017.

Polymorph	Crystal System	Space Group	Unit Cell Parameters			
			a/nm	*b*/nm	*c*/nm	β/degree
anatase	tetragonal	$I4_1/amd$	0.379	-	0.951	-
rutile	tetragonal	$P4_2/mnm$	0.459	-	0.296	-
brookite	orthorhombic	$Pbca$	0.918	0.545	0.515	-
TiO$_2$ (B)	monoclinic	$C2/m$	1.216	0.374	0.651	107.3

Once considered inactive, brookite is now gaining more and more attention in the field of photocatalysis due to its peculiar performance. A very comprehensive review of brookite was published some years ago by Di Paola et al. [10]. while some detailed reviews on TiO$_2$ for photocatalytic applications can be found in the literature, focusing on electronic properties [16], exposure of tailored facets [17] and nanostructural control [18]. The present review is focused on the advancement in brookite synthesis for photocatalytic applications, a research field in rapid expansion. First, the most advanced synthetic methodologies to produce pure brookite and well-defined brookite-containing composites are presented, together with some guidelines for thorough characterization of the materials. Then, some applications of brookite as a photocatalyst are outlined in comparison with the performance of other TiO$_2$ polymorphs. Finally, structure/activity relations are summarized and a perspective on the future development of brookite nanostructured materials is given.

2. Characterization

TiO$_2$ polymorphs are usually identified by X-ray diffraction (XRD) or Raman spectroscopy. The presence of brookite in a TiO$_2$ sample can be evidenced by the presence of the characteristic (121) peak at 2θ = 30.8° in the XRD pattern (Figure 1a). While rutile is also straightforward to identify from XRD patterns, thanks to the (110) peak at 27.43°, the detection of anatase in an apparently pure brookite sample is complicated by the overlapping of the (101) main diffraction peak of anatase at 25.28° with the (120) and (111) peaks of brookite, at 25.34° and 25.28°, respectively (Figure 1a). In order to quantitatively assess the phase composition of a TiO$_2$ sample, the whole XRD pattern should be fitted by a structure refinement method, such as the Rietveld method [19]. The method is very accurate as it allows us to take into account the broadening of the XRD peaks at higher Bragg diffraction angles and the possible preferred crystal orientations in plate- or rod-like crystallites, for which the reflex intensities vary from that predicted for a completely random distribution. In a study by some of the present authors on the synthesis of TiO$_2$ nanocrystals with precise and tunable exposed facets, the Rietveld method was used in conjunction with simulations and electron microscopy results to determine the phase and the average nanocrystal dimensions [20].

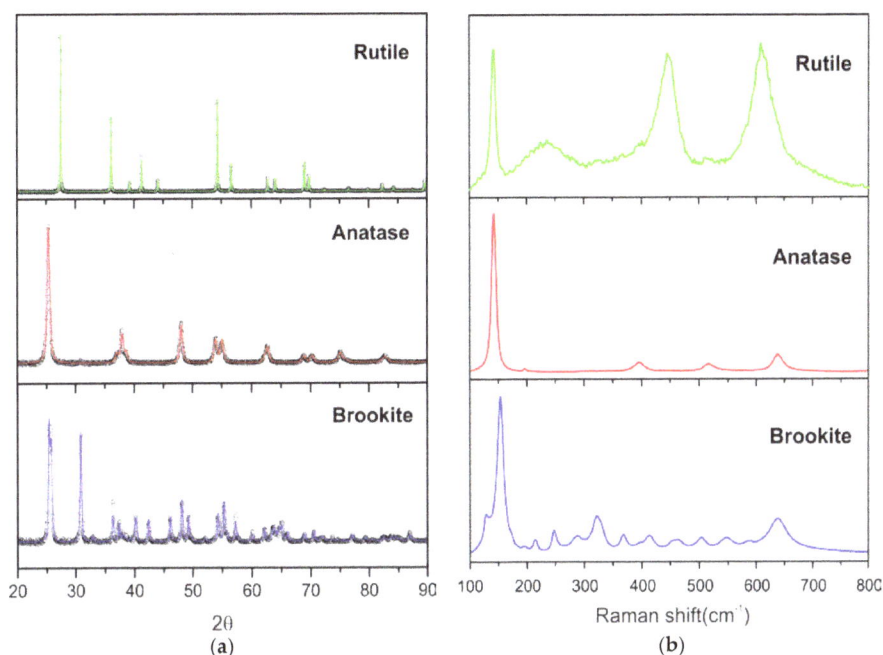

Figure 1. (a) XRD patterns (colored dots) and Rietveld analysis (black lines) of pure rutile, anatase and brookite (reproduced from [21] with the permission of the Elsevier, 2017) and (b) corresponding Raman spectra.

Raman spectroscopy can be used in combination with XRD in order to gain more insight into the phase composition of TiO_2 samples. Raman is very sensitive to minor phase impurities which might go undetected by XRD, and allows spatial mapping in the 10×10 micrometer resolution in confocal microscope setups. Since the vibrational spectrum of brookite presents more bands than the other two polymorphs, small amounts of brookite can be detected (Figure 1b) [22]. The more symmetric anatase and rutile structure gives rise to much simpler Raman spectra, with characteristic features evidenced in Figure 1b [23]. Insights into the surface/bulk distribution of phases in a TiO_2 sample can be gained by a combination of common visible/IR Raman with UV Raman spectroscopy, which is more surface-sensitive due to the adsorption of UV light by TiO_2 [24,25]. This approach was used to demonstrate that, in the anatase → rutile transition, rutile forms in the bulk, while anatase can persist on the surface of the sample to higher temperatures than previously thought. Applying this method to the study of brookite-based samples could lead to a deeper understanding of brookite formation and phase transitions.

Transmission electron microscopy (TEM) can provide insights in the phase composition of small areas of the material (hundreds of nanometers in size) by electron diffraction over a selected area electron diffraction pattern (SAED) [26]. In this regard, it can be used in combination with XRD, which gives bulk-sensitive results. Moreover, high-resolution TEM (HRTEM) can reveal the presence of amorphous material, which can be also quantified quite precisely by electron energy loss spectroscopy (EELS), provided that a crystalline and amorphous standard are available [27]. HRTEM can be used to study preferential exposure of facets, crystal morphology, grain boundaries and presence of defects, which can all influence the photocatalytic activity of a material and are therefore very important in structure/activity relation studies. Nonetheless, in some cases it is hard to distinguish between TiO_2 phases, making a definitive assignment difficult if nanocrystals are not suitably aligned with respect to the electron beam [28]. Moreover, the presence of defects and lattice deformation can lead to slight changes of diffractograms, so intensive characterization is needed in this kind of HRTEM study [26].

Quite interestingly, X-ray absorption spectroscopy (XAS) techniques have seldom been used to study TiO_2 polymorphism [29–31], despite the crystal structure sensitivity of such methods. A reason for this is that XAS techniques are not for routine characterization, since they require access to synchrotron facilities. However, Ti L_3-edge and O K-edge XAS can be used to study the local Ti and O symmetry and to investigate the crystal structure in great detail. For instance, the pre-edge region of the Ti L_3-edge allows accurate quantification of the degree of crystalline nature of a sample, in a more sensitive way than is performed by XRD [30]. Finally, in-situ studies would give new insights into phase transition processes and in structure–activity relations.

Other characterization techniques, not sensitive to phase composition, are nonetheless essential in order to fully characterize and study TiO_2 materials, especially for their use in photocatalysis and to study structure/activity relations. A comprehensive discussion of all such techniques is out of the scope of this paper, but recent reviews on the topic can be found in the literature [32,33].

3. Synthesis

The first synthesis of pure brookite dates back to the 1950s [34]. Since then, a lot of progress has been made in controlling the phase composition of TiO_2 materials, and now a number of synthetic methods exist to produce pure and mixed-phase brookite, rutile and anatase. The effect of the preparation parameters of TiO_2 materials on the final phase composition has also been widely investigated and was recently reviewed by Kumar and Rao [35]. However, the development of new phase-specific synthetic routes remains a trial-and-error process, due to the many thermodynamic and kinetic factors influencing the final phase composition. Indeed, the thermodynamic stability of polymorphs in nanocrystal systems is strongly affected by surface energy contributions, which in turn depend on the nanoparticles size, shape, exposed facets and surface adsorbates, as recently reviewed [36]. Kinetic processes such as reactant diffusion and ripening can be even more influential in particle formation and growth regimes, which usually occur far from equilibrium. The fact that brookite synthesis is more challenging than that of anatase and rutile is due to the narrower energy stability window of brookite compared to other polymorphs.

Moreover, brookite, anatase and monoclinic TiO_2 (B) are all metastable phases, and are transformed to the most thermodynamically stable polymorph—rutile—by annealing at high temperatures. Therefore, post-synthetic temperature treatments, usually employed to improve the degree of crystallization, to reduce the sample or to remove organic precursors, should be carefully controlled in order to avoid the phase transformation of metastable polymorphs. The process of phase transformation has been widely investigated, especially on anatase, but its mechanism is not yet consolidated, to the point that all these transitions were observed: anatase to brookite to rutile [37], brookite to anatase to rutile [38], anatase to rutile [39,40], and brookite to rutile [41]. This is due to the fact that the phase transformation process depends on many factors, such as starting material, temperature, particle size, surface area [37], surface defect sites and presence of adsorbates [42]. The transformation sequence was rationalized in terms of surface enthalpy crossover upon coarsening [37], but further studies are needed in order to shed some light on such a complex process.

Hydrothermal methods are the most commonly employed to synthesize brookite. Briefly, a TiO_2 precursor is mixed with water or organic solvents, an acid or base and additives such as capping/chelating agents or salts, and heated in an autoclave to moderately high temperatures and pressures. Typical Ti precursors are chlorides ($TiCl_3$, $TiCl_4$), $Ti(SO_4)_2$, alkoxides (e.g., isopropoxide, butoxide), alkaline titanates and amorphous titania. In general, there is no one-fit-all strategy in order to obtain brookite; acidic or basic conditions can be employed to obtain pure brookite, depending on the set of conditions, and additives which are necessary in one case can be avoided by changing other synthetic parameters. Nonetheless, varying the reaction parameters in a systematic way, different polymorphs can be obtained [43,44].

Adapting the procedure reported by Zhao et al. [45], some of us showed that pure brookite, anatase and rutile with a well-defined shape and dimension can be obtained using titanium (IV) bis(ammonium lactate) dihydroxide, $Ti(NH_4C_3H_4O_3)_2(OH)_2$, and varying the concentration of urea in aqueous solution (0 to 7.0 M) at 160 °C for 24 h (Figure 2). Pure brookite nanorod formation was favored by higher urea concentrations, corresponding to higher pH values due to urea decomposition, in accordance with the literature [45,46]. Interestingly, the use of comparable urea concentrations to those reported by Zhao et al. resulted in slightly different phase compositions of the final material, probably because of the influence of setup parameters such as autoclave free volume and heating/cooling method [21].

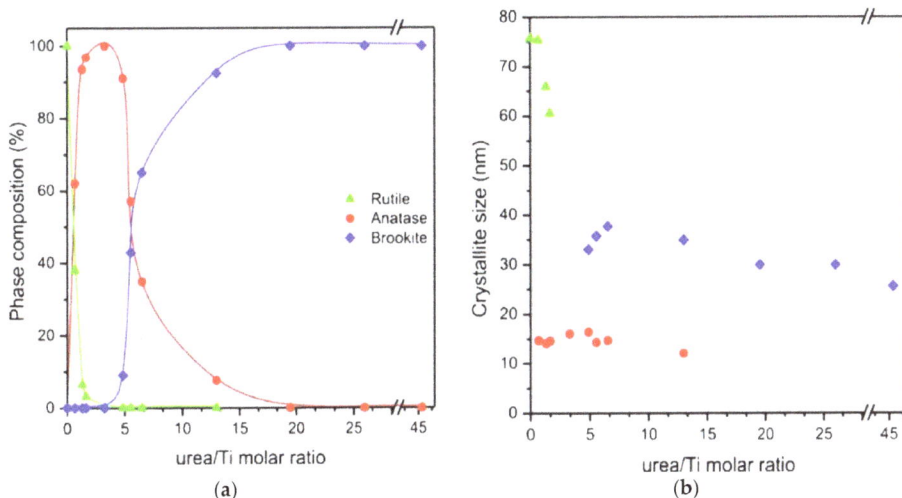

Figure 2. (a) Evolution of phase composition and (b) mean crystallite sizes as a function of the initial urea concentration. Reproduced from reference [21] with the permission of the Elsevier, 2017.

The drawbacks of hydrothermal synthesis are high energy consumption, low throughput and of difficult scalability due to the high pressure and temperature conditions of the sealed synthesis vessel, which also makes in-situ monitoring of the synthesis a challenging task. However, to the best of our knowledge, pure brookite has not been synthesized so far by other wet-chemistry methods, such as sol-gel, ionic liquids assisted synthesis, ultrasound and microwave-assisted synthesis and microemulsion [35,47]. The reason for this is most probably related to the aforementioned narrower stability window of brookite compared to other polymorphs, but this aspect surely deserves more investigation in the future.

Mechanochemical activation methods, such as ball milling (BM), are emerging as more sustainable synthetic methods because of the absence of bulk solvents and the possibility to use metal oxides or hydroxides as starting materials [35,48]. While pure brookite synthesis by BM has not been reported so far, future studies could clarify the role of the reaction parameters on the phase composition of the final material. Moreover, brookite could be used as a starting material in BM synthesis in order to study its stability (and therefore the feasibility of the synthetic method) and the phases formed under different conditions.

Modification of TiO_2 by doping (H, N, C, metal ions), co-doping and self-doping (Ti^{3+}, black titania [49]) are general strategies to enhance TiO_2 light harvesting and photocatalytic properties. Doped brookite synthesis by the introduction of dopants or their precursors in brookite hydrothermal synthesis was recently reported [10]. As suggested by a computational study of TiO_2 polymorphs (H,N)-co-doping, the effect of doping on the electric properties of anatase, rutile and brookite can be different. For instance, (H,N)-co-doping reduced the band gap of anatase and brookite, but had little effect on rutile [50]. Combined experimental and theoretical results on this topic could greatly extend our knowledge of TiO_2, and some efforts have been recently spent in this sense [51].

Self-doping of TiO_2 by Ti^{3+} can be accomplished by various reduction methods, using plasma or thermal, chemical or electrochemical activation [49]. With the intent of avoiding harsh, harmful and expensive methods, recent studies focused on the development of yet other synthetic routes [52,53]. For instance, pure black brookite synthesis was recently reported, in which melted Al was used as a reducing agent in a two-zone vacuum furnace, instead of employing more dangerous H_2 [52]. In this case, the Al reduction reportedly starts from the brookite surface and causes lattice disorders and oxygen vacancies in the surface layer. Xi et al. reported a hydrothermal synthesis of black brookite using post-annealing treatment in order to introduce large amount of Ti^{3+} defects in the bulk of the nanoparticles [53]. Interestingly, no rutile or anatase were detected after N_2 annealing treatment at elevated temperatures (up to 700 °C), suggesting that brookite can be stabilized by reduction. Systematic studies comparing different synthetic routes are greatly desired, since the synthetic method can affect the properties and performances of black TiO_2 materials, leading to the diverse results rationalization found in the literature [49]. Controlling the amount and spatial distribution of surface lattice disorder, oxygen vacancies, Ti^{3+} ions, Ti–OH and Ti–H groups present in black TiO_2 would also open the way to band-gap engineering in order to enhance photocatalytic performances in targeted reactions.

Finally, in order to scrupulously study structure/activity relations in TiO_2 materials, the synthesized samples should not only have controlled phase composition, but also well-defined crystallite dimensions and morphology. Indeed, different facets of TiO_2 crystals have distinct photocatalytic properties as a result of different atomic and electronic structure [17,54]. The nanostructure also plays a pivotal role in photocatalysis. For instance, the rate of electron–hole recombination in brookite nanorods can be controlled by engineering their length, as recently reported by some of us [28]. The synthesis of nanostructured TiO_2 has been widely investigated in the literature [18,20,55], so that a wide library of morphology-controlled materials can now be produced. Structure/activity studies on TiO_2 are thriving, as they can provide great insights into reaction mechanisms and give an indication of how to enhance the performance of state-of-the-art photocatalysts.

4. Photocatalytic Studies

The phase composition of TiO_2 nanomaterials is one of the factors strongly influencing photocatalytic reaction activity and selectivity [56]. This is mainly due to the phase-dependency of electronic structures (band gap, charge-carrier lifetime and mobility, exciton diffusion length, trapping sites) and of preferentially exposed surfaces with different energy, atom arrangements and adsorption sites. While anatase- and rutile-based nanomaterials have been widely investigated and their photocatalytic performances are well established [9,17], brookite has slowly gained the status of an efficient photocatalyst only in the last couple of decades, as recently reviewed [10]. Here, we report the latest advances in photocatalysis over brookite-based materials, focusing on comparative and in-depth studies, to give a critical view of the present challenges in structure/activity relation studies.

One of the reasons for the different photocatalytic performances of TiO_2 polymorphs is the trapping of photogenerated charge carriers (electrons and holes) in defect sites, resulting in slower electron-hole recombination kinetics and phase-dependent charge carrier stabilization. The density of such occupied mid-gap trap states (DOTS) can be studied by time-resolved vis-IR absorption spectroscopy (TRAS) [57,58]. In a recent experimental study, the depth of electron-traps in brookite was estimated to be 0.4 eV, which is deeper than that of anatase (<0.1 eV), but shallower than that of rutile (~0.9 eV) [57]. These results are in good accordance with the calculated stabilization energies for rutile (0.8–1 eV) and anatase (0–0.2 eV), in which oxygen vacancies or interstitial Ti atoms were considered as trap site defects [59,60]. Notably, to the best of our knowledge, no theoretical study has been reported on electron trapping at the defects of brookite, even if similar results to anatase and rutile are expected. Some of the authors recently reported a study on trapped carriers in different TiO_2 polymorphs in the ms–s timescales, typical of photocatalytic reactions [58]. Also in these slow timescales, anatase and brookite exhibit dispersive power law recombination dynamics, in accordance with shallow charge trapping, while rutile exhibits logarithmic decay kinetics, indicative of deeper charge trapping.

The stabilization of electrons in trap states has two opposite effects: on the one side, it reduces the reactivity of electrons, and on the other, it causes an increase in the lifetime of electrons and holes, because the probability of recombination is decreased. Therefore, an appropriate electron trap depth can help in maximizing the yield of long-lived, but still reactive, charge carriers. In the presence of traps of moderate depth, such as in brookite, electrons can take part in reduction reactions, while long-lived holes can participate in photocatalytic oxidation reactions, such as water oxidation (Figure 3). In anatase, the trap-depth is too shallow to extend the lifetime of holes, while in rutile it is too deep for electron-consuming reactions (Figure 3). Notably, deep electron trapping in rutile extends the lifetime of holes, which could promote water oxidation. However, the low reactivity of deeply trapped electrons causes the overall activity to be low [61].

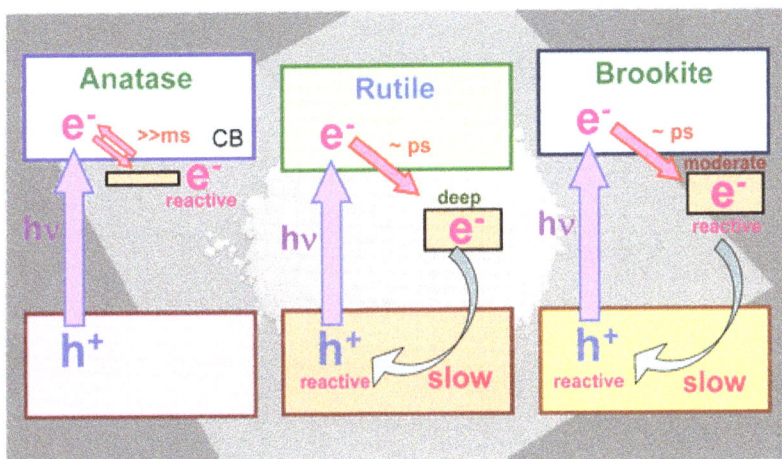

Figure 3. Schematic representation of electron trapping sites in anatase, rutile and brookite. CB: conduction band. Reproduced from reference [57] with permission of the American Chemical Society, 2017.

Particle size can also have a great influence on photocatalytic activity due to geometric and electronic effects. While a smaller particle size results in larger surface areas (which means more active sites per volume) and shorter mean distances for carriers to migrate to the surface, it also brings about some drawbacks such as less volume to generate and separate charge carriers, an increase of band gap due to quantum confinement and incomplete band relaxation to the bulk level (which means a smaller potential drop and electric field in the space charge region, and hindered charge separation) [62]. Therefore, a compromise in particle size between the positive influence of having more reactive sites and the negative influence of increased recombination is wanted. Nonetheless, electron-hole separation can be improved by controlling the morphology of nanocrystals, while maintaining high specific surface areas.

As recently demonstrated by some of the authors by a combination of electron paramagnetic resonance (EPR) spectroscopy and in-situ TRAS [28], H_2 production rates in the photoreforming of ethanol, glucose and glycerol increases with the increasing length of Pt-decorated brookite nanorods due to the longer lifetime of charge-separated states (Figure 4). A broad photoinduced absorption feature in the visible range is obtained in TRAS when pumping above the brookite band gap, which was assigned to trapped holes lying energetically within the band gap and physically on or near the surface of the TiO$_2$ nanoparticles [63]. The TRAS normalized kinetics reported in Figure 4 show an ultrafast rise due to the trapping of photoexcited holes of the order of a few hundred fs, followed by a decay due to the recombination of holes with electrons or reaction with adsorbates. The increase in the lifetime of the trapped holes with nanorod length was attributed to the different surface energy of crystal faces exposed on the tips and the sides of the nanorods, which drives the electrons and holes to different crystal faces, leading to enhanced charge carrier separation in longer nanorods [64]. Studies of photocatalytic coupling of methanol to methyl formate and the photo-oxidation of acetaldehyde to acetate on brookite nanorods revealed similar length-dependent rate enhancement, corroborating the proposed model [65].

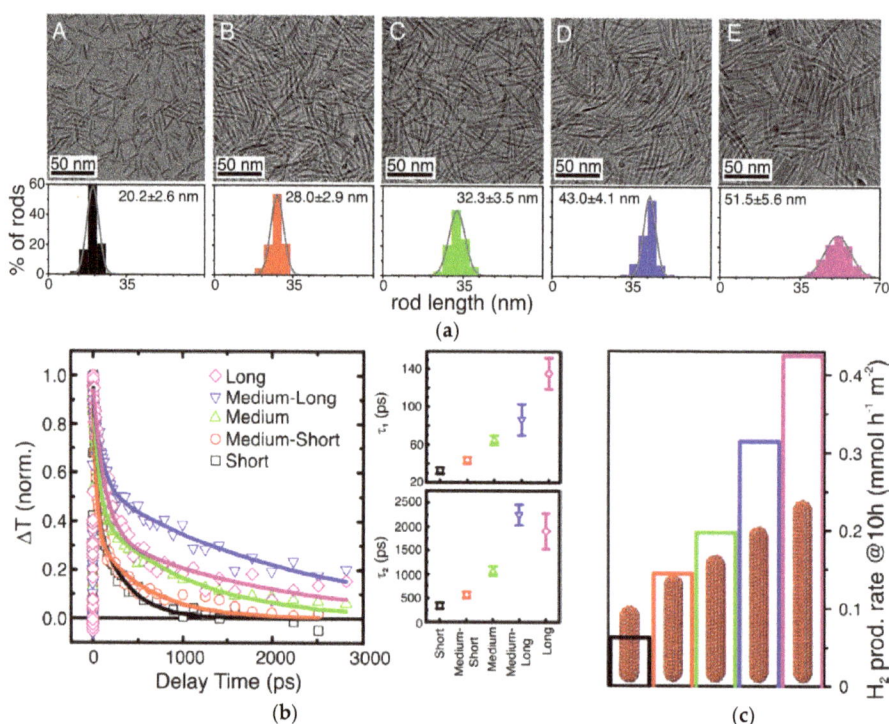

Figure 4. (a) TEM images and size analysis for the brookite nanorods with different lengths; (b) Normalized TRAS (time-resolved vis-IR absorption spectroscopy) kinetic traces of Pt-decorated nanorods in ethanol. Insets show fitted time constants (τ_1 (Top) and τ_2 (Bottom)); (c) H$_2$ production rates from ethanol photoreforming on 1 wt. % Pt−brookite nanorods of different length after 10 h under illumination, normalized by the surface area of the photocatalysts. Reproduced from reference [28] with the permission of the PNAS, 2017.

The EPR results further indicated that electrons in longer nanorods transfer to Pt faster than electrons in shorter nanorods. The capture of electrons by Pt increases the lifetime of the charge-separated state, as indicated by the much faster decays observed for nanorods without Pt in ethanol [28]. A similar effect was observed for Au surface decoration of brookite nanorods [66], which enabled electrons a lifetime of four orders of magnitude longer due to efficient hopping on brookite lateral facets. Conversely, when Au nanoparticles are introduced in the bulk of nanorods, they act as recombination centers for plasmonic carriers in the fs timescale.

From the above discussion, it is clear that the control over exposed crystal faces is an essential aspect in fundamental photocatalytic studies, as the surface energy, chemical surface state, number/type of defects and reactivity strongly depend on the atomic arrangement of the exposed surfaces [17,54,67,68]. In the last decade, the synthesis of TiO_2 anatase, rutile and brookite nanocrystals with controlled exposed faces has been reported by several group, employing specific precursors/conditions or suitable structure-directing agents. All polymorphs show pronounced structure/activity relations, with distinct crystal faces showing dramatically different reactivity and specifically promoting reduction or oxidation [64]. Ohno et al. reported clear evidence of such face-dependent reactivity by studying the photodeposition of Pt (by reduction of $PtCl_6^-$) and of PbO_2 (by oxidation of Pb^{2+} nitrate) on rutile, anatase [69], and brookite [68,70]. On brookite TiO_2 nanorods, reduction and oxidation reactions were observed to proceed predominantly on {210} and {212} exposed crystal faces, respectively [68,70]. In brookite nanosheets, {201} facets acted as oxidation sites, while {210} and {101} facets acted as reduction sites [64]. Face-specific reactivity was also employed to selectively modify the tips of brookite nanorods with Fe^{3+} ions, to enhance the activity in acetaldehyde photocatalytic oxidation [68].

Theoretical studies also evidenced the importance of the energetics of crystal faces. The sequence of surface energies of ten stoichiometric 1×1 low-index surfaces of brookite was calculated by first-principle density functional theory (DFT) simulations, showing that the electronic and chemical properties of brookite and the other TiO_2 phases can be significantly different [67]. Recently, combined theoretical and experimental studies of brookite nanomaterials demonstrated that preferential exposure of {121} faces, with under-coordinated atoms and lower VB potential, led to higher performance in photodegradation, while preferential exposure of {211} surface, with higher CB potential, resulted in enhanced H_2 productivity [71,72]. Such calculations also suggested that electrons struggle to migrate from bulk to {121} faces, in accordance for their poor H_2 production efficiency.

While pure phase TiO_2 materials are extensively investigated for fundamental studies, mixed phase TiO_2 materials have been shown to provide enhanced photocatalytic performances, due to charge transfer across the interface of different phases [73]. For instance, the widely investigated TiO_2 Degussa P25, used as a standard in the majority of TiO_2-related studies, is a mixture of anatase and rutile particles (anatase/rutile ratio ~70:30) in which a synergistic effect of the two phases is ascribed to the spatial separation of photogenerated charge carriers [74]. Despite the general consensus in such mechanisms, there is disagreement about the direction of charge transfer, which depends on the relative positions of the conduction bands and trap states in the two polymorphs. Electron transfer from anatase to rutile was proposed by Kawahara et al., based on results of the photodeposition of Ag on a patterned bilayer-type TiO_2 photocatalyst consisting of anatase and rutile phases [75]. SEM images clearly showed silver particles mostly deposited on the anatase surface, except at the interfacial region, in which silver deposition occurred preferentially at the rutile layer boundary, suggesting electron transfer from anatase to rutile. On the other hand, the opposite mechanism was proposed by other studies basing on EPR spectroscopy results, in which photogenerated electrons migrated from rutile to anatase trapping sites [74,76].

Recently, a similar enhancement of photocatalytic activity in the photodegradation of organic molecules and CO_2 photoreduction has been also observed for brookite/anatase [45,77,78] and brookite/rutile mixtures [79,80]. Moreover, anatase-rich/brookite mixture (75:25) was shown to be much more active for CO_2 photoreduction than Degussa P25, an anatase-rich/rutile mixture with similar anatase fraction [45]. These results suggest that electron transfer may take pace also in brookite-based TiO_2 composites, which may be a new direction for the development of efficient photocatalysts. However, to the best of our knowledge, no direct evidence for electron transfer, e.g., by EPR, has been reported so far. DFT calculations on pairs of rutile, anatase and brookite TiO_2 slabs showed that in most cases highest occupied molecular orbital (HOMO) and lowest unoccupied molecular orbital (LUMO) states were separated in the two phases, indicating that charge transfer may take place upon photoexcitation [81]. Such separation was observed to be dependent on the native HOMO–LUMO states of the two components used to build the composites and on the lattice match between the surfaces of the two phases. Future experimental and theoretical investigations focused on the structural and electronic properties of brookite TiO_2 composite interfaces are needed to gain further insights into efficient photocatalyst preparation. On the other hand, since the presence of interfaces can drastically improve the activity of TiO_2 materials, careful characterization is essential to rule out contamination in fundamentals studies aimed to assess the differences in the activity of polymorphs.

It should be noted that the relative activity of polymorphs is dependent on the considered photocatalytic reaction, so that a general activity trend among brookite, anatase and rutile (and their composites) does not exist [82,83]. For instance, two distinct activity trends were observed for anatase, brookite and their composites in MeOH photoreforming and in dichloroacetic acid (DCA) degradation [82]. This was explained by two different rate-determining factors: H_2 production was favored by the cathodically-shifted flatband potentials of brookite materials with respect to anatase, while DCA degradation rates were highest for anatase, and observed to increase with surface area (Figure 5a). Organic degradation studies can be further complicated by a non-trivial effect of calcination temperature and the sorption capacity of ions and of dissolved gases (e.g., O_2) in water on the relative activity of brookite and anatase [83]. Normalizing the photocatalytic activity by a certain value may help in rationalizing the results, as observed by normalization to initial absorbed Ag(I) or Cr(VI) amount [83] and by normalization to the surface area in anatase/brookite materials used in photocatalytic reforming of ethanol and glycerol [21]. In the latter case, the activity normalized to the surface area continuously increased with the brookite content, indicating that the exposed facets of brookite nanorods possess an intrinsic higher activity in H_2 production than that of the other polymorphs (Figure 5b,c) [21].

Some insights in the factors determining the high intrinsic activity of brookite surfaces come from DFT calculations on the structure and adsorption properties of brookite (210) and anatase (101) surfaces, the most commonly exposed facets in nanocrystals [84,85]. The two surfaces are structurally similar, but the brookite (210) surface presents shorter interatomic distances and a different block arrangement. These features result in enhanced reactivity toward strong dissociative adsorption of H_2O and HCOOH, and generate highly active junction sites, all factors entailing higher specific activity of brookite (210) surfaces [84]. On the other hand, charge analysis calculations revealed that $CO_2^{\bullet-}$ radical anion (the first intermediate in CO_2 reduction) formed on the negatively charged anatase (101) surface, but not on the negatively charged brookite (210) surface, indicating that brookite is not a suitable catalyst for CO_2 reduction [85]. Nonetheless, the study confirmed the previous results on the favorable adsorption energetics of brookite (210), and revealed that the presence of oxygen vacancies on the brookite (210) surface enhanced the charge transfer to the CO_2 molecule, promoting $CO_2^{\bullet-}$ radical formation and CO_2 reduction.

These results were supported by in-situ diffuse reflectance infrared Fourier transform spectroscopy (DRIFTS) and CO_2 photoreduction photocatalytic experiments on defect-free and oxygen-deficient brookite, anatase and rutile TiO_2 nanocrystals [86]. Oxygen vacancies (V_O) and Ti^{3+} sites were created by helium pretreatment of the as-prepared TiO_2 at a moderate temperature on anatase and brookite, but not on rutile. The production of CO and CH_4 from CO_2 photoreduction was remarkably enhanced on defective anatase and brookite TiO_2 (up to 10-fold enhancement) as compared to the defect-free surfaces, thanks to enhanced light harvesting and CO_2 adsorption on defect sites. Defective brookite was the most active photocatalyst among the investigated TiO_2 polymorphs, which was tentatively explained by the favored formation of oxygen vacancies, faster reaction rate of CO_2^- with adsorbed H_2O or surface OH groups, and an additional reaction route involving an HCOOH intermediate. High CO_2 photoreduction rates were also observed for defective brookite nanosheets with Ti^{3+} self-doping, synthesized by an innovative hydrothermal method combined with post-annealing treatment, avoiding commonly employed harsh and costly physical methods [53]. Defective brookite nanomaterials are surely worthy of more extensive investigation, in terms of the development of new synthetic routes and of in-depth characterization (e.g., by EPR [87,88]) and photocatalytic studies.

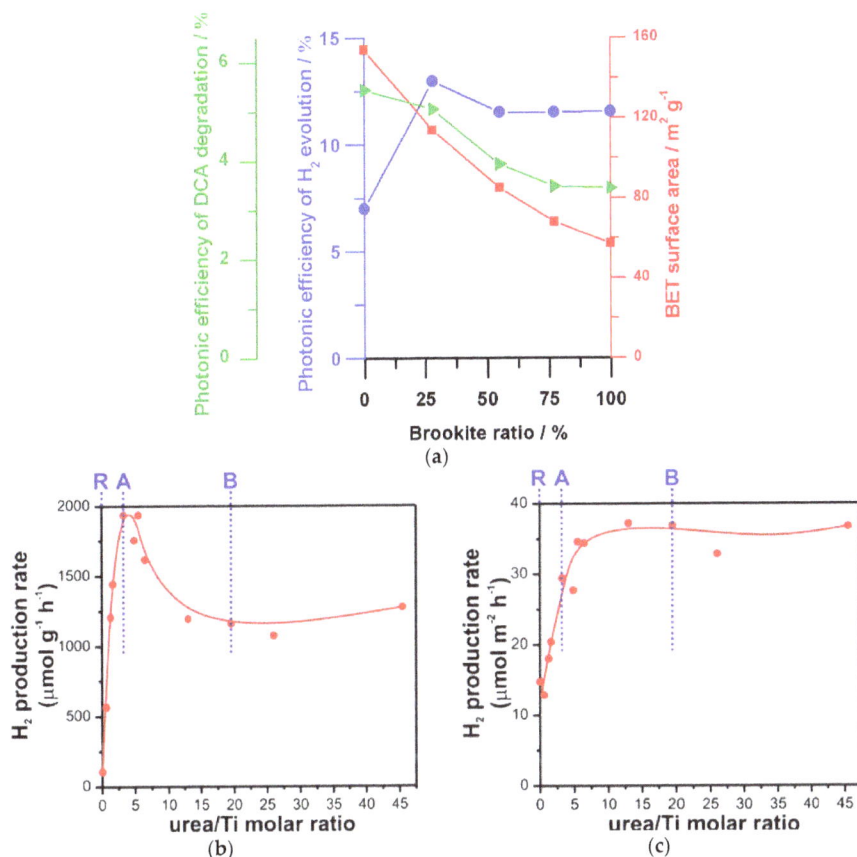

Figure 5. (a) Effect of brookite ratio on the photonic efficiency of TiO_2 composites in dicholoracetic acid (DCA) degradation (green triangles) and H_2 production (blue circles). BET surface area (red squares) is reported for comparison with the efficiency trends. Reprinted with permission of reference [82]; (b) H_2 production rate by photocatalytic ethanol dehydrogenation over Pt(0.2 wt. %)/TiO_2 prepared using different urea/Ti molar ratios, normalized with respect to the mass of catalysts and (c) with respect the surface area. Markers correspond to the urea/Ti molar ratios that allow the preparation of pure phase materials: rutile (R), anatase (A) and brookite (B). Reproduced from reference [21] with permission of the American Chemical Society, 2017.

Selectivity studies in photocatalysis have long been lacking, but are now gaining momentum due to the significant advantages presented by photocatalytic reactions as alternative, green routes in organic and inorganic chemistry [89]. The selectivity of photocatalytic systems can be enhanced by two main strategies: modification of the photocatalyst and optimization of reaction conditions (e.g., aeration, type of solvent, concentration and type of anions [90], use of membranes) [91]. Not surprisingly, the TiO_2 phase composition was demonstrated to have an effect on the selectivity of many photocatalytic reactions, such as ammonia oxidation [92], photoinduced decomposition of acetone, oxygenate photoreforming [6], and selective oxidation of alcohols to aldehydes [91,93]. For instance, the product distribution of ammonia oxidation was different in the case of rutile, yielding nitrates as a major product, and of anatase and brookite, yielding mild-oxidation products such as nitrites and N_2 [92]. In a study by some of the authors, brookite was observed to be less active in the complete mineralization of glycerol during photocatalytic reforming, leading to higher H_2/CO_2 and opening interesting opportunities in selective oxidation of sacrificial agents. On rutile, on the other hand, a selective dehydrogenation of glycerol through the secondary OH group was observed (although with very low activity) [6]. Furthermore, brookite nanorods were observed to selectively produce H_2O_2, a high value-added and green oxidant, during photoelectrochemical water splitting, also showing a very low onset potential for water oxidation ($E_{onset} \sim -0.2$ V vs. reversible hydrogen electrode (RHE)) [66]. These findings suggest that the effect of TiO_2 phase composition should be investigated more in detail in the highly novel field of selective photocatalysis.

Future theoretical and experimental studies on well-defined nanostructured TiO_2 materials will be essential to unravel the conundrum regarding the relative photocatalytic performances of different polymorphs and their mixtures, considering the discrepancies observed in the literature. We strongly advise the pursuit of rigorous studies of the electronic structure of TiO_2 nanomaterials and of their facet-specific properties, which requires the development of precise and tunable synthesis methods, yielding size and morphology-controlled materials. The task is arduous, but can be now undertaken using the tools of simulations, high-throughput synthesis and nanotechnology.

5. Conclusions and Perspectives

Brookite-based materials are receiving increasing attention in photocatalytic and related applications, due to their novelty and peculiar properties. This "brookite rush" is rapidly producing new evidence of the profound influence of phase composition on photocatalytic activity and selectivity, unraveling the mechanistic aspects and structure/activity relations in photocatalytic processes. Nonetheless, we are still fumbling around in the dark regarding the paramount electronic properties of brookite (e.g., its energy gap), so further theoretical and experimental studies are needed in order to reach a new level of understanding and a wide consensus in the scientific community on this topic. However, the solid research work previously published on anatase and rutile seems to have discouraged similar fundamental investigation of brookite. A prime example of this is the case of the postulated electron transfer in the brookite/anatase and brookite/rutile interfaces, which was not supported by direct evidence, since similar work was done for anatase/rutile composites [45,77–80]. While such electron transfer is reasonably expected, more caution should be taken in future studies of brookite.

An intrinsic burden to the development of a wide range of morphology-controlled brookite nanomaterials is posed by the limited understanding of solvothermal synthetic mechanisms. This is a common issue in the synthesis of nanomaterials, usually carried out by a trial-and-error approach. Nonetheless, nanostructured brookite (e.g., nanorods [28], nanosheets [64], and nanocubes [94]) has been successfully synthesized, and alternative strategies are being proposed to produce highly-oriented brookite and TiO_2 nanoparticles with the desirable specific external crystal facet (e.g., templating with graphene [95] or 2D oxides [96]). The playground is now open to new areas of investigation, such as the development of scalable and sustainable brookite synthesis, modification by doping [10,50–53], controlled introduction of defects, and synthesis of brookite nanocomposites [94,97] and hybrid organic-inorganic systems [98]. In particular, the effect of TiO_2 phase composition in nanocomposites with metal organic frameworks (MOFs), quantum dots and nanocarbons is still to be assessed. The formation of phase-dependent charged interfaces in such hierarchically-structured materials could be a promising strategy for the design of improved photocatalysts [62].

Investigation of new composites could lead to a great enhancement of photocatalytic performance, stimulating further advances in synthetic methods, characterization techniques and elucidation of catalytic mechanisms. For instance, a non-trivial influence of co-catalyst loading on CO_2 photoreduction activity and selectivity was recently reported for brookite nanocubes loaded with Ag nanoparticles [94]. A maximum yield of CO was observed for 0.5 wt. % Ag loading, which corresponded to the case of the preferential deposition of Ag nanoparticles on {210} faces, while for intermediate loadings (1 wt. %) of Ag, the CH_4 selectivity increased, due to an agglomeration of Ag on {210} faces and the deposition of small nanoparticles on {001} faces (Figure 6). These fascinating results were tentatively rationalized in terms of an interplay of charge transfers, adsorption properties and light harvesting effects, revealing how the synergistic effects in semiconductors/co-catalyst systems are much more complex than expected, and how they can be used to finely tune the performances of photocatalytic systems.

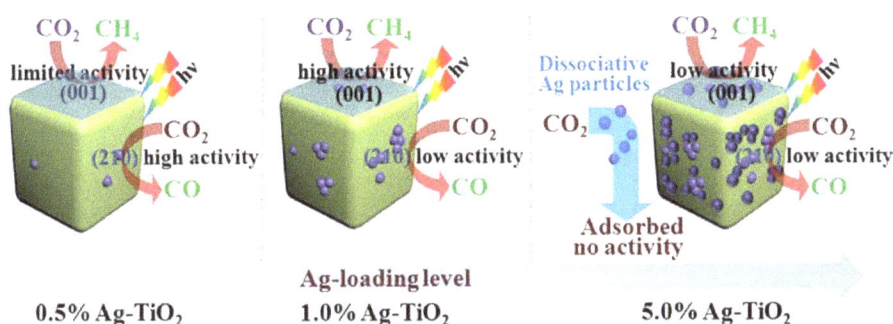

Figure 6. Graphical representation of the influence of Ag co-catalyst wt. % content on the CO_2 photoreduction activity of brookite nanocubes. Reproduced from reference [94] with permission of the Elsevier, 2017.

In-depth studies concerning different band gap and trap states, face reactivity, the effects of crystal morphology and dimension, the presence of vacancies, and the generation of adsorbates and radicals are of great interest, and recently great progress has been made in the study of these properties. Photo-induced heterogeneous electron transfer (ET) across the semiconductor and adsorbed molecules, leading to the formation of radicals (e.g., •OH) is of particular relevance for any photocatalytic reaction, but studying such highly reactive species is challenging [91,99]. An emerging approach is based on the use of organic dye probes in microscopic fluorescence imaging, for the sensitive detection of reactive oxygen species, their diffusion in solution or air, and the identification of photocatalytic active facets on semiconductor surfaces [100,101]. Recently, a mass spectrometry-based approach was used to investigate the ultrafast ET of photoelectrons generated by ultraviolet irradiation on the surfaces of semiconductor nanoparticles or crystalline facets, providing a new technique for studying the photo-electric properties of various materials [102]. Similar studies of brookite would lead to a deeper understanding of the reaction mechanism and provide valuable insights into strategies to enhance activity and selectivity in many photocatalytic reactions.

The phase-dependent activity and selectivity observed in photocatalytic reactions over brookite, rutile and anatase is probably the most exciting and yet still not well investigated aspect concerning TiO_2 polymorphs. Brookite nanomaterials were found to perform better than other polymorphs in various photocatalytic reactions (e.g., CO_2 photoreduction [86,103], photoreforming of oxygenates [21,82], ammonia oxidation [92]), and showed promising properties for future applications also in other fields, such as photoelectrochemical devices [66], dye-sensitized solar cells (DSSC) [104,105], bio-applications and self-cleaning materials [106,107]. The days in which brookite was considered just an undesirable byproduct are long gone, it is now time to dig deeper and find out even more about this and other less familiar TiO_2 polymorphs.

Acknowledgments: Matteo Monai, Tiziano Montini and Paolo Fornasiero acknowledge financial support from the University of Trieste through the FRA2015 project and ICCOM CNR.

Author Contributions: Matteo Monai, Tiziano Montini and Paolo Fornasiero wrote the paper.

Conflicts of Interest: The authors declare no conflict of interest. The founding sponsors had no role in the design of the study; in the collection, analyses, or interpretation of data; in the writing of the manuscript, and in the decision to publish the results.

References

1. Sood, S.; Gouma, P. Polymorphism in nanocrystalline binary metal oxides. *Nanomater. Energy* **2013**, *2*, 82–96. [CrossRef]
2. Machala, L.; Tuček, J.; Zbořil, R. Polymorphous Transformations of Nanometric Iron(III) Oxide: A Review. *Chem. Mater.* **2011**, *23*, 3255–3272. [CrossRef]
3. Laarif, A.; Theobald, F. The lone pair concept and the conductivity of bismuth oxides Bi_2O_3. *Solid State Ion.* **1986**, *21*, 183–193. [CrossRef]
4. Quintana, A.; Varea, A.; Guerrero, M.; Suriñach, S.; Baró, M.D.; Sort, J.; Pellicer, E. Structurally and mechanically tunable molybdenum oxide films and patterned submicrometer structures by electrodeposition. *Electrochim. Acta* **2015**, *173*, 705–714. [CrossRef]
5. Prasad, A.K.; Kubinski, D.J.; Gouma, P.I. Comparison of sol–gel and ion beam deposited MoO_3 thin film gas sensors for selective ammonia detection. *Sens. Actuators B Chem.* **2003**, *93*, 25–30. [CrossRef]
6. Beltram, A.; Romero-Ocaña, I.; Josè Delgado Jaen, J.; Montini, T.; Fornasiero, P. Photocatalytic valorization of ethanol and glycerol over TiO_2 polymorphs for sustainable hydrogen production. *Appl. Catal. A Gen.* **2016**, *518*, 167–175. [CrossRef]
7. Carraro, G.; Maccato, C.; Gasparotto, A.; Montini, T.; Turner, S.; Lebedev, O.I.; Gombac, V.; Adami, G.; Van Tendeloo, G.; Barreca, D.; et al. Enhanced Hydrogen Production by Photoreforming of Renewable Oxygenates Through Nanostructured Fe_2O_3 Polymorphs. *Adv. Funct. Mater.* **2014**, *24*, 372–378. [CrossRef]
8. Schneider, J.; Matsuoka, M.; Takeuchi, M.; Zhang, J.; Horiuchi, Y.; Anpo, M.; Bahnemann, D.W. Understanding TiO_2 Photocatalysis: Mechanisms and Materials. *Chem. Rev.* **2014**, *114*, 9919–9986. [CrossRef] [PubMed]
9. Ma, Y.; Wang, X.; Jia, Y.; Chen, X.; Han, H.; Li, C. Titanium Dioxide-Based Nanomaterials for Photocatalytic Fuel Generations. *Chem. Rev.* **2014**, *114*, 9987–10043. [CrossRef] [PubMed]
10. Di Paola, A.; Bellardita, M.; Palmisano, L. Brookite, the Least Known TiO_2 Photocatalyst. *Catalysts* **2013**, *3*, 36–73. [CrossRef]
11. Cai, J.; Wang, Y.; Zhu, Y.; Wu, M.; Zhang, H.; Li, X.; Jiang, Z.; Meng, M. In Situ Formation of Disorder-Engineered TiO_2(B)-Anatase Heterophase Junction for Enhanced Photocatalytic Hydrogen Evolution. *ACS Appl. Mater. Interfaces* **2015**, *7*, 24987–24992. [CrossRef] [PubMed]
12. Li, B.-H.; Lin, Y.; Wang, J.-L.; Zhang, X.; Wang, Y.-R.; Jiang, Y.; Li, T.-S.; Liu, L.-M.; Chen, L.; Zhang, W.-L.; et al. Formation of New Phases to Improve the Visible-Light Photocatalytic Activity of Tio_2 (B) via Introducing Alien Elements. *J. Phys. Chem. C* **2017**, *121*, 52–59. [CrossRef]
13. Buckeridge, J.; Butler, K.T.; Catlow, C.R.A.; Logsdail, A.J.; Scanlon, D.O.; Shevlin, S.A.; Woodley, S.M.; Sokol, A.A.; Walsh, A. Polymorph Engineering of TiO_2: Demonstrating How Absolute Reference Potentials Are Determined by Local Coordination. *Chem. Mater.* **2015**, *27*, 3844–3851. [CrossRef]
14. Samat, M.H.; Taib, M.F.M.; Hassan, O.H.; Yahya, M.Z.A.; Ali, A.M.M. Structural, electronic and optical properties of brookite phase titanium dioxide. *Mater. Res. Express* **2017**, *4*, 44003. [CrossRef]
15. De Angelis, F.; Di Valentin, C.; Fantacci, S.; Vittadini, A.; Selloni, A. Theoretical Studies on Anatase and Less Common TiO_2 Phases: Bulk, Surfaces, and Nanomaterials. *Chem. Rev.* **2014**, *114*, 9708–9753. [CrossRef] [PubMed]
16. Linsebigler, A.L.; Lu, G.; Yates, J.T. Photocatalysis on TiO_2 Surfaces: Principles, Mechanisms, and Selected Results. *Chem. Rev.* **1995**, *95*, 735–758. [CrossRef]
17. Liu, G.; Yang, H.G.; Pan, J.; Yang, Y.Q.; Lu, G.Q.; Cheng, H.-M. Titanium Dioxide Crystals with Tailored Facets. *Chem. Rev.* **2014**, *114*, 9559–9612. [CrossRef] [PubMed]
18. Wang, X.; Li, Z.; Shi, J.; Yu, Y. One-Dimensional Titanium Dioxide Nanomaterials: Nanowires, Nanorods, and Nanobelts. *Chem. Rev.* **2014**, *114*, 9346–9384. [CrossRef] [PubMed]
19. Rietveld, H.M. A profile refinement method for nuclear and magnetic structures. *J. Appl. Crystallogr.* **1969**, *2*, 65–71. [CrossRef]

20. Gordon, T.R.; Cargnello, M.; Paik, T.; Mangolini, F.; Weber, R.T.; Fornasiero, P.; Murray, C.B. Nonaqueous Synthesis of TiO$_2$ Nanocrystals Using TiF4 to Engineer Morphology, Oxygen Vacancy Concentration, and Photocatalytic Activity. *J. Am. Chem. Soc.* **2012**, *134*, 6751–6761. [CrossRef] [PubMed]

21. Romero Ocaña, I.; Beltram, A.; Delgado Jaén, J.J.; Adami, G.; Montini, T.; Fornasiero, P. Photocatalytic H$_2$ production by ethanol photodehydrogenation: Effect of anatase/brookite nanocomposites composition. *Inorg. Chim. Acta* **2015**, *431*, 197–205. [CrossRef]

22. Iliev, M.N.; Hadjiev, V.G.; Litvinchuk, A.P. Raman and infrared spectra of brookite (TiO$_2$): Experiment and theory. *Vib. Spectrosc.* **2013**, *64*, 148–152. [CrossRef]

23. Nikodemski, S.; Dameron, A.A.; Perkins, J.D.; O'Hayre, R.P.; Ginley, D.S.; Berry, J.J. The Role of Nanoscale Seed Layers on the Enhanced Performance of Niobium doped TiO$_2$ Thin Films on Glass. *Sci. Rep.* **2016**, *6*, 32830. [CrossRef] [PubMed]

24. Zhang, J.; Li, M.; Feng, Z.; Chen, J.; Li, C. UV Raman Spectroscopic Study on TiO$_2$. I. Phase Transformation at the Surface and in the Bulk. *J. Phys. Chem. B* **2006**, *110*, 927–935. [CrossRef] [PubMed]

25. Zhang, J.; Xu, Q.; Li, M.; Feng, Z.; Li, C. UV Raman Spectroscopic Study on TiO$_2$. II. Effect of Nanoparticle Size on the Outer/Inner Phase Transformations. *J. Phys. Chem. C* **2009**, *113*, 1698–1704. [CrossRef]

26. Buonsanti, R.; Grillo, V.; Carlino, E.; Giannini, C.; Kipp, T.; Cingolani, R.; Cozzoli, P.D. Nonhydrolytic Synthesis of High-Quality Anisotropically Shaped Brookite TiO$_2$ Nanocrystals. *J. Am. Chem. Soc.* **2008**, *130*, 11223–11233. [CrossRef] [PubMed]

27. Bertoni, G.; Beyers, E.; Verbeeck, J.; Mertens, M.; Cool, P.; Vansant, E.F.; Van Tendeloo, G. Quantification of crystalline and amorphous content in porous TiO$_2$ samples from electron energy loss spectroscopy. *Ultramicroscopy* **2006**, *106*, 630–635. [CrossRef]

28. Cargnello, M.; Montini, T.; Smolin, S.Y.; Priebe, J.B.; Delgado Jaén, J.J.; Doan-Nguyen, V.V.T.; McKay, I.S.; Schwalbe, J.A.; Pohl, M.-M.; Gordon, T.R.; et al. Engineering titania nanostructure to tune and improve its photocatalytic activity. *Proc. Natl. Acad. Sci. USA* **2016**, *113*, 3966–3971. [CrossRef] [PubMed]

29. Schneider, K.; Zajac, D.; Sikora, M.; Kapusta, C.; Michalow-Mauke, K.; Graule, T.; Rekas, M. XAS study of TiO$_2$-based nanomaterials. *Radiat. Phys. Chem.* **2015**, *112*, 195–198. [CrossRef]

30. Ruus, R.; Kikas, A.; Saar, A.; Ausmees, A.; Nõmmiste, E.; Aarik, J.; Aidla, A.; Uustare, T.; Martinson, I. Ti 2p and O 1s X-ray absorption of TiO$_2$ polymorphs. *Solid State Commun.* **1997**, *104*, 199–203. [CrossRef]

31. Angelomé, P.C.; Andrini, L.; Calvo, M.E.; Requejo, F.G.; Bilmes, S.A.; Soler-Illia, G.J.A.A. Mesoporous Anatase TiO$_2$ Films: Use of Ti K XANES for the Quantification of the Nanocrystalline Character and Substrate Effects in the Photocatalysis Behavior. *J. Phys. Chem. C* **2007**, *111*, 10886–10893. [CrossRef]

32. Luo, C.; Ren, X.; Dai, Z.; Zhang, Y.; Qi, X.; Pan, C. Present Perspectives of Advanced Characterization Techniques in TiO$_2$-Based Photocatalysts. *ACS Appl. Mater. Interfaces* **2017**, *9*, 23265–23286. [CrossRef] [PubMed]

33. Morra, E.; Giamello, E.; Chiesa, M. EPR approaches to heterogeneous catalysis. The chemistry of titanium in heterogeneous catalysts and photocatalysts. *J. Magn. Reson.* **2017**, *280*, 89–102. [CrossRef] [PubMed]

34. Keesmann, I. Zur hydrothermalen Synthese von Brookit. *Z. Anorg. Allg. Chem.* **1966**, *346*, 30–43. [CrossRef]

35. Kumar, S.G.; Rao, K.S.R.K. Polymorphic phase transition among the titania crystal structures using a solution-based approach: From precursor chemistry to nucleation process. *Nanoscale* **2014**, *6*, 11574–11632. [CrossRef] [PubMed]

36. Zhang, H.; Banfield, J.F. Structural Characteristics and Mechanical and Thermodynamic Properties of Nanocrystalline TiO$_2$. *Chem. Rev.* **2014**, *114*, 9613–9644. [CrossRef] [PubMed]

37. Zhang, H.; Banfield, J.F. Understanding Polymorphic Phase Transformation Behavior during Growth of Nanocrystalline Aggregates: Insights from TiO$_2$. *J. Phys. Chem. B* **2000**, *104*, 3481–3487. [CrossRef]

38. Ye, X.; Sha, J.; Jiao, Z.; Zhang, L. Thermoanalytical characteristic of nanocrystalline brookite-based titanium dioxide. *Nanostruct. Mater.* **1997**, *8*, 919–927. [CrossRef]

39. Su, W.; Zhang, J.; Feng, Z.; Chen, T.; Ying, P.; Li, C. Surface Phases of TiO$_2$ Nanoparticles Studied by UV Raman Spectroscopy and FT-IR Spectroscopy. *J. Phys. Chem. C* **2008**, *112*, 7710–7716. [CrossRef]

40. Kumar, K.-N.P.; Keizer, K.; Burggraaf, A.J.; Okubo, T.; Nagamoto, H.; Morooka, S. Densification of nanostructured titania assisted by a phase transformation. *Nature* **1992**, *358*, 48–51. [CrossRef]

41. Kominami, H.; Kohno, M.; Kera, Y. Synthesis of brookite-type titanium oxide nano-crystals in organic media. *J. Mater. Chem.* **2000**, *10*, 1151–1156. [CrossRef]

42. Ma, Y.; Xu, Q.; Chong, R.; Li, C. Photocatalytic H_2 production on TiO_2 with tuned phase structure via controlling the phase transformation. *J. Mater. Res.* **2013**, *28*, 394–399. [CrossRef]
43. Li, J.-G.; Ishigaki, T.; Sun, X. Anatase, Brookite, and Rutile Nanocrystals via Redox Reactions under Mild Hydrothermal Conditions: Phase-Selective Synthesis and Physicochemical Properties. *J. Phys. Chem. C* **2007**, *111*, 4969–4976. [CrossRef]
44. Katsumata, K.; Ohno, Y.; Tomita, K.; Taniguchi, T.; Matsushita, N.; Okada, K. Synthesis of Amphiphilic Brookite Nanoparticles with High Photocatalytic Performance for Wide Range of Application. *ACS Appl. Mater. Interfaces* **2012**, *4*, 4846–4852. [CrossRef] [PubMed]
45. Zhao, H.; Liu, L.; Andino, J.M.; Li, Y. Bicrystalline TiO_2 with controllable anatase–brookite phase content for enhanced CO_2 photoreduction to fuels. *J. Mater. Chem. A* **2013**, *1*, 8209. [CrossRef]
46. Zhao, B.; Lin, L.; He, D. Phase and morphological transitions of titania/titanate nanostructures from an acid to an alkali hydrothermal environment. *J. Mater. Chem. A* **2013**, *1*, 1659–1668. [CrossRef]
47. Manjumol, K.A.; Jayasankar, M.; Vidya, K.; Mohamed, A.P.; Nair, B.N.; Warrier, K.G.K. A novel synthesis route for brookite rich titanium dioxide photocatalyst involving organic intermediate. *J. Sol-Gel Sci. Technol.* **2015**, *73*, 161–170. [CrossRef]
48. Salari, M.; Mousavi khoie, S.M.; Marashi, P.; Rezaee, M. Synthesis of TiO_2 nanoparticles via a novel mechanochemical method. *J. Alloys Compd.* **2009**, *469*, 386–390. [CrossRef]
49. Chen, X.; Liu, L.; Huang, F. Black titanium dioxide (TiO_2) nanomaterials. *Chem. Soc. Rev.* **2015**, *44*, 1861–1885. [CrossRef] [PubMed]
50. Pan, H.; Zhang, Y.-W.; Shenoy, V.B.; Gao, H. Effects of H-, N-, and (H, N)-Doping on the Photocatalytic Activity of TiO_2. *J. Phys. Chem. C* **2011**, *115*, 12224–12231. [CrossRef]
51. Choi, M.; Lee, J.H.; Jang, Y.J.; Kim, D.; Lee, J.S.; Jang, H.M.; Yong, K. Hydrogen-doped Brookite TiO_2 Nanobullets Array as a Novel Photoanode for Efficient Solar Water Splitting. *Sci. Rep.* **2016**, *6*, 36099. [CrossRef] [PubMed]
52. Zhu, G.; Lin, T.; Lu, X.; Zhao, W.; Yang, C.; Wang, Z.; Yin, H.; Liu, Z.; Huang, F.; Lin, J. Black brookite titania with high solar absorption and excellent photocatalytic performance. *J. Mater. Chem. A* **2013**, *1*, 9650–9653. [CrossRef]
53. Xin, X.; Xu, T.; Wang, L.; Wang, C. Ti3+-self doped brookite TiO_2 single-crystalline nanosheets with high solar absorption and excellent photocatalytic CO_2 reduction. *Sci. Rep.* **2016**, *6*, 23684. [CrossRef] [PubMed]
54. Liu, G.; Yu, J.C.; Lu, G.Q. (Max); Cheng, H.-M. Crystal facet engineering of semiconductor photocatalysts: motivations, advances and unique properties. *Chem. Commun.* **2011**, *47*, 6763. [CrossRef] [PubMed]
55. Wen, J.; Li, X.; Liu, W.; Fang, Y.; Xie, J.; Xu, Y. Photocatalysis fundamentals and surface modification of TiO_2 nanomaterials. *Cuihua Xuebao Chin. J. Catal.* **2015**, *36*, 2049–2070. [CrossRef]
56. Kumar, S.G.; Rao, K.S.R.K. Comparison of modification strategies towards enhanced charge carrier separation and photocatalytic degradation activity of metal oxide semiconductors (TiO_2, WO_3 and ZnO). *Appl. Surf. Sci.* **2017**, *391*, 124–148. [CrossRef]
57. Vequizo, J.J.M.; Matsunaga, H.; Ishiku, T.; Kamimura, S.; Ohno, T.; Yamakata, A. Trapping-Induced Enhancement of Photocatalytic Activity on Brookite TiO_2 Powders: Comparison with Anatase and Rutile TiO_2 Powders. *ACS Catal.* **2017**. [CrossRef]
58. Moss, B.; Lim, K.K.; Beltram, A.; Moniz, S.; Tang, J.; Fornasiero, P.; Barnes, P.; Durrant, J.; Kafizas, A. Comparing photoelectrochemical water oxidation, recombination kinetics and charge trapping in the three polymorphs of TiO_2. *Sci. Rep.* **2017**, *7*, 2938. [CrossRef] [PubMed]
59. Na-Phattalung, S.; Smith, M.F.; Kim, K.; Du, M.-H.; Wei, S.-H.; Zhang, S.B.; Limpijumnong, S. First-principles study of native defects in anatase TiO_2. *Phys. Rev. B* **2006**, *73*, 125205. [CrossRef]
60. Mattioli, G.; Filippone, F.; Alippi, P.; Amore Bonapasta, A. Ab initio study of the electronic states induced by oxygen vacancies in rutile and anatase TiO_2. *Phys. Rev. B* **2008**, *78*, 241201. [CrossRef]
61. Yamakata, A.; Vequizo, J.J.M.; Matsunaga, H. Distinctive Behavior of Photogenerated Electrons and Holes in Anatase and Rutile TiO_2 Powders. *J. Phys. Chem. C* **2015**, *119*, 24538–24545. [CrossRef]
62. Li, L.; Salvador, P.A.; Rohrer, G.S. Photocatalysts with internal electric fields. *Nanoscale* **2014**, *6*, 24–42. [CrossRef] [PubMed]
63. Henderson, M.A. A surface science perspective on TiO_2 photocatalysis. *Surf. Sci. Rep.* **2011**, *66*, 185–297. [CrossRef]

64. Lin, H.; Li, L.; Zhao, M.; Huang, X.; Chen, X.; Li, G.; Yu, R. Synthesis of High-Quality Brookite TiO_2 Single-Crystalline Nanosheets with Specific Facets Exposed: Tuning Catalysts from Inert to Highly Reactive. *J. Am. Chem. Soc.* **2012**, *134*, 8328–8331. [CrossRef] [PubMed]

65. Pepin, P.A.; Diroll, B.T.; Choi, H.J.; Murray, C.B.; Vohs, J.M. Thermal and Photochemical Reactions of Methanol, Acetaldehyde, and Acetic Acid on Brookite TiO_2 Nanorods. *J. Phys. Chem. C* **2017**, *121*, 11488–11498. [CrossRef]

66. Naldoni, A.; Montini, T.; Malara, F.; Mróz, M.M.; Beltram, A.; Virgili, T.; Boldrini, C.L.; Marelli, M.; Romero-Ocaña, I.; Delgado, J.J.; et al. Hot Electron Collection on Brookite Nanorods Lateral Facets for Plasmon-Enhanced Water Oxidation. *ACS Catal.* **2017**, *7*, 1270–1278. [CrossRef]

67. Gong, X.-Q.; Selloni, A. First-principles study of the structures and energetics of stoichiometric brookite TiO_2. *Phys. Rev. B* **2007**, *76*, 235307. [CrossRef]

68. Ohno, T.; Higo, T.; Saito, H.; Yuajn, S.; Jin, Z.; Yang, Y.; Tsubota, T. Dependence of photocatalytic activity on aspect ratio of a brookite TiO_2 nanorod and drastic improvement in visible light responsibility of a brookite TiO_2 nanorod by site-selective modification of Fe^{3+} on exposed faces. *J. Mol. Catal. A Chem.* **2015**, *396*, 261–267. [CrossRef]

69. Ohno, T.; Sarukawa, K.; Matsumura, M. Crystal faces of rutile and anatase TiO_2 particles and their roles in photocatalytic reactions. *New J. Chem.* **2002**, *26*, 1167–1170. [CrossRef]

70. Ohno, T.; Higo, T.; Murakami, N.; Saito, H.; Zhang, Q.; Yang, Y.; Tsubota, T. Photocatalytic reduction of CO_2 over exposed-crystal-face-controlled TiO_2 nanorod having a brookite phase with co-catalyst loading. *Appl. Catal. B Environ.* **2014**, *152*, 309–316. [CrossRef]

71. Zhao, M.; Xu, H.; Chen, H.; Ouyang, S.; Umezawa, N.; Wang, D.; Ye, J. Photocatalytic reactivity of {121} and {211} facets of brookite TiO_2 crystals. *J. Mater. Chem. A* **2015**, *3*, 2331–2337. [CrossRef]

72. Xu, Y.; Lin, H.; Li, L.; Huang, X.; Li, G. Precursor-directed synthesis of well-faceted brookite TiO_2 single crystals for efficient photocatalytic performances. *J. Mater. Chem. A* **2015**, *3*, 22361–22368. [CrossRef]

73. Kapilashrami, M.; Zhang, Y.; Liu, Y.-S.; Hagfeldt, A.; Guo, J. Probing the Optical Property and Electronic Structure of TiO_2 Nanomaterials for Renewable Energy Applications. *Chem. Rev.* **2014**, *114*, 9662–9707. [CrossRef] [PubMed]

74. Li, G.; Gray, K.A. The solid–solid interface: Explaining the high and unique photocatalytic reactivity of TiO_2-based nanocomposite materials. *Chem. Phys.* **2007**, *339*, 173–187. [CrossRef]

75. Kawahara, T.; Konishi, Y.; Tada, H.; Tohge, N.; Nishii, J.; Ito, S. A Patterned TiO_2(Anatase)/TiO_2(Rutile) Bilayer-Type Photocatalyst: Effect of the Anatase/Rutile Junction on the Photocatalytic Activity. *Angew. Chem.* **2002**, *15*, 2935–2937. [CrossRef]

76. Hurum, D.C.; Agrios, A.G.; Gray, K.A.; Rajh, T.; Thurnauer, M.C. Explaining the Enhanced Photocatalytic Activity of Degussa P25 Mixed-Phase TiO_2 Using EPR. *J. Phys. Chem. B* **2003**, *107*, 4545–4549. [CrossRef]

77. Shen, X.; Zhang, J.; Tian, B.; Anpo, M. Tartaric acid-assisted preparation and photocatalytic performance of titania nanoparticles with controllable phases of anatase and brookite. *J. Mater. Sci.* **2012**, *47*, 5743–5751. [CrossRef]

78. Ardizzone, S.; Bianchi, C.L.; Cappelletti, G.; Gialanella, S.; Pirola, C.; Ragaini, V. Tailored Anatase/Brookite Nanocrystalline TiO_2. The Optimal Particle Features for Liquid- and Gas-Phase Photocatalytic Reactions. *J. Phys. Chem. C* **2007**, *111*, 13222–13231. [CrossRef]

79. Xu, H.; Zhang, L. Controllable One-Pot Synthesis and Enhanced Photocatalytic Activity of Mixed-Phase TiO_2 Nanocrystals with Tunable Brookite/Rutile Ratios. *J. Phys. Chem. C* **2009**, *113*, 1785–1790. [CrossRef]

80. Boppella, R.; Basak, P.; Manorama, S.V. Viable Method for the Synthesis of Biphasic TiO_2 Nanocrystals with Tunable Phase Composition and Enabled Visible-Light Photocatalytic Performance. *ACS Appl. Mater. Interfaces* **2012**, *4*, 1239–1246. [CrossRef] [PubMed]

81. Li, W.-K.; Hu, P.; Lu, G.; Gong, X.-Q. Density functional theory study of mixed-phase TiO_2: Heterostructures and electronic properties. *J. Mol. Model.* **2014**, *20*, 2215. [CrossRef] [PubMed]

82. Kandiel, T.A.; Feldhoff, A.; Robben, L.; Dillert, R.; Bahnemann, D.W. Tailored Titanium Dioxide Nanomaterials: Anatase Nanoparticles and Brookite Nanorods as Highly Active Photocatalysts. *Chem. Mater.* **2010**, *22*, 2050–2060. [CrossRef]

83. Li, Z.; Cong, S.; Xu, Y. Brookite vs Anatase TiO_2 in the Photocatalytic Activity for Organic Degradation in Water. *ACS Catal.* **2014**, *4*, 3273–3280. [CrossRef]

84. Li, W.-K.; Gong, X.-Q.; Lu, G.; Selloni, A. Different Reactivities of TiO$_2$ Polymorphs: Comparative DFT Calculations of Water and Formic Acid Adsorption at Anatase and Brookite TiO$_2$ Surfaces. *J. Phys. Chem. C* **2008**, *112*, 6594–6596. [CrossRef]

85. Rodriguez, M.M.; Peng, X.; Liu, L.; Li, Y.; Andino, J.M. A Density Functional Theory and Experimental Study of CO$_2$ Interaction with Brookite TiO$_2$. *J. Phys. Chem. C* **2012**, *116*, 19755–19764. [CrossRef]

86. Liu, L.; Zhao, H.; Andino, J.M.; Li, Y. Photocatalytic CO$_2$ Reduction with H$_2$O on TiO$_2$ Nanocrystals: Comparison of Anatase, Rutile, and Brookite Polymorphs and Exploration of Surface Chemistry. *ACS Catal.* **2012**, *2*, 1817–1828. [CrossRef]

87. Livraghi, S.; Rolando, M.; Maurelli, S.; Chiesa, M.; Paganini, M.C.; Giamello, E. Nature of Reduced States in Titanium Dioxide as Monitored by Electron Paramagnetic Resonance. II: Rutile and Brookite Cases. *J. Phys. Chem. C* **2014**, *118*, 22141–22148. [CrossRef]

88. Livraghi, S.; Chiesa, M.; Paganini, M.C.; Giamello, E. On the Nature of Reduced States in Titanium Dioxide As Monitored by Electron Paramagnetic Resonance. I: The Anatase Case. *J. Phys. Chem. C* **2011**, *115*, 25413–25421. [CrossRef]

89. Palmisano, G.; Garcia-Lopez, E.; Marci, G.; Loddo, V.; Yurdakal, S.; Augugliaro, V.; Palmisano, L. Advances in selective conversions by heterogeneous photocatalysis. *Chem. Commun.* **2010**, *46*, 7074–7089. [CrossRef] [PubMed]

90. Zhang, X.; Xiong, X.; Xu, Y. Brookite TiO$_2$ photocatalyzed degradation of phenol in presence of phosphate, fluoride, sulfate and borate anions. *RSC Adv.* **2016**, *6*, 61830–61836. [CrossRef]

91. Kou, J.; Lu, C.; Wang, J.; Chen, Y.; Xu, Z.; Varma, R.S. Selectivity Enhancement in Heterogeneous Photocatalytic Transformations. *Chem. Rev.* **2017**, *117*, 1445–1514. [CrossRef] [PubMed]

92. Altomare, M.; Dozzi, M.V.; Chiarello, G.L.; Di Paola, A.; Palmisano, L.; Selli, E. High activity of brookite TiO$_2$ nanoparticles in the photocatalytic abatement of ammonia in water. *Catal. Today* **2015**, *252*, 184–189. [CrossRef]

93. Addamo, M.; Augugliaro, V.; Bellardita, M.; Di Paola, A.; Loddo, V.; Palmisano, G.; Palmisano, L.; Yurdakal, S. Environmentally Friendly Photocatalytic Oxidation of Aromatic Alcohol to Aldehyde in Aqueous Suspension of Brookite TiO$_2$. *Catal. Lett.* **2008**, *126*, 58–62. [CrossRef]

94. Li, K.; Peng, T.; Ying, Z.; Song, S.; Zhang, J. Ag-loading on brookite TiO$_2$ quasi nanocubes with exposed {210} and {001} facets: Activity and selectivity of CO$_2$ photoreduction to CO/CH$_4$. *Appl. Catal. B Environ.* **2016**, *180*, 130–138. [CrossRef]

95. Karimipour, M.; Sanjari, M.; Molaei, M. The synthesis of highly oriented brookite nanosheets using graphene oxide as a sacrificing template. *J. Mater. Sci. Mater. Electron.* **2017**, *28*, 9410–9415. [CrossRef]

96. Yuan, H.; Han, K.; Dubbink, D.; Mul, G.; ten Elshof, J.E. Modulating the External Facets of Functional Nanocrystals Enabled by Two-Dimensional Oxide Crystal Templates. *ACS Catal.* **2017**, 6858–6863. [CrossRef]

97. Li, K.; Peng, B.; Jin, J.; Zan, L.; Peng, T. Carbon nitride nanodots decorated brookite TiO$_2$ quasi nanocubes for enhanced activity and selectivity of visible-light-driven CO$_2$ reduction. *Appl. Catal. B Environ.* **2017**, *203*, 910–916. [CrossRef]

98. Shang, Q.; Huang, X.; Tan, X.; Yu, T. High Activity Ti3+-Modified Brookite TiO$_2$/Graphene Nanocomposites with Specific Facets Exposed for Water Splitting. *Ind. Eng. Chem. Res.* **2017**, *56*, 9098–9106. [CrossRef]

99. Anpo, M.; Yamashita, H.; Ichihashi, Y.; Fujii, Y.; Honda, M. Photocatalytic Reduction of CO$_2$ with H$_2$O on Titanium Oxides Anchored within Micropores of Zeolites: Effects of the Structure of the Active Sites and the Addition of Pt. *J. Phys. Chem. B* **1997**, *101*, 2632–2636. [CrossRef]

100. Chen, T.; Dong, B.; Chen, K.; Zhao, F.; Cheng, X.; Ma, C.; Lee, S.; Zhang, P.; Kang, S.H.; Ha, J.W.; et al. Optical Super-Resolution Imaging of Surface Reactions. *Chem. Rev.* **2017**, *117*, 7510–7537. [CrossRef] [PubMed]

101. Tachikawa, T.; Yonezawa, T.; Majima, T. Super-Resolution Mapping of Reactive Sites on Titania-Based Nanoparticles with Water-Soluble Fluorogenic Probes. *ACS Nano* **2013**, *7*, 263–275. [CrossRef] [PubMed]

102. Zhong, H.; Zhang, J.; Tang, X.; Zhang, W.; Jiang, R.; Li, R.; Chen, D.; Wang, P.; Yuan, Z. Mass spectrometric monitoring of interfacial photoelectron transfer and imaging of active crystalline facets of semiconductors. *Nat. Commun.* **2017**, *8*, 14524. [CrossRef] [PubMed]

103. Li, K.; Peng, B.; Peng, T. Recent Advances in Heterogeneous Photocatalytic CO$_2$ Conversion to Solar Fuels. *ACS Catal.* **2016**, *6*, 7485–7527. [CrossRef]

104. Magne, C.; Cassaignon, S.; Lancel, G.; Pauporté, T. Brookite TiO$_2$ Nanoparticle Films for Dye-Sensitized Solar Cells. *ChemPhysChem* **2011**, *12*, 2461–2467. [CrossRef] [PubMed]

105. Xu, J.; Wu, S.; Jin, J.; Peng, T. Preparation of brookite TiO$_2$ nanoparticles with small sizes and the improved photovoltaic performance of brookite-based dye-sensitized solar cells. *Nanoscale* **2016**, *8*, 18771–18781. [CrossRef] [PubMed]
106. Liu, K.; Cao, M.; Fujishima, A.; Jiang, L. Bio-Inspired Titanium Dioxide Materials with Special Wettability and Their Applications. *Chem. Rev.* **2014**, *114*, 10044–10094. [CrossRef] [PubMed]
107. Shibata, T.; Irie, H.; Ohmori, M.; Nakajima, A.; Watanabe, T.; Hashimoto, K. Comparison of photochemical properties of brookite and anatase TiO$_2$ films. *Phys. Chem. Chem. Phys.* **2004**, *6*, 1359–1362. [CrossRef]

© 2017 by the authors. Licensee MDPI, Basel, Switzerland. This article is an open access article distributed under the terms and conditions of the Creative Commons Attribution (CC BY) license (http://creativecommons.org/licenses/by/4.0/).

catalysts

MDPI

Review

On the Origin of Enhanced Photocatalytic Activity of Copper-Modified Titania in the Oxidative Reaction Systems

Marcin Janczarek [1,2,*] and Ewa Kowalska [1,*]

[1]   Institute for Catalysis, Hokkaido University, N21, W10, Sapporo 001-0021, Japan
[2]   Department of Chemical Technology, Gdansk University of Technology, Narutowicza Str. 11/12, 80-233 Gdansk, Poland
*   Correspondence: marcin@cat.hokudai.ac.jp (M.J.); kowalska@cat.hokudai.ac.jp (E.K.)

Received: 12 October 2017; Accepted: 24 October 2017; Published: 27 October 2017

Abstract: Modification of titania with copper is a promising way to enhance the photocatalytic performance of TiO_2. The enhancement means the significant retardation of charge carriers' recombination ratio and the introduction of visible light activity. This review focuses on two main ways of performance enhancement by copper species—i.e., originated from plasmonic properties of zero-valent copper (plasmonic photocatalysis) and heterojunctions between semiconductors (titania and copper oxides). The photocatalytic performance of copper-modified titania is discussed for oxidative reaction systems due to their importance for prospective applications in environmental purification. The review consists of the correlation between copper species and corresponding variants of photocatalytic mechanisms including novel systems of cascade heterojunctions. The problem of stability of copper species on titania, and the methods of its improvement are also discussed as important factors for future applications. As a new trend in the preparation of copper-modified titania photocatalyst, the role of particle morphology (faceted particles, core-shell structures) is also described. Finally, in the conclusion section, perspectives, challenges and recommendations for future research on copper-modified titania are formulated.

Keywords: photocatalysis; copper-modified titania; oxidative reaction systems; heterojunction

1. Introduction

Titanium dioxide (TiO_2, titania) has played a crucial role in the development of semiconductor photocatalysis over the last 40 years [1]. The applicative potential of TiO_2-photocatalysis is presently focused on the areas such as environmental remediation (water treatment [2–4] and air purification [5]), renewable energy processes—water splitting for hydrogen production [6], conversion of CO_2 to hydrocarbons [7], solar cells [8]—and self-cleaning surfaces [9]. TiO_2 is characterized by those properties that are indispensable to fulfill the requirements of efficient, stable, and green photocatalytic material (long term stability, chemical inertness, corrosion resistance, non-toxicity) [1].

Despite the above-mentioned benefits, the application of titania is still limited to regions with a high intensity of solar radiation due to its wide bandgap (ca. 3.0 to 3.2 eV). Therefore, there is a necessity to incorporate visible light absorption with TiO_2. Various strategies of titania performance improvement towards visible light responses have been proposed in recent years such as doping, modification, semiconductor coupling, and dye sensitization [10]. As a consequence of such modifications, TiO2 showed an absorption band in the visible light region and, in most cases, also photocatalytic activity under visible light irradiation for different reaction systems.

The photocatalytic activity of titania is also limited by the recombination of the photogenerated electron-hole pairs [11] as typical for all semiconducting materials. Titania exhibits only weak bandgap emission upon the recombination of conduction band electrons with valence band holes, and the irradiative recombination involving trap states is optically allowed [1,12]. Recombination is generally caused by impurities, defects or other factors, which introduce bulk or surface imperfections into the crystal and, depending on titania properties, it may occur either on the surface or in the bulk (usually a decrease in particle size results in an increase in surface recombination) [13,14]. The incorporation of species capable of promoting charge separation may reduce this phenomenon, and thus result in improvement of overall quantum efficiency of photocatalytic system. Promoted charge separation has been reported for titania modified with various modifiers, e.g., ions [14,15], noble metals [16,17], other semiconductors (heterojunction coupling) [18,19], and doped with metal ions [14].

Considering the above-mentioned limitations (vis inactivity and charge carriers' recombination), and similar ways of their removal (surface modification, doping, heterojunction), it is thought that effective materials based on TiO_2, which meet those requirements are highly designable. It must be pointed out that modification of titania towards visible light photocatalytic activity could significantly influence photocatalytic performance under UV irradiation and in some cases a decrease in photocatalytic activity has been observed since modifiers/dopants could also work as recombination centers for electrons and holes, therefore a proper selection procedure for titania modification should consider both limitation issues.

Among different metallic candidates for TiO_2 modification, copper (Cu) is a very promising material. Cu-based nanocatalysts have significant applicability in nanotechnology including catalytic organic transformations, electrocatalysis, and photocatalysis [20–22]. In comparison with other noble metals (gold, platinum and silver) recognized as very efficient co-catalysts of titania, copper—as a consequence of its abundance in the Earth's crust—is an inexpensive material, 100 times and 6000 times cheaper than silver and gold, respectively [21]. Furthermore, owing to the location in the same group of the periodic table as gold and silver, copper has similar properties due to its electronic configuration and the face centered cubic (FCC) structure of the atom's location. Therefore, one can expect the comparable potential to improve the photocatalytic activity of TiO_2, simultaneously with the higher possibility of successful application. Copper can exist in a wide range of accessible oxidation states: Cu^0, Cu^I, Cu^{II} and Cu^{III}. In this connection, the active copper species in TiO_2 photocatalytic system are copper oxides (Cu_2O, CuO) and metallic copper. Cu may also co-exist in both forms (the oxide and zero-valent) and this is the most probable case when titania is modified with metallic copper under anaerobic conditions, and thus-obtained photocatalyst is subsequently kept under aerobic conditions (surface oxidation of metallic copper deposits). The variety of copper forms can provide difficulty in understanding of their role in considered reaction systems. Therefore, a statement that the nature of Cu species in TiO_2 photocatalysis is still not clearly understood may be made. There is also some possibility of copper ions doping into TiO_2 crystal lattice. The radius of Cu^{2+} is 0.087 nm, which is larger than the 0.0745-nm radius of Ti^{4+}. Moreover, a huge difference in valence state suggests that Cu^{2+} should not replace Ti^{4+} to enable displacement doping at the crystal lattice site [23]. There is only the possibility of incorporating Cu^{2+} ions into interstitial positions in the lattice [24–27].

Various papers on copper-modified titania have been already published, including comprehensive reviews [22,28] about copper species in photocatalysis. However, these reviews are mainly focused on the photocatalytic reduction reactions, i.e., hydrogen evolution and/or CO_2 to methane conversion. Of course, the Cu/TiO_2 system is most known for the mentioned reaction systems—and the number of papers in this field is the highest—but the application of Cu/TiO_2 for the photocatalytic oxidation of organic compounds is also very promising both under ultraviolet (UV) and visible/solar irradiation, and the role of Cu species in this system needs clarification. Therefore, in the present work, an overview of papers has been carried out to compare different mechanisms resulting from an existence of various forms of copper combined with titania in relation to efficiencies of oxidative reaction systems.

2. Photocatalytic Systems Based on Cu/TiO$_2$

2.1. Presence of Copper Ions in the TiO$_2$ Suspension

The first papers about Cu/TiO$_2$ systems were mainly connected with photodeposition of copper on titania surface as the efficient method for Cu removal from water environments [29–32]. Cu^{II} ions were reduced to metallic copper with the participation of photogenerated electrons (e$^-$) and holes (h$^+$) according to the following Reactions (1)–(3) [29]:

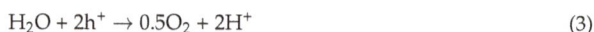

$$TiO_2 + h\nu \rightarrow e^- + h^+ \tag{1}$$

$$Cu^{2+} + 2e^- \rightarrow Cu \tag{2}$$

$$H_2O + 2h^+ \rightarrow 0.5O_2 + 2H^+ \tag{3}$$

In opposition to the above results of Bard's group [29], Hermann et al. suggested that Cu^{2+} ions in TiO$_2$ suspension were photoreduced to Cu$^+$ ions and that metallic copper could not be obtained [33]. However, Jacobs et al. showed that photodeposition of zero-valent Cu particles was possible and was preceded by photodeposition of Cu$_2$O particles [32].

Another type of study focused on the influence of the presence of copper ions on the efficiency of photocatalytic oxidation of organic compounds. It was found that addition of dissolved copper ions to TiO$_2$ reaction system improved significantly the rate of photocatalytic oxidation [34–39]. For example, Bard et al. observed [29] that Cu^{2+} ions retarded the electron-hole recombination by trapping of photogenerated electrons (Equation (2)). The reduced forms, in turn, could prevent the recombination by hole trapping (Equation (4)).

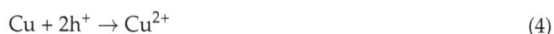

$$Cu + 2h^+ \rightarrow Cu^{2+} \tag{4}$$

Generally, the small concentration of Cu^{2+} ions prevents charge recombination. In the case of large amounts of Cu^{2+} ions, their influence on the reaction rate can be detrimental considering that Reactions (2) and (4) are short-circuiting. In 1985, Okamoto et al. showed that a high concentration of Cu^{2+} (50 mM) inhibited the efficiency of phenol photooxidation (the first study using copper for photocatalytic oxidation of organic compounds) [34]. The detrimental effect of a high amount of Cu^{2+} ions was also explained by absorbance of UV/vis radiation by copper species [40]. Precipitation of dissolved copper as a hydroxide could also decrease the rate of photocatalytic oxidation by reflecting UV irradiation through increased solution turbidity (shielding effect) [39]. Brezova et al. observed an inhibition effect of dissolved Cu^{2+} ions even at low concentrations (>1 mM) in the phenol oxidation system. The key point is that the observed photodeposition of metallic copper and Cu$_2$O on titania may significantly influence the process of charge carriers and radical intermediate generation and recombination and finally, along with Equation (2), result in the inhibition of photodegradation [40]. Butler and Davis observed negligible adsorption of copper (and other metals) on titania surface and suggested that mainly dissolved copper (and other metals) increased the reaction rate via a homogeneous pathway rather than surface reactions on copper-modified titania [39,40]. They proposed a mechanism involving formation of a ternary reactive complex between copper, organic compound or its oxidation intermediate and oxygen-containing species such as O$_2$, H$_2$O$_2$. Bideau et al. analyzed TiO$_2$ system in the presence of Cu^{2+} ions for the oxidation of three organic compounds: formic acid, acetic acid and propionic acid, and monocarboxylate complexes of Cu^{2+} were identified as important species in the reaction kinetics [35–37,41]. The observed increase in the reaction rate in the mentioned studies was explained in terms of the effects of Cu^{2+}/Cu$^+$ redox couple, which inhibited the electron-hole recombination and the inner sphere mechanism of Cu^{2+} ions with the organic compounds, forming organo-metallic intermediates.

2.2. Copper Species Deposited on TiO$_2$ Surface

2.2.1. Metallic Copper—Plasmonic Photocatalysis

Copper (similar to gold and silver) nanoparticles (Cu NPs)–titania couple can be a very promising photocatalytic system for both UV and visible light-induced reactions. Cu NPs as surface modifiers of TiO$_2$ can inhibit electron-hole recombination under UV irradiation working as an electron sink [42,43], because Fermi energy level of metallic Cu lies below the conduction band of TiO$_2$, and therefore photogenerated electrons in TiO$_2$ can be easily transferred to Cu NPs. Other important advantage of Cu NPs is that they can activate TiO$_2$ towards visible light due to localized surface plasmon resonance (LSPR) of Cu, thus being so-called "plasmonic photocatalysts" [44–49]. The LSPR is excited when light interacts with the free electrons of a metallic nanostructure, which results in the collective excitations (oscillations) that lead to significant improvement of the local electromagnetic fields surrounding the nanoparticles [50]. Spherical metallic nanoparticles, such as gold, silver and copper, are sufficiently small (in comparison with the wavelength of the light) to be resonant with the light coming from all directions and indicate coloration as well as a local electromagnetic field enhancement on the particle surface even without the condition of total internal reflection [44]. The existence of characteristic surface plasmon resonance band absorbing light in the visible region results in visible-light activity of noble metal-modified titania, probably due to an efficient transfer of the photoexcited electrons from metal particles to the conduction band of TiO$_2$ (energy transfer and plasmonic heating have been also proposed as possible reasons of vis response [51]). This would result in electron-deficiency in metal and electron-richness in TiO$_2$ and therefore, the direct photocatalytic oxidation occurs on the metal surface rather than on TiO$_2$ surfaces as shown in Figure 1 [52] or indirect oxidation proceeds by reactive oxygen species (initiated by superoxide anion radical formation).

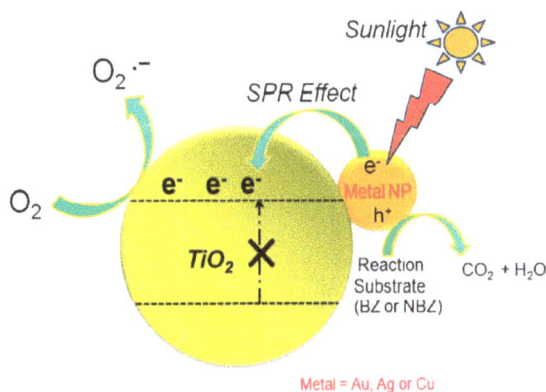

Figure 1. Surface plasmon resonance (SPR) effect of metal nanoparticles for TiO$_2$ photocatalysis under sunlight irradiation. Reprinted with permission from [52]. Copyright Royal Society of Chemistry, 2015.

Based on the unique properties of LSPR, many studies focused mainly on Au and Ag as plasmonic metals, which improve photocatalytic properties of TiO$_2$ [53]. However, very few studies have been reported about photocatalytic performance of Cu NPs in connection with titania [47–49]. The main problem with utilizing Cu NPs as a "plasmonic sensitizer" is the known fact that zero-valent copper is easily oxidized and lose plasmon resonance properties gradually under ambient conditions [54–56]. For example, (i) although, titania modification by strong radiolytic reduction of Cu^{2+} or photodeposition under anaerobic conditions resulted in formation of zero-valent copper, Cu^0 was subsequently oxidized under ambient conditions forming CuO/TiO$_2$ and Cu/Cu$_2$O/Cu/TiO$_2$, respectively [57–59] and (ii) even Cu NPs immersed in water were oxidized by dissolved oxygen [60].

The reported main solutions of this problem consider the necessity of preventing copper oxidation by maintaining Cu in anoxic environments [55,61–63] or protecting its surface with chemical corrosion inhibitors [64], polymeric layers [47,55], and by oxide encapsulation [65]. The optimal Cu NPs based plasmonic photocatalysts should work without necessity of addition of surface-obscuring chemical stabilizers and would be storable indefinitely under ambient conditions [49]. Most of these studies were focused on Cu NPs alone but not on Cu NPs-TiO$_2$, where it is necessary to consider interactions between Cu NPs and TiO$_2$ during preparation procedure [66].

Most studies about Cu NPs-TiO$_2$ were connected with unprotected copper under reductive, oxygen free conditions. Zhang et al. prepared TiO$_2$ nanotube arrays with copper NPs by pulsed electrochemical deposition method [48]. They reported visible light activity of this material for hydrogen evolution. Similarly, Kum et al. investigated the properties of the visible light active Cu/TiO$_2$ photocatalyst for hydrogen production [67]. This material showed high stability and photocatalytic activity in the evacuated chamber. However, it is difficult to distinguish whether the photocatalyst was activated by excitation of copper LSPR or band gap heterojunction with copper oxides since photocatalysis experiments were driven by broadband visible light irradiation.

In order to apply the Cu NPs-TiO$_2$ system for an oxidative system, the persistence of metallic copper to oxidative environment is necessary. Yamaguchi et al. prepared plasmonic Cu NPs deposited on TiO$_2$ electrode and protected by polyvinyl alcohol, which resulted in efficient photocurrent response due to LSPR [47]. In the latest research, De Sario et al. studied titania aerogels supporting ca. 5-nm Cu NPs [49]. They found that plasmonic properties of Cu NPs were preserved by offering an extended interfacial contact with reduced TiO$_2$ support (characteristic design issue for aerogels). An arrangement of Cu NPs at high-surface area TiO$_2$ (with multiple Cu-TiO$_2$ contacts per Cu nanoparticle) allowed the obtaining of a Cu/TiO$_2$ system with good stability and a high content of Cu NPs. Figure 2 shows DRS spectra of Cu/TiO$_2$[aerogel] (with characteristic peaks for copper LSPR at ca. 770 nm) before and after photoelectrochemical oxidation of methanol confirming high stability of this material (almost the same plasmonic properties of Cu/TiO$_2$).

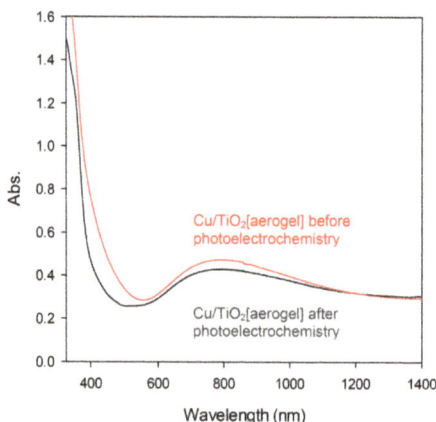

Figure 2. Diffuse-reflectance UV–visible spectroscopy of Cu nanoparticle-modified TiO$_2$ aerogels fabricated by photocatalytic reduction of Cu^{2+} at the TiO$_2$ surface before (red) and after (black) photoelectrochemical oxidation of methanol. Reprinted with permission from [49]. Copyright Royal Society of Chemistry, 2017.

2.2.2. Copper Oxides—Heterojunction Systems with TiO$_2$

The Concept of Heterojunction between Copper Oxides and Titania

It is well known that coupling TiO$_2$ with other semiconductors with different redox energy levels can lead to an increase in photocatalytic efficiency by the enhancement of the charge carrier separation process and thus an increase in the lifetime of the charge carriers. Moreover, the proper selection of coupled semiconductors can also activate the heterojunction system towards a visible light response. To create the efficient heterojunction system, both semiconductors must possess different energy levels from their corresponding conduction and valence bands. In such a configuration, several advantages can be obtained: (a) an improvement of charge carriers' separation; (b) an increase in the lifetime of the charge carrier; and (c) an enhancement of the interfacial charge transfer efficiency to adsorbed substrates [68,69].

In the case of an efficient interparticle electron transfer between the semiconductor and TiO$_2$, the conduction band of TiO$_2$ must be more anodic than the corresponding band of the sensitizer. Under visible light irradiation, only the semiconductor-sensitizer is excited and the electrons photoexited to its conduction band are injected into the TiO$_2$ conduction band. If the valence band of the sensitizer is more cathodic than the valence band of TiO$_2$, the hole generated in the semiconductor remains there (enable to migrate to TiO$_2$). These thermodynamic conditions favor the phenomenon of electron injections, as shown in Figure 3a. When the system of coupled semiconductors works under UV–vis irradiation, both semiconductors are excited. For heterojunction system described above, two origins of electrons in the conduction band (CB) of titania are considered: (i) injected into TiO$_2$ (the same as under vis excitation) from coupled semiconductor; and (ii) photoexcited electrons from titania valence band under UV irradiation, resulting in a high concentration of electrons in CB of TiO$_2$. Whereas, holes left in the valence band (VB) of TiO$_2$ may migrate to the VB of coupled semiconductor and influence a high concentration of holes in the couple semiconductor/electrolyte interface (Figure 3b) [70].

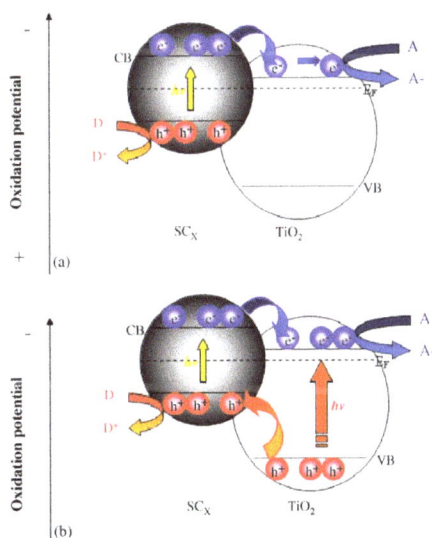

Figure 3. (**a**) Energy diagram illustrating the coupling of two semiconductors (SC) in which vectorial electron transfer occurs from the light-activated SC to the nonactivated TiO2; (**b**) Diagram depicting the coupling of SC in which vectorial movement of electrons and holes is possible. Reprinted with permission from [70]. Copyright Elsevier, 2005.

Cu$_2$O and CuO are *p*-type semiconductors with band gap energies of 2.1 eV and 1.7 eV, respectively. The band positions of copper oxides in relation to TiO$_2$ are shown in Figure 4a.

Depending on the band positions of two semiconductors forming heterojunction, three different types of heterojunction could be distinguished: I, II and III (Figure 4b). Type II, which provides the optimum band positions for efficient charge carrier separation, is considered the most probable for the Cu$_x$O/TiO$_2$ system. Photoexcited electrons are transferred from CB(B) to CB(A) and this transfer can occur directly between semiconductors due to favorable energetics of the relative positions of CBs, or due to band bending at the interface inducing and internal electric field. Whereas, holes are transferred simultaneously from VB(A) to VB(B) and as a result photogenerated electrons and holes are separated from each other, reducing the recombination probability and increasing the lifetimes of the charge carriers [71].

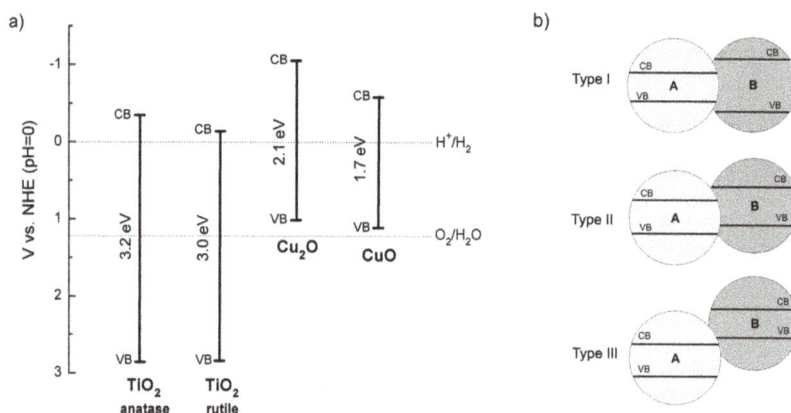

Figure 4. (a) Band gap energies and band positions of titania (anatase and rutile) and copper oxides—values of band positions were taken from the reference. Data collected from [71]; (b) Types of heterojunction system of coupled semiconductors.

Both copper oxides, especially Cu$_2$O, are very promising semiconductors for photocatalytic hydrogen production [72–75]. The main limitation of their application results from not good photostability, which is very important issue for oxidative photocatalytic systems. Generally, Cu$_2$O possesses higher photocatalytic activity than CuO for degradation of organic compounds [75,76]. Although CuO has much smaller band gap than Cu$_2$O, and thus is able to absorb more vis photons, the positions of CB and VB for CuO are insufficient to catalyze the production of hydroxyl and superoxide radicals, which are primary initiators for the photocatalytic oxidation of organic compounds [75]. Deng et al. found that combined CuO/Cu$_2$O nanostructures can be efficient photocatalysts for the photodegradation of organic compounds and possess better resistance against photocorrosion [76]. They proposed that the co-existence of CuO could inhibit the photocorrosion of Cu$_2$O, and furthermore the CuO/Cu$_2$O system can be more photocatalytically efficient than single CuO or Cu$_2$O.

CuO-TiO$_2$ Heterojunction

Taking into consideration above-mentioned issues, the heterojunction between *p*-type semiconductors–copper oxides and *n*-type one–titania could be a very promising way to improve the photocatalytic activity of titania under UV or/and visible light irradiation and to obtain more active and stable photocatalytic material than copper oxides alone. In the middle of eighties Okamoto et al. prepared Cu-deposited TiO$_2$ by UV-irradiation of deaerated suspension of slightly reduced (in hydrogen) anatase powder in the presence of CuSO$_4$ [34]. They assumed that only metallic copper was obtained on titania surface (but without experimental evidence). It was found that Cu-deposited TiO_2 was significantly more active in phenol photocatalytic oxidation than single TiO_2 in the presence of Cu^{2+} ions. However, the useless of this photocatalyst for wastewater treatment was suggested because of dissolution of deposited Cu into the reaction solution transforming again to Cu^+ and Cu^{2+} ions. This research expressed two main problems with the discussion on Cu/TiO_2: careful characterization of copper forms connected with titania, and the stability of prepared photocatalyst. To understand the mechanism of improvement of photocatalytic activity and to analyze perspectives of application, thorough research considering above issues is necessary.

The first research including aspects of different forms of copper in connection with titania and photooxidation activity was performed by Song et al. in 1999 [77]. They investigated the effect of oxidation state of copper loaded on titania surface on photocatalytic oxidation reaction. Cu was photodeposited on TiO_2 surface in the presence of $CuSO_4$ and ethanol as a hole scavenger. Moreover, to obtain Cu/TiO_2 photocatalysts with different oxidation states of surface-loaded copper, samples were annealed at 200 °C and 300 °C. They reported that non-annealed samples contained only metallic copper, whereas annealing at 200 °C and 300 °C resulted in copper oxidation to mainly Cu_2O and CuO, respectively. It was found that metallic copper deposited on titania improved the photocatalytic activity of TiO_2 for photooxidation of 1,4-dichlorobenzene. In contrast, the use of oxidized form of copper on titania resulted in a significant decrease in photocatalytic activity and it was proposed that copper oxide clusters on TiO_2 were not effective in the transfer of photoexited electrons [77].

Another report, which confirmed detrimental effect of CuO presence, was done by Chiang et al. for photocatalytic oxidation of cyanides in the CuO/TiO_2 system [78]. CuO/TiO_2 was prepared by photodeposition method and subsequent thermal treatment in air at 110 °C for various copper contents (0.05–10 at. % (atomic percentage) Cu). Only slight improvement of photocatalytic activity was observed for 0.1 at. % Cu, due to suggested electron trapping by CuO. It was proposed that a decrease in photocatalytic activity with a further increase in CuO content was caused by reduced photon absorption since well dispersed nanosized CuO particles covered the surface of TiO_2 (shielding effect). Moreover, it was suggested that higher concentration of CuO could promote recombination of photogenerated holes with the trapped electrons resulting in a decrease in available holes for redox reactions. Chiang et al. also examined photocatalytic activity in cyanide oxidation system under the presence of dissolved Cu^{2+} ions. The presence of copper ions was detrimental for the rate of reaction. This decrease could be explained in terms of the competition reaction of copper(I) cyanide complexes for the photogenerated hydroxyl group [78].

Shun-Xin et al. proposed another preparation method of Cu/TiO_2, i.e., impregnation of titania precursor in copper salt during sol-gel synthesis [79]. They suggested the co-existence of both copper oxides on titania surface and copper ions (Cu^{2+} and Cu^+) in the lattice of titania (doping). Authors proposed that Cu^{2+} could trap the excited electron in CB of titania, inhibiting electron-hole pair recombination, whereas Cu^+ could transfer electron to oxygen adsorbed on the surface of photocatalyst accelerating interfacial electron transfer [79].

Arana et al. studied methyl tert-butyl ether (MTBE) photocatalytic oxidation on CuO-TiO_2 under UV irradiation [80]. They observed a significant improvement of photocatalytic activity of copper-modified titania in comparison with pure TiO_2. Considering the reduction potential for Cu^{2+}/Cu^+ (+0.17 V vs. NHE (normal hydrogen electrode)), CuO deposits on TiO_2 surface could react with the photogenerated electrons through Reaction (5), and Cu^+ ions could be re-oxidized to Cu^{2+} by oxygen, H_2O_2 or other oxidizing species present in the Medium (6):

$$Cu^{2+} + e^-{}_{CB} \rightarrow Cu^+ \tag{5}$$

$$Cu^+ + (O_2, H_2O_2, \text{other oxidants}) \rightarrow Cu^{2+} + e^- \tag{6}$$

Therefore, the Reaction (5) of Cu^{2+} ions with the photogenerated electrons should slow down electron-hole recombination resulting in activity enhancement [80].

The mechanism for the Cu/TiO_2 system under visible light irradiation was proposed by Irie et al. for 2-propanol oxidation [81,82]. They prepared Cu(II)-grafted TiO_2 photocatalyst by impregnation method using $CuCl_2 \cdot 2H_2O$ as the source of copper. Authors suggested that Cu^{2+} ions existed in the form of amorphous CuO clusters. They proposed that visible light irradiation initiated interfacial charge transfer (IFCT): electrons in the VB of TiO_2 were directly transferred to Cu^{2+} and Cu^+, and the holes created in the VB of TiO_2 decomposed organic compounds (Figure 5) [82]. It was proposed that for efficient IFCT, Cu^{2+} ions had to be atomically isolated on the TiO_2 surface. Authors suggested that produced Cu^+ ions could reduce O_2 by a multielectron reduction, being re-oxidized to Cu^{2+}. This multi-step reduction process could be initiated by a two-electron reduction of oxygen to peroxide (Equation (7)), and the following four-electron reductions (Equations (8) and (9)).

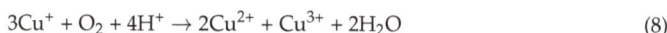

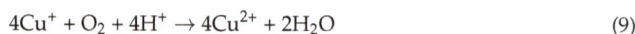

$$2Cu^+ + O_2 + 2H^+ \rightarrow 2Cu^{2+} + H_2O_2 \tag{7}$$

$$3Cu^+ + O_2 + 4H^+ \rightarrow 2Cu^{2+} + Cu^{3+} + 2H_2O \tag{8}$$

$$4Cu^+ + O_2 + 4H^+ \rightarrow 4Cu^{2+} + 2H_2O \tag{9}$$

It was concluded that Cu^{2+} ion in the amorphous form affords the smooth reversibility between Cu^{2+} and Cu^+ forms [82]. The possibility of this mechanism was also confirmed by this group for other material, i.e., Fe(III)-grafted TiO_2 [83]. Similar mechanism was proposed for CuO/WO_3 system by Arai et al. [84]. Photoexcited electrons in WO_3 were transferred to CuO and Cu^{2+} was reduced to Cu^+, and reduced surface was re-oxidized (Cu^+ to Cu^{2+}) by oxygen.

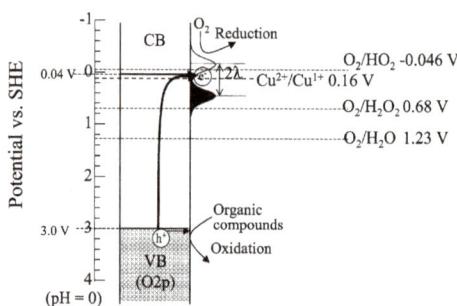

Figure 5. Mechanism for the generation of photocatalytic activity of Cu(II)/TiO_2 under visible light. Visible light irradiation induces interfacial charge transfer (IFCT) from the valence band (VB) of TiO_2 to the Cu^{2+} ion. Reprinted with permission from [82]. Copyright American Chemical Society, 2009.

The properties of photocatalytic material prepared by Irie et al. [82] were optimized in the field of crystallinity, the interfacial junction between TiO_2 and CuO nanoclusters, and the amount of CuO nanoclusters. The synthesized CuO-TiO_2 nanocomposites exhibited efficient interfacial charge transfer (IFCT) for decomposition of volatile organic compounds and strong anti-pathogenic effects under indoor conditions. The mechanism of those dual type properties is presented in Figure 6. It was proposed that Cu^+ species were very efficient for enhancing the antibacterial and anti-viral properties of TiO_2 photocatalysts (even under dark conditions), i.e., holes in the valence band (VB) of TiO_2, generated under irradiation with visible light, in combination with Cu^+ species, could attack the outer membrane, proteins, and nucleic acid (DNA and RNA) of viruses and bacteria, resulting in their death and inactivation [85].

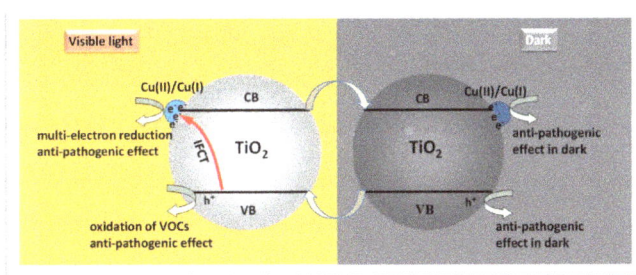

Figure 6. Proposed processes of photocatalysis and inactivation of viruses and bacteria under visible-light irradiation and dark conditions. Reprinted from [85] under CC BY-NC 3.0 license, 2009.

To understand the mechanism of coupled photocatalysts, it is necessary to know the exact positions of the valence band (VB) and conductance band (CB) of semiconductors. The ambiguity over the positions of the conduction and valence bands in CuO (and Cu_2O) results in different explanations for the mechanistic aspects of heterojunction systems. For example, Li et al. prepared CuO/TiO_2 (rutile) nanocomposites by impregnation and subsequent calcination in air [86]. They tested photocatalytic properties of this material using methylene blue as a model compound under UV/vis irradiation. The authors proposed that CuO/TiO_2 system could be classified as type I of heterojunction, as presented in the inset of Figure 7 (different CB and VB positions for CuO than presented in Figure 4a). In this system, since the band edges of TiO_2 bracketed those of CuO, coupling between CuO and TiO_2 could lead to decreased photocatalytic activity due to transfer of both electrons and holes simultaneously from TiO_2 to CuO resulting in their recombination in CuO. However, when the copper content was very low (0.1 wt %) significant improvement of photocatalytic activity was observed in comparison to unmodified titania (Figure 7). Therefore, it was proposed that copper existed in this sample as highly dispersed CuO clusters and substitional Cu^{2+} (Ti-O-Cu linkages), and UV illumination resulted in trapping of electrons in TiO_2 lattice. Photogenerated electrons were localized at Ti sites in the bulk of titania lattice and holes at interfacial sites of TiO_2-CuO (by EPR spectroscopy). However, at higher copper loadings, no photogenerated electrons were observed at titania lattice trapping sites. Slamet et al. explained this phenomenon as a "shading effect" [87] (also known as "shielding effect" or "inner filter effect" and observed for various titania surface modifications, e.g., with Au NPs [88], and Pt NPs [89]): colored CuO absorbed light and reduced the photoexcitation capacity of TiO_2 (competition about photon absorption). Li et al. concluded that efficient electron transfer from the CB of TiO_2 to CuO led to the absence of Ti^{3+} (trapped electrons) in CuO-TiO_2 at high CuO loadings. They formulated a hypothesis that the form of CuO (clusters or nanocrystallites) had a significant influence on the type of charge transfer kinetics between TiO_2 and CuO [86].

Moniz et al. prepared a CuO/TiO_2 (P25) composite using microwave co-precipitation technique [90]. It was found that 5% CuO/P25 exhibited near 1.6 times higher efficiency than pure P25 in mineralization of a model organic pollutant (2,4-dichlorophenoxyacetic acid (herbicide)) under UV/Vis. To analyze the mechanism of photocatalytic activity, the authors determined the band edge positions of CuO and TiO_2. They experimentally determined those positions by the interpretation of impedance (Mott-Schottky) plots. The difference in conduction band energies of the two materials suggested that electron transfer from TiO_2 to CuO was feasible, given the 0.34 V offset which could provide sufficient driving force for charge separation (Figure 8). It was also evidenced by the fact that pure CuO was inactive for photo-decomposition, and thus photocatalytic oxidation by CuO did not occur [90]. The concept of this junction was not connected with the increase of visible light absorption of P25 but only with more efficient charge separation. Photogenerated electrons were transferred from TiO_2 to the conductions band of CuO, followed by transfer of photoexcited holes from TiO_2 to the surface for the expected oxidation reactions.

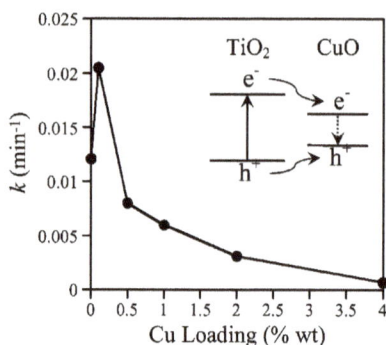

Figure 7. First-order rate constants (k in min^{-1}) as measured by methylene blue degradation under UV irradiation. Synthesized CuO-TiO$_2$ nanocomposites with different copper loadings were used as the photocatalysts. Inset: a schematic diagram showing the photoinduced charge separation (e$^-$ in conduction band and h$^+$ in valence band) in TiO$_2$, charge transfer from TiO$_2$ to CuO and subsequent charge recombination in CuO (dotted arrow). The difference between the conduction band edges of TiO$_2$ and CuO is estimated to be 0.75 eV. Reprinted with permission from [86]. Copyright American Chemical Society, 2008.

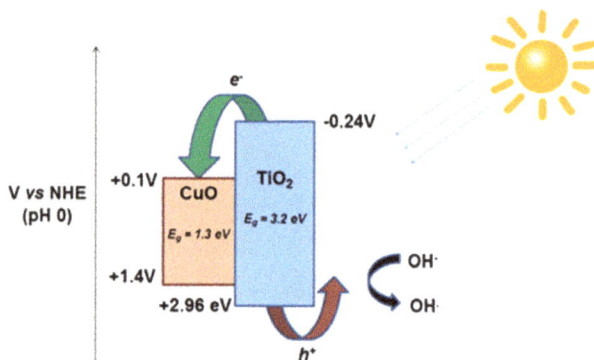

Figure 8. Mechanism of charge transfer in CuO/TiO$_2$ heterojunction. Reprinted with permission from [90]. Copyright John Wiley & Sons, Inc., 2015.

Siah et al. prepared CuO photodeposited on TiO$_2$ [91]. Under UV irradiation, the photocatalytic activity for 2,4-dichlorophenoxyacetic acid decomposition increased by a maximum factor of 4.3 compared with unmodified TiO$_2$, when TiO$_2$ was loaded with 0.75 mol % CuO. In contrast, under visible light and solar simulator irradiation, the optimum loading of CuO was much lower (0.1 mol %), and enhancements of 22.5 and 2.4 times were observed. In the case of samples loaded with higher amounts of CuO (1–5 mol %), CuO clusters can mask the surface of titania, which prevents the light source from reaching the active sites. CuO clusters in CuO-high loaded TiO$_2$ may act also as recombination centers. Therefore, in addition to the increased charge separation and improved visible light absorption, the masking effect will also be considered in designing efficient heterojunctions in the copper oxides–titania system [91].

Another strategy to obtain efficient visible light active photocatalyst, based on titania–copper(II) oxide, was proposed for self-doped TiO$_2$ (Ti^{3+}) grafted with amorphous CuO [92]. Cu-modified self-doped TiO$_2$ was obtained by thermal treatment of mixture of titanium oxides (Ti$_2$O$_3$ and TiO$_2$ mixture) in air, and their subsequent impregnation with copper salt solution. As it was reported in earlier Irie's works [81,82], in the case of titania modified with amorphous nanoclusters of Cu^{2+} oxide, electrons in the valence band of TiO_2 could be directly excited to CuO nanoclusters, which could serve as an efficient oxygen reduction site through the multi-electron process. Significant improvement of visible light activity was achieved for CuO-modified-self-doped titania (CuO/Ti^{3+}-TiO_2) in comparison with CuO/TiO_2 in the oxidation of gaseous 2-propanol. Under visible light irradiation, induced electrons on an isolated Ti^{3+} band could be transferred to the surface CuO nanoclusters efficiently, in addition to the direct charge transfer from the titania VB to the CuO nanoclusters, as shown in Figure 9. Therefore, for CuO/Ti^{3+}-TiO_2, under visible light irradiation, there were holes in the VB decomposed 2-propanol, while photoinduced electrons were consumed (via oxygen reduction) on CuO nanoclusters. It was proposed that amorphous Cu(II) oxide nanoclusters grafted on titania surface suppressed the recombination of electron—hole pairs (at the isolated Ti^{3+} band), and acted as a co-catalyst for efficient oxygen reduction to consume the photoinduced electrons. Therefore, the photocatalytic activity of CuO/Ti^{3+}-TiO_2 under both UV and visible light irradiation was very high and stable [92].

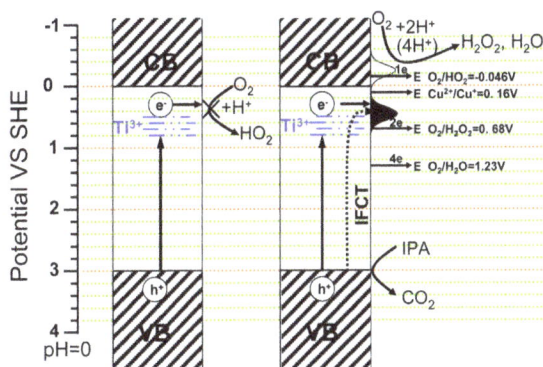

Figure 9. Proposed photocatalytic mechanism of CuO/Ti^{3+}-TiO_2 visible light activation. Reprinted with permission from [92]. Copyright American Chemical Society, 2011.

Cu_2O-TiO_2 Heterojunction

For the first time, a Cu_2O/TiO_2 heterojunction system without the presence of other forms of copper was prepared by electrochemical deposition of Cu_2O on Ti foil by Siripala et al. as a thin film heterojunction photocathode [93]. They observed a photoresponce demonstrating efficient light-induced charge carriers' separation at Cu_2O-TiO_2 junction. Moreover, it was proposed that TiO_2 film successfully protected the Cu_2O layer against photocorrosion. In the next study, Li et al. prepared composites of Cu_2O-TiO_2 (P25) by electrochemical method for UV-decomposition of dye brilliant red [94]. Prepared material was more active than pure P25, but unfortunately the mechanism was not discussed.

Bessekhouad et al. studied photocatalytic efficiency of Cu_2O/TiO_2, prepared by direct mixing of both semiconductors [70]. The activity of this heterojunction system was checked by using model organic pollutant Orange II under both UV and visible light irradiation. It was assumed that Brownian motion sufficed to permit charge transfers between the particles of the two solids. They observed that the amount of Cu_2O in the heterojunction played an important role, since vis photocatalytic efficiency increased with an increase in Cu_2O amount reaching saturation at ca. 30% (Figure 10a). It was proposed that under visible light irradiation, electrons from Cu_2O could be injected into the CB of TiO_2, and at titania surface react with dissolved oxygen molecules and induce a formation of oxygen peroxide radicals ($O_2^{\bullet-}$). Similarly, in the case of the UV system, an increase in content of Cu_2O resulted in enhanced efficiency, but resultant activities at high content of Cu_2O were only slightly higher than that of pure TiO$_2$, whereas for low content of Cu$_2$O a significant deactivation of TiO$_2$ was noticed (Figure 10b). This fact was attributed to the increase of charge trapping mechanism contributions when both semiconductors are excited. According to the authors, the CB of Cu$_2$O was more negative than the CB of TiO$_2$ favoring the electron transfer from Cu$_2$O to TiO$_2$. Other conditions, such as the kinetics of reactions occurring at each surface, should be also considered. For example, the holes generated in Cu$_2$O had to react faster to induce an efficient charge separation [70]. Moreover, it should be reminded that lack of holes' consumption (oxidation reaction by holes) could result in photocorrosion of the sensitizer (here Cu$_2$O) [69].

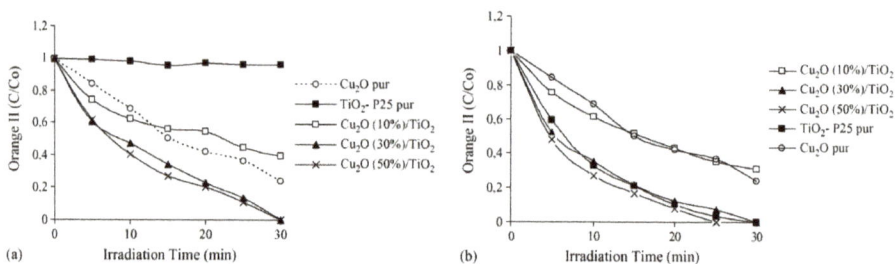

Figure 10. Photocatalytic degradation of Orange II in Cu$_2$O/TiO$_2$ system under: (**a**) visible irradiation and (**b**) UV–vis irradiation. Reprinted with permission from [70]. Copyright Elsevier, 2005.

Huang et al. performed similar investigations on Cu$_2$O/TiO$_2$ system under UV and visible light irradiation for decomposition of Orange II [95]. They used titania P25, Cu(CH$_3$COO)$_2$·H$_2$O as a copper source and alcohol-aqueous based chemical precipitation method to prepare heterojunction photocatalysts, in which copper was present in only one form-Cu$_2$O (2–3-nm particle size). Similar to Bessekhouad work, it was found, that an increase in content of Cu$_2$O resulted in higher photocatalytic activity (70%-Cu$_2$O content with highest activity). They reported 6 and 27 times higher photocatalytic activity for Cu$_2$O/TiO$_2$ system than that for pure P25 for UV–Vis and visible light irradiation, respectively. The weak point of this and Bessekhouad's research under visible light irradiation is the fact that vis-response should not be tested for dyes due to possibility of semiconductor sensitization [96,97]. Although Huang et al. justified that the excited electrons could not migrate from Orange II to Cu$_2$O, since the reduction potential of Acid Orange II was lower (-1.25 V) than CB of Cu$_2$O (-1.5 V), the possibility of electron migration to CB of titania could not be omitted. According to Figure 11, under UV light irradiation, both Cu$_2$O and TiO$_2$ could be excited (process (1) and (3) corresponding to Equations (10) and (11), respectively), the generated electrons in Cu$_2$O and holes in TiO$_2$ could migrate to CB of TiO$_2$ (process (2)) and VB of Cu2O (process (4)), respectively. This transfer process is thermodynamic favorable since both CB and VB of Cu$_2$O lie above that of TiO$_2$ (heterojunction type II), which results in prolongation of the lifetime of excited electrons and holes, inducing higher quantum efficiency. Meanwhile, the generated electrons react with dissolved oxygen molecules and produce O$_2$$^{\bullet-}$ (Equations (12) and (13)) to decompose organic pollutants (Equation (14)), while the holes could induce some oxidation process directly (Equation (15)). Under visible light irradiation, process (1) can also occur due to the narrow band gap of Cu$_2$O (2.0 eV), and thus Orange II still can be photocatalytically decomposed by the redox reactions (Equations (11)–(14)), where Equation (12) is possible due to the transfer of electrons from Cu$_2$O (process (2)):

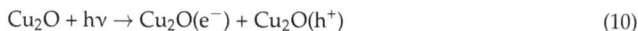

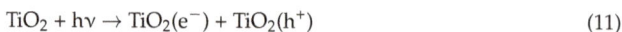

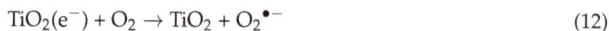

$$Cu_2O + h\nu \rightarrow Cu_2O(e^-) + Cu_2O(h^+) \tag{10}$$

$$TiO_2 + h\nu \rightarrow TiO_2(e^-) + TiO_2(h^+) \tag{11}$$

$$TiO_2(e^-) + O_2 \rightarrow TiO_2 + O_2{}^{\bullet-} \tag{12}$$

$$Cu_2O(e^-) + O_2 \rightarrow Cu_2O + O_2{}^{\bullet-} \tag{13}$$

$$OII^*{}_{ads} + O_2{}^{\bullet-} \rightarrow products \tag{14}$$

$$OII^*{}_{ads} + TiO_2(h^+) \rightarrow products \tag{15}$$

Similar conclusions were proposed by Zhang et al., who prepared TiO_2 film covered by a Cu_2O microgrid by the microsphere lithography method [98]. The underlying TiO_2 film was composed of nanosized particles covered by microgrids of Cu_2O, which were composed of particles smaller than 20 nm. By comparison of three types of thin films—TiO_2, Cu_2O and TiO_2-Cu_2O for methylene blue degradation—it was found that under both UV and visible light irradiation the heterojunction system was the most active. The authors suggested the same mechanism of heterojunction as that explained above (Figure 11).

A modified explanation of the enhanced photocatalytic activity of Cu_2O/TiO_2 heterojunction system considering the role of Ti^{3+} was proposed by Xiong et al. [99]. TiO_2 and Cu_2O were prepared separately and final coupled photocatalyst was obtained by suspension mixing. It was reported that Cu_2O/TiO_2 photocatalyst had much better vis activity in degradation of brilliant red X-3B and photocatalytic hydrogen evolution than TiO_2 and Cu_2O alone. It was proposed that photogenerated electrons from Cu_2O were captured by Ti^{4+} ions in TiO_2 and thus being reduced to Ti^{3+} ions, as shown in Figure 12. Whereas, left holes in the valence band of Cu_2O formed "hole centers" hampering the charge carriers' recombination. Therefore, trapped electrons (in Ti^{3+} ions with prolonged lifetime) could be transferred to the interface between the composite and solution, resulting in retarded recombination between electrons in TiO^2 (Ti^{3+}) and holes in Cu_2O [99].

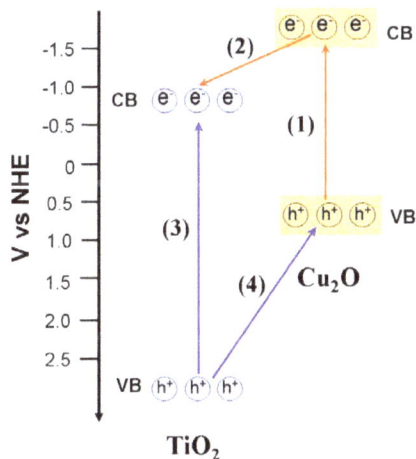

Figure 11. The scheme of excitation and separation of electrons and holes for Cu_2O/TiO_2 heterostructures under irradiation. Transformations (1) and (3) represent the excitation process in Cu_2O and TiO_2, respectively; process (2) and (4) stand for the transfer of electrons and holes between Cu_2O and TiO_2). Transformations (1) and (2) occur both under UV and visible light, while process (3) and (4) under UV light. Reprinted with permission from [95]. Copyright Elsevier, 2009.

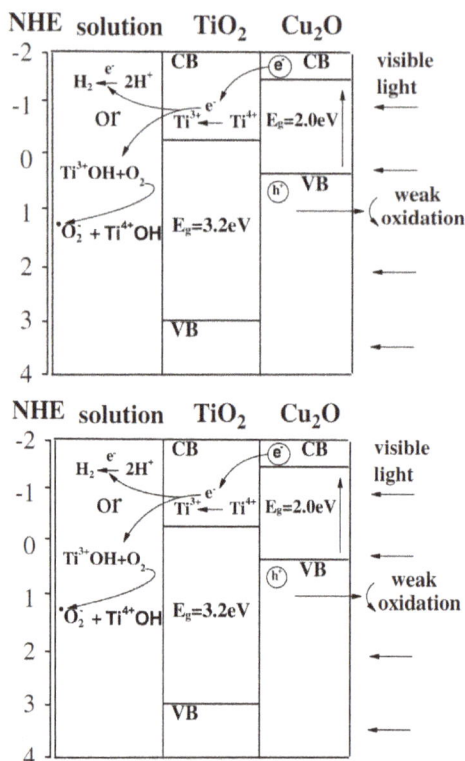

Figure 12. Interfacial electron transfer in Cu_2O/TiO_2 composite in the presence of Ti^{3+}. Reprinted with permission from [99]. Copyright Elsevier, 2011.

Wang et al. prepared Cu_2O-loaded TiO_2 nanotube arrays (NTAs) by electrochemical anodization of Ti foil and subsequent Cu_2O deposition by an ultrasonication-assisted sequential chemical bath deposition (S-CBD) method [100]. They check both visible light photocatalytic (Rhodamine B as model pollutant) and photoelectrocatalytic activities of Cu_2O/TiO_2 NTAs. The photocatalysts with small content of Cu_2O exhibited the largest photocurrent and photoconversion efficiency under visible light irradiation, as well as the highest visible light photocatalytic degradation rate of RhB. In particular, when 0.5 V bias potential was applied, Cu_2O/TiO_2 NTA photoelectrodes were found to possess superior photoelectrocatalytic efficiency, due to a synergistic effect of electricity and visible light irradiation [100].

However, in the case of heterojunctions for photocatalysts obtained according to standard methods an improved charge separation was observed, some limitations were also suggested, due to the incompact contact and small interface between two semiconductors. Therefore, the preparation of interconnected heterostructures with a large interface between two semiconductors was proposed for improvement of the charge transfer efficiency. For example, $Cu_2O@TiO_2$ core-shell photocatalyst was prepared by in situ hydrolysis and crystallization method (Figure 13), and tested under solar radiation for 4-nitrophenol degradation [101]. It was found that $Cu_2O@TiO_2$ had absorption properties similar to pure Cu_2O: strong visible light absorption in the range of 400–600 nm (Figure 14a), and was the most active among other coupled photocatalysts, e.g., prepared by physical mixing $Cu_2O/TiO_{2(PM)}$ and by chemical method denoted to Cu_2O/TiO_2 (Figure 14b). Moreover, weaker photoluminescence signals for $Cu_2O@TiO_2$ sample than for pure Cu_2O suggested retardation of charge carriers' recombination, probably because introduction of TiO_2 shell could reduce the surface oxygen vacancies and defect in Cu_2O core [101].

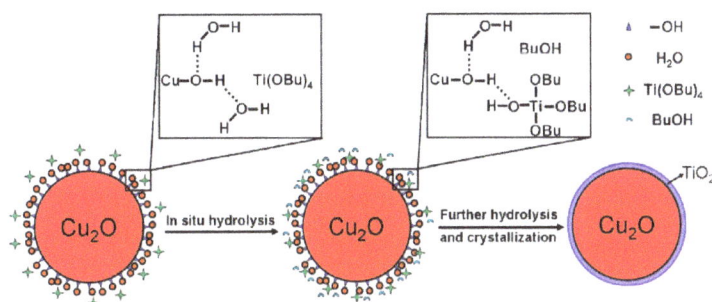

Figure 13. The formation of compact interface between two components ascribed as in situ hydrolysis and crystallization synthetic method of core-shell $Cu_2O@TiO_2$. Reprinted with permission from [101]. Copyright Elsevier, 2011.

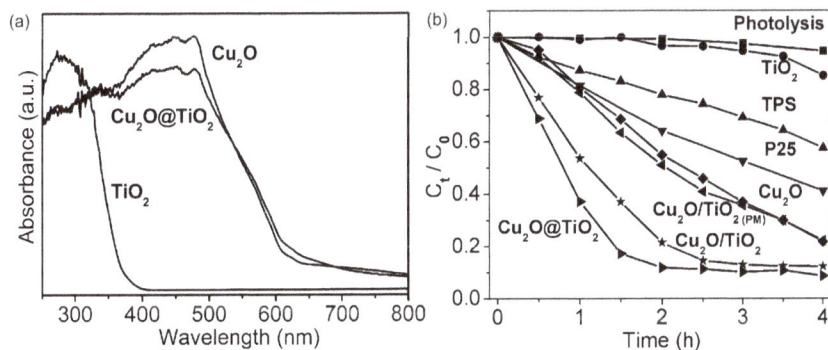

Figure 14. (a) UV–Vis diffuse reflection spectra of pure Cu_2O and $Cu_2O@TiO_2$ sample; (b) comparison of photocatalytic degradation of 4-nitrophenol under solar light in the absence of the photocatalyst and over different photocatalytic systems. Reprinted with permission from [101]. Copyright Elsevier, 2011.

Another strategy for obtaining efficient copper oxide–titania coupled photocatalyst is based on consideration of the morphology of copper oxide and titania. For example, Liu et al. synthesized Cu_2O nanospheres decorated with TiO_2 nanoislands (Cu_2O-NS/TiO_2-NI) by a facile hydrolyzation reaction followed by a solvent-thermal process. [102]. It was found that Cu_2O-NS/TiO_2-NI demonstrated superior photocatalytic performance under visible light illumination for methyl orange degradation, *E. coli* bacteria disinfection, and also a better stability during the photocatalysis process from their specific structure. The design of a partial coverage of Cu_2O nanosphere with TiO_2 nano-islands enabled the reaction between photo-generated holes and water to produce hydroxyl radicals or directly with organic pollutants/microorganisms in water. Therefore, the accumulation of holes on the underlying Cu_2O films prevented photocorrosion, and subsequently made this photocatalyst stable during the photocatalysis process (Figure 15). It was shown that part of Ti^{4+} was reduced to Ti^{3+} in Cu_2O-NS/TiO_2-NI, suggesting that under visible light irradiation photo-generated electrons in Cu_2O were transferred to TiO_2 and trapped there. After shutting off the irradiation, these trapped electrons could be gradually released from TiO_2 to react with O_2 producing reactive radicals, which imparted the photocatalytic memory effect to Cu_2O-NS/TiO_2-NI photocatalyst (Figure 15) [102].

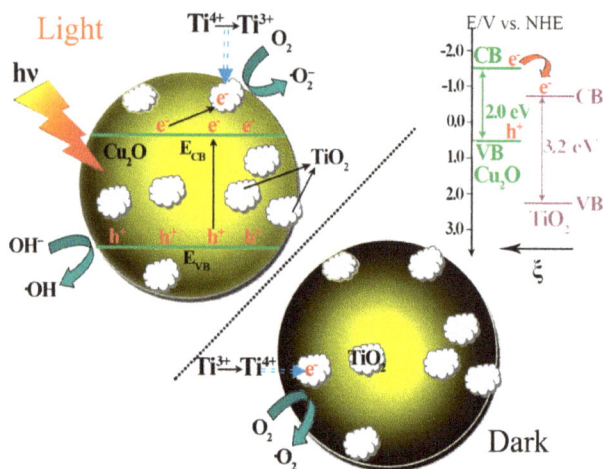

Figure 15. Proposed energy band structure of the Cu2O/TiO2 *p-n* heterojunction, the photocatalytic activity enhancement mechanism under visible light illumination, and the post-illumination catalytic memory mechanism in the dark. Reprinted with permission from [102]. Copyright American Chemical Society, 2014.

It is well known that controlled morphology by application of exposed facets (faceted semiconductors) significantly enhances photocatalytic performance, due to excellent crystallinity and low content of defects, e.g., octahedral and decahedral anatase particles [103,104]. Therefore, it is not surprising that similar concept was applied for copper oxide–titania heterojunction systems. For example, Liu et al. prepared $Cu_2O@TiO_2$ photocatalysts using Cu_2O with exposed facets [105]. First, Cu_2O nanocrystals with different types of morphology: cubes, cuboctahedra and octahedra were synthesized, and then $Cu_2O@TiO_2$ core-shell polyhedral nanostructures were prepared by hydrothermal reaction. It was proven then the morphology of Cu_2O cores was well preserved during hydrothermal process (Figure 16A). Well-crystallized anatase nanoparticles created a uniform and rough layer covering the smooth surface of Cu_2O core. The photocatalytic activity of $Cu_2O@TiO_2$ in the visible light was checked in relation to methylene blue and 4-nitrophenol. The activity of all coupled samples was higher than that of single Cu_2O and TiO_2, and increased with the following order regarding Cu_2O morphology: cubes < cuboctahedra < octahedra. The surface photovoltage spectroscopy (SPS) was used to analyze photoinduced carrier separation and charge transfer behavior, where the magnitude of SPS response peak depended on the amount of net charge accumulated on the material surface. It was shown that SPS response correlated with the type of facet configuration, and $Cu_2O@TiO_2$ octahedra had the strongest SPS response, $Cu_2O@TiO_2$ cuboctahedra—medium, whereas $Cu_2O@TiO_2$ cubes—the lowest (Figure 16B), showing the same tendency as the photocatalytic activity. Moreover, it was found that $Cu_2O@TiO_2$ core-shell photocatalysts had different band offset values (Figure 16C), demonstrating a clear facet-dependent activity. The photocatalytic performance of $Cu_2O@TiO_2$ polyhedra was different with that of pure Cu_2O polyhedra without TiO_2 shells, which could be attributed to the different driving forces for the charge carrier separation [105].

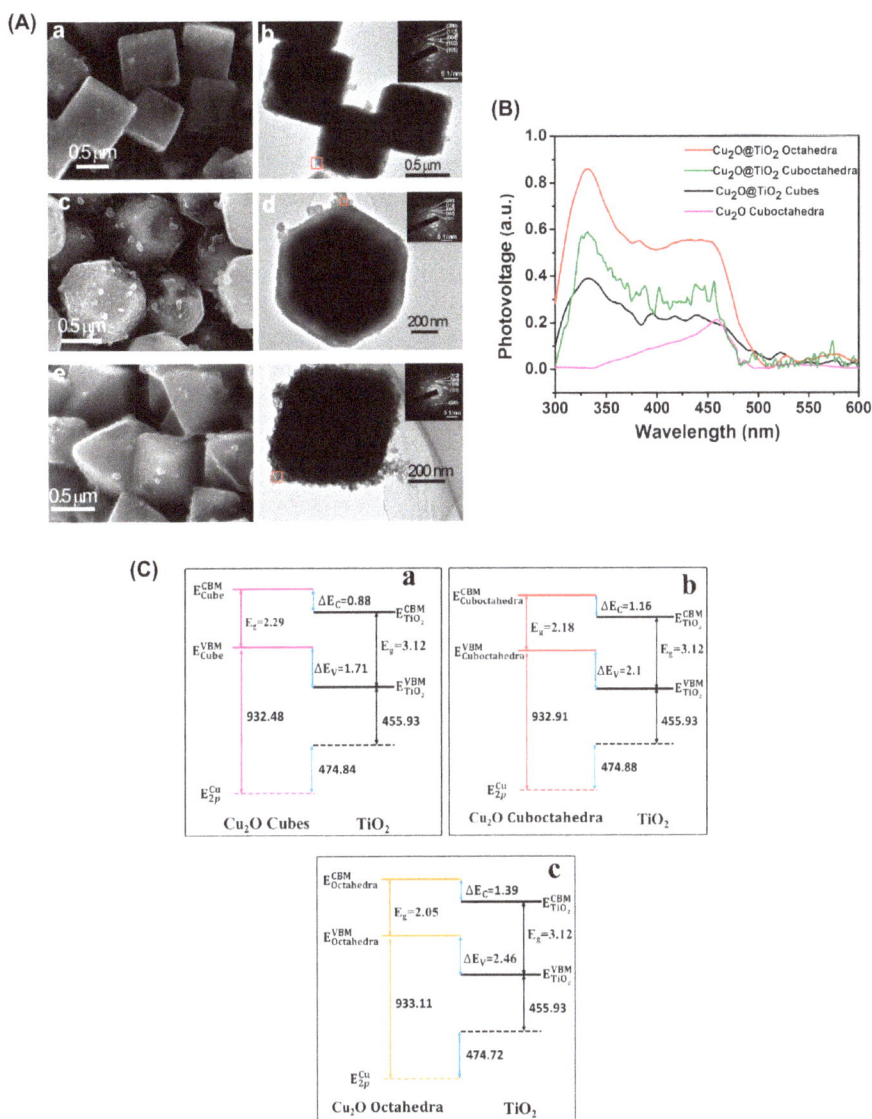

Figure 16. (A) SEM and TEM images of $Cu_2O@TiO_2$, Cu_2O nanocrystals in the shape of cubes—(**a,b**) and, cuboctahedra—(**c,d**), and octahedra—(**e,f**); (B) Surface photovoltage spectroscopy (SPS) spectra of $Cu_2O@TiO_2$ core-shell polyhedra, compared with that of the as-synthesized Cu_2O cuboctahedra; (C) Energy band diagrams for different types of morphology. Reprinted with permission from [105]. Copyright American Chemical Society, 2015.

Mixed Copper Species Deposited on Titania Surface

Frequently, during preparation of copper-modified TiO_2, a mixture of copper species is obtained [58,59], and such systems are named as "cascade heterojunction systems." For example, Helaili et al. prepared Cu_2O/TiO_2, $Cu/Cu_2O/TiO_2$ and $Cu/Cu_2O/CuO/TiO_2$ heterojunctions using chemical methods, and examined their potential application as photocatalysts for oxidative reactions

(Orange II as a model dye pollutant) under visible light irradiation [106]. The obtained results suggested the following rules, which could govern the electrons transfer mechanism: (a) metallic Cu improved the photocatalytic activity of single semiconductor by the formation of apparent Ohmic junction enhancing the charges transfer kinetics; (b) electromotive forces developed from semiconductor/semiconductor heterojunction suppressed the effect of metallic Cu due to the formation of an apparent Schottky type junction; (c) diffusion potential developed from multiple steps electrons transfer had more impact for improving the efficiency than that developed from one step transition; and (d) heterojunction cascade obtained from monobloc of photosensitizers was more efficient than that obtained from direct mixture of the semiconductors [106]. Xing et al. prepared $Cu/Cu_2O/TiO_2$ photocatalyst as an efficient heterojunction system for terephthalic acid degradation under UV/Vis irradiation [107]. According to the mechanism presented in Figure 17, under UV/vis irradiation electrons and holes are photogenerated on both Cu_2O and TiO_2. Because the Fermi energy level of Cu is lower than that of $_p$-type Cu_2O, electrons from the CB of Cu_2O migrate to Cu until the two Fermi levels are aligned. Holes in the VB of Cu_2O recombine with electrons in the CB of TiO_2. Therefore, holes in the VB of TiO_2 are efficiently separated from electrons, and engage in hydroxyl radicals' formation. In the case of visible light irradiation, only Cu_2O is photoexcited. Electrons in the CB of Cu_2O are trapped by highly conductive Cu, whereas migration of holes from the VB of Cu_2O to VB of TiO_2 is impossible, and thus, the negligible amount of hydroxyl radicals can be detected. This transmission pathway (under UV) is analogous to that of the Z-scheme system instead of a type II heterojunction system. It means that $Cu/Cu_2O/TiO_2$ system with very intensive charge separation is much more efficient in the case of UV/vis system than for visible light system, where Z-scheme model dominates [107].

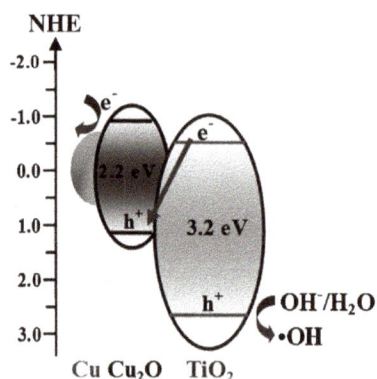

Figure 17. Reaction pathway for photoexcited e–h pairs in $Cu/Cu_2O/TiO_2$ nanojunctions. Reprinted with permission from [107]. Copyright John Wiley & Sons, Inc., 2013.

Analogically, it is thought that the morphology of titania should be also important issue for the efficiency of copper species–titania heterojunction system. Janczarek et al. prepared copper-modified titania based on decahedral anatase particles (DAP) with eight equivalent (101) facets and two (001) facets [58]. Copper species were photodeposited on DAP surface. It was estimated that Cu_2O was a predominant form of copper (82.0%), whereas other copper species existed in minority, CuO (12.8%) and Cu(0) (5.2%). Copper species formed small nanoclusters of ca. 2 nm, which were uniformly distributed on all facets of DAP. This material $Cu/Cu_2O/CuO/TiO_2$(DAP) was especially active in UV/vis system for acetic acid oxidation but very low improvement of activity for 2-propanol oxidation was observed under visible light irradiation, confirming mechanistic issues described in the previously mentioned studies [58].

Qiu et al. prepared Cu_xO-TiO_2 nanocomposites dedicated to indoor environments for degradation of volatile organic compounds and pathogens [108]. The structure of copper species deposited on titania surface was nanocluster mixture of Cu_2O and CuO. The balance between $Cu(I)$ and $Cu(II)$ states in Cu_xO was critical to achieving efficient VOC decomposition and antipathogenic activity. It was found that the optimum $Cu(I)/Cu(II)$ ratio in Cu_xO nanoclusters was 1.3. This material had also good antiviral and antibacterial activity under dark conditions. The ratio of $Cu(I)/Cu(II)$ did not markedly change after long-term visible-light irradiation, indicating that the multielectron reduction reaction was catalytic in air with a turnover number greater than 22 [108]. Luna et al. studied Cu_2O-CuO-TiO_2(P25) semiconductors for gallic acid oxidation under UV/vis LEDs irradiation [109]. By using UV light, CuO addition to TiO_2 did not enhance the photoactivity and acted like an impurity or a defect at the surface, favoring the fast recombination of charge carriers, or like an electron fast scavenger. In the case of visible light system (Figure 18), CuO addition evidenced the injection of charge carriers from Cu_2O to TiO_2 and CuO, as suggested by TRMC study. Gallic acid degradation proceeded by a sequence of reactions involving the formation of intermediates such as, maleic, fumaric, oxalic and formic acids, before reaching its mineralization [109].

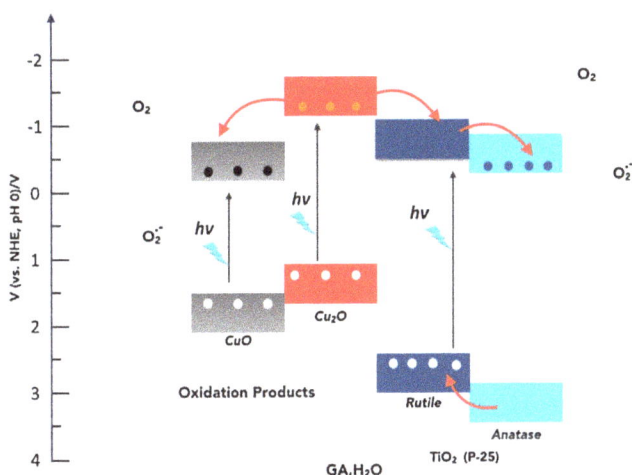

Figure 18. A proposed scheme showing the photoredox processes during degradation of gallic acid (GA) in presence of Cu_2O–CuO–TiO_2 (P25) photocatalysts, air and visible light (LEDs). Reprinted with permission from [109]. Copyright Elsevier, 2016.

3. Conclusions

The design of an efficient and stable photocatalyst for oxidative reaction systems with high activity under UV and visible light is still a challenge. Among various metals considered as titania modifiers, copper is one of the most promising because of its abundancy, low cost and similar properties to expensive noble metals such as gold and silver. The enhancement of TiO_2 photocatalytic activity with copper modification can be realized by two main mechanistic concepts: preparation of plasmonic photocatalyst with metallic copper on titania surface and formation of heterojunctions between titania and copper oxides. Direct doping of titania lattice by copper ions is rather difficult because of different sizes of Cu^{2+} and Ti^{4+}. The chemical stability of copper species is a very important issue for copper-modified TiO_2. This issue should be always considered in design thinking about copper-modified titania, primarily for metallic copper. In this review, the successful attempts to obtain stable metallic Cu were presented, and this breakthrough research may cause an increase in the number of research in the copper-focused, visible light-oriented plasmonic photocatalysis.

This review also highlights various heterojunction concepts according to different copper oxides–titania configurations. Owing to the authors' knowledge there is still no comprehensive research paper considering all main variants of copper oxides–titania heterojunctions to perform a comparable study. To analyze the mechanism of heterojunction, it is necessary to calculate exact positions of VB and CB of semiconductors. The problem is that these values for copper oxides can significantly differ in various papers, resulting in change of the possible type of heterojunction. Therefore, an accurate photoelectrochemical characterization of each component of heterojunction system is required. Among copper oxides in relation to heterojunction with titania, Cu_2O has the promising application potential to wastewater treatment including its very good antipathogenic properties, even better that metallic copper [108]. For oxidative reaction systems, cascade heterojunction with different forms of copper is also a perspective option. Additional presence of metallic copper in such heterojunction system is beneficial under UV irradiation, but is detrimental in the case of visible light irradiation. The presence of Ti^{3+} ions can also positively influence the efficiency of heterojunction system. Moreover, the content of copper oxides (Cu_2O/CuO) in cascade heterojunction system is a very significant parameter influencing photocatalytic activity. The new research areas are opening with the development of methods, which allow to obtain copper oxides and titania with specific particle morphology. The influence of different types of crystal facets of Cu_2O on the efficiency of Cu_2O/TiO_2 system has been confirmed and is still under debate. Furthermore, the core-shell configurations for copper oxide–titania heterojunctions can increase a photocatalytic efficiency and stability of whole photocatalytic system. Additionally, the post-illumination catalytic memory phenomenon, observed in Cu_2O(nanospheres)/TiO_2(nanoislands) can be important extension for photocatalytic properties under dark conditions.

Acknowledgments: M.J. acknowledges Hokkaido University for guest lecture position (2016–2017).

Author Contributions: M.J. wrote the paper; E.K. corrected the manuscript. All authors read and approved the final manuscript.

Conflicts of Interest: The authors declare no conflict of interest.

References

1. Schneider, J.; Matsuoka, M.; Takeuchi, M.; Zhang, J.; Horiuchi, Y.; Anpo, M.; Bahnemann, D.W. Understanding TiO_2 photocatalysis: Mechanisms and materials. *Chem. Rev.* **2014**, *114*, 9919–9986. [CrossRef] [PubMed]
2. Bahnemann, D.W. Photocatalytic water treatment: Solar energy applications. *Sol. Energy* **2004**, *77*, 445–459. [CrossRef]
3. Lee, S.A.; Choo, K.H.; Lee, H.I.; Hyeon, T.; Choi, W.; Kwon, H.H. Use of ultrafiltration membranes for the separation of TiO_2 photocatalysts in drinking water treatment. *Ind. Eng. Chem. Res.* **2001**, *40*, 1712–1719. [CrossRef]
4. Politano, A.; Cupolillo, A.; Di Profio, G.; Arafat, H.A.; Chiarello, G.; Curcio, E. When plasmonics meets membrane technology. *J. Phys. Condens. Matter* **2016**, *28*, 363003. [CrossRef] [PubMed]
5. Hay, S.O.; Obee, T.; Luo, Z.; Jiang, T.; Meng, Y.T.; He, J.K.; Murphy, S.C.; Suib, S. The viability of photocatalysis for air purification. *Molecules* **2015**, *20*, 1319–1356. [CrossRef] [PubMed]
6. Chen, X.B.; Shen, S.H.; Guo, L.J.; Mao, S.S. Semiconductor-based photocatalytic hydrogen generation. *Chem. Rev.* **2010**, *110*, 6503–6570. [CrossRef] [PubMed]
7. Li, K.; Peng, B.S.; Peng, T.Y. Recent advances in heterogeneous photocatalytic CO_2 conversion to solar fuels. *ACS Catal.* **2016**, *6*, 7485–7527. [CrossRef]
8. Roose, B.; Pathak, S.; Steiner, U. Doping of TiO_2 for sensitized solar cells. *Chem. Soc. Rev.* **2015**, *44*, 8326–8349. [CrossRef] [PubMed]
9. Parkin, I.P.; Palgrave, R.G. Self-cleaning coatings. *J. Mater. Chem.* **2005**, *15*, 1689–1695. [CrossRef]
10. Pelaez, M.; Nolan, N.T.; Pillai, S.C.; Seery, M.K.; Falaras, P.; Kontos, A.G.; Dunlop, P.S.M.; Hamilton, J.W.J.; Byrne, J.A.; O'Shea, K.; et al. A review on the visible light active titanium dioxide photocatalysts for environmental applications. *Appl. Catal. B Environ.* **2012**, *125*, 331–349. [CrossRef]
11. Ohtani, B. Titania photocatalysis beyond recombination: A critical review. *Catalysts* **2013**, *3*, 942–953. [CrossRef]

12. Emeline, A.V.; Ryabchuk, V.K.; Serpone, N. Dogmas and misconceptions in heterogeneous photocatalysis. Some enlightened reflections. *J. Phys. Chem. B* **2005**, *109*, 18515–18521. [CrossRef] [PubMed]

13. Herrmann, J.M. Heterogeneous photocatalysis: State of the art and present applications. *Top. Catal.* **2005**, *34*, 49–65. [CrossRef]

14. Choi, W.Y.; Termin, A.; Hoffmann, M.R. The role of metal-ion dopants in quantum-sized TiO$_2$—Correlation between photoreactivity and charge-carrier recombination dynamics. *J. Phys. Chem.* **1994**, *98*, 13669–13679. [CrossRef]

15. Yu, J.C.; Yu, J.G.; Ho, W.K.; Jiang, Z.T.; Zhang, L.Z. Effects of F-doping on the photocatalytic activity and microstructures of nanocrystalline TiO$_2$ powders. *Chem. Mater.* **2002**, *14*, 3808–3816. [CrossRef]

16. Ohtani, B.; Kakimoto, M.; Nishimoto, S.; Kagiya, T. Photocatalytic reaction of neat alcohols by metal-loaded titanium(IV) oxide particles. *J. Photochem. Photobiol. A Chem.* **1993**, *70*, 265–272. [CrossRef]

17. Sclafani, A.; Herrmann, J.M. Influence of metallic silver and of platinum-silver bimetallic deposits on the photocatalytic activity of titania (anatase and rutile) in organic and aqueous media. *J. Photochem. Photobiol. A Chem.* **1998**, *113*, 181–188. [CrossRef]

18. Engweiler, J.; Harf, J.; Baiker, A. WO$_x$/TiO$_2$ catalysts prepared by grafting of tungsten alkoxides: Morphological properties and catalytic behavior in the selective reduction of NO by NH$_3$. *J. Catal.* **1996**, *159*, 259–269. [CrossRef]

19. Vinodgopal, K.; Kamat, P.V. Enhanced rates of photocatalytic degradation of an azo-dye using SnO$_2$/TiO$_2$ coupled semiconductor thin-films. *Environ. Sci. Technol.* **1995**, *29*, 841–845. [CrossRef] [PubMed]

20. Allen, S.E.; Walvoord, R.R.; Padilla-Salinas, R.; Kozlowski, M.C. Aerobic copper-catalyzed organic reactions. *Chem. Rev.* **2013**, *113*, 6234–6458. [CrossRef] [PubMed]

21. Bhanushali, S.; Ghosh, P.; Ganesh, A.; Cheng, W.L. 1D copper nanostructures: Progress, challenges and opportunities. *Small* **2015**, *11*, 1232–1252. [CrossRef] [PubMed]

22. Clarizia, L.; Spasiano, D.; Di Somma, I.; Marotta, R.; Andreozzi, R.; Dionysiou, D.D. Copper-modified-TiO$_2$ catalysts for hydrogen generation through photoreforming of organics. A short review. *Int. J. Hydrogen Energy* **2014**, *39*, 16812–16831. [CrossRef]

23. Yan, H.H.; Zhao, T.J.; Li, X.J.; Hun, C.H. Synthesis of Cu-doped nano-TiO$_2$ by detonation method. *Ceram. Int.* **2015**, *41*, 14204–14211. [CrossRef]

24. Sreekantan, S.; Zaki, S.M.; Lai, C.W.; Tzu, T.W. Copper-incorporated titania nanotubes for effective lead ion removal. *Mater. Sci. Semicond. Process.* **2014**, *26*, 620–631. [CrossRef]

25. Janczarek, M.; Zielińska-Jurek, A.; Markowska, I.; Hupka, J. Transparent thin films of Cu–TiO$_2$ with visible light photocatalytic activity. *Photochem. Photobiol. Sci.* **2014**, *14*, 591–596. [CrossRef] [PubMed]

26. Xin, B.; Wang, P.; Ding, D.; Liu, J.; Ren, Z.; Fu, H. Effect of surface species on Cu-TiO$_2$ photocatalytic activity. *Appl. Surf. Sci.* **2008**, *254*, 2569–2574. [CrossRef]

27. Liu, Z.; Wang, Y.; Peng, X.; Li, Y.; Liu, Z.; Liu, C.; Ya, J.; Huang, Y. Photoinduced superhydrophilicity of TiO$_2$ thin film with hierarchical Cu doping. *Sci. Technol. Adv. Mater.* **2012**, *13*, 025001. [CrossRef] [PubMed]

28. Gawande, M.B.; Goswami, A.; Felpin, F.X.; Asefa, T.; Huang, X.; Silva, R.; Zou, X.; Zboril, R.; Varma, R.S. Cu and Cu-based nanoparticles: Synthesis and applications in catalysis. *Chem. Rev.* **2016**, *116*, 3722–3811. [CrossRef] [PubMed]

29. Reiche, H.; Dunn, W.W.; Bard, A.J. Heterogeneous photocatalytic and photosynthetic deposition of copper on TiO$_2$ and WO$_3$ powders. *J. Phys. Chem.* **1979**, *83*, 2248–2251. [CrossRef]

30. Foster, N.S.; Noble, R.D.; Koval, C.A. Reversible photoreductive deposition and oxidative dissolution of copper ions in titanium dioxide aqueous suspensions. *Environ. Sci. Technol.* **1993**, *27*, 350–356. [CrossRef]

31. Foster, N.S.; Lancaster, A.N.; Noble, R.D.; Koval, C.A. Effect of Organics on the Photodeposition of Copper in Titanium-Dioxide Aqueous Suspensions. *Ind. Eng. Chem. Res.* **1995**, *34*, 3865–3871. [CrossRef]

32. Jacobs, J.W.M.; Kampers, F.W.H.; Rikken, J.M.G.; Bullelieuwma, C.W.T.; Koningsberger, D.C. Copper photodeposition on TiO$_2$ studied with HREM and EXAFS. *J. Electrochem. Soc.* **1989**, *136*, 2914–2923. [CrossRef]

33. Herrmann, J.M.; Disdier, J.; Pichat, P. Photoassisted platinum deposition on TiO$_2$ powder using various platinum complexes. *J. Phys. Chem.* **1986**, *90*, 6028–6034. [CrossRef]

34. Okamoto, K.; Yamamoto, Y.; Tanaka, H.; Tanaka, M.; Itaya, A. Heterogeneous photocatalytic decomposition of phenol over TiO$_2$ powder. *Bull. Chem. Soc. Jpn.* **1985**, *58*, 2015–2022. [CrossRef]

35. Bideau, M.; Claudel, B.; Faure, L.; Rachimoellah, M. Photo-oxidation of formic acid by oxygen in the presence of titanium dioxde and dissolved copper ions: Oxygen transfer and reaction kinetics. *Chem. Eng. Commun.* **1990**, *93*, 167–179. [CrossRef]

36. Bideau, M.; Claudel, B.; Faure, L.; Kazouan, H. The photo-oxidation of acetic acid by oxygen in the presence of titanium dioxide and dissolved copper ions. *J. Photochem. Photobiol. A Chem.* **1991**, *61*, 269–280. [CrossRef]

37. Bideau, M.; Claudel, B.; Faure, L.; Kazouan, H. The photo-oxidation of propionic acid by oxygen in the presence of TiO₂ and dissolved copper ions. *J. Photochem. Photobiol. A Chem.* **1992**, *67*, 337–348. [CrossRef]

38. Brezova, V.; Blazkova, A.; Borosova, E.; Ceppan, M.; Fiala, R. The influence of dissolved metal ions on the photocatalytic degradation of phenol in aqueous TiO₂ suspensions. *J. Mol. Catal. A Chem.* **1995**, *98*, 109–116. [CrossRef]

39. Butler, E.C.; Davis, A.P. Photocatalytic oxidation in aqueous titanium dioxide suspensions: The influence of dissolved transition metals. *J. Photochem. Photobiol. A Chem.* **1993**, *70*, 273–283. [CrossRef]

40. Sykora, J. Photochemistry of copper complexes and their environmental aspects. *Coord. Chem. Rev.* **1997**, *159*, 95–108. [CrossRef]

41. Bideau, M.; Claudel, B.; Faure, L.; Kazouan, H. Metallic complexes as intermediates in homogeneously and heterogeneously photocatalysed reactions. *J. Photochem. Photobiol. A Chem.* **1994**, *84*, 57–67. [CrossRef]

42. Kraeutler, B.; Baard, A.J. Heterogeneous photocatalytic preparation of supported catalysts. Photodeposition of platinum on titanium dioxide powder and other substrates. *J. Am. Chem. Soc.* **1978**, *100*, 4317–4318. [CrossRef]

43. Koudelka, M.; Sanchez, J.; Augustynski, J. Electrochemical and surface characteristics of the photocatalytic platinum deposits on titania. *J. Phys. Chem.* **1982**, *86*, 4277–4280. [CrossRef]

44. Ueno, K.; Misawa, H. Surface plasmon-enhanced photochemical reactions. *J. Photochem. Photobiol. C Photochem. Rev.* **2013**, *15*, 31–52. [CrossRef]

45. Kowalska, E.; Prieto Mahaney, O.O.; Abe, R.; Ohtani, B. Visible-light-induced photocatalysis through surface plasmon excitation of gold on titania surfaces. *Phys. Chem. Chem. Phys.* **2010**, *12*, 2344–2355. [CrossRef] [PubMed]

46. Ingram, D.B.; Linic, S. Water splitting on composite plasmonic-metal/semiconductor photoelectrodes: Evidence for selective plasmon-induced formation of charge carriers near the semiconductor surface. *J. Am. Chem. Soc.* **2011**, *133*, 5202–5205. [CrossRef] [PubMed]

47. Yamaguchi, T.; Kazuma, E.; Sakai, N.; Tatsuma, T. Photoelectrochemical responses from polymer-coated plasmonic copper nanoparticles on TiO₂. *Chem. Lett.* **2012**, *41*, 1340–1342. [CrossRef]

48. Zhang, S.; Peng, B.; Yang, S.; Wang, H.; Yu, H.; Fang, Y.; Peng, F. Non-noble metal copper nanoparticles-decorated TiO₂ nanotube arrays with plasmon-enhanced photocatalytic hydrogen evolution under visible light. *Int. J. Hydrogen Energy* **2015**, *40*, 303–310. [CrossRef]

49. DeSario, P.A.; Pietron, J.J.; Brintlinger, T.H.; McEntee, M.; Parker, J.F.; Baturina, O.; Stroud, R.M.; Rolison, D.R. Oxidation-stable plasmonic copper nanoparticles in photocatalytic TiO₂ nanoarchitectures. *Nanoscale* **2017**, *9*, 11720–11729. [CrossRef] [PubMed]

50. Chan, G.H.; Zhao, J.; Hicks, E.M.; Schatz, G.C.; Van Duyne, R.P. Plasmonic properties of copper nanoparticles fabricated by nanosphere lithography. *Nano Lett.* **2007**, *7*, 1947–1952. [CrossRef]

51. Kowalska, E. Plasmon-assisted catalysis. In *Gold Nanoparticles for Physics, Chemistry and Biology*; Pluchery, C.L.O., Ed.; World Scientific: Singapore, 2017; pp. 319–364.

52. Kaur, R.; Pal, B. Plasmonic coinage metal—TiO₂ hybrid nanocatalysts for highly efficient photocatalytic oxidation under sunlight irradiation. *New J. Chem.* **2015**, *39*, 5966–5976. [CrossRef]

53. Hou, W.; Cronin, S.B. A review of surface plasmon resonance-enhanced photocatalysis. *Adv. Funct. Mater.* **2012**, *23*, 1612–1619. [CrossRef]

54. Kanninen, P.; Johans, C.; Merta, J.; Kontturi, K. Influence of ligand structure on the stability and oxidation of copper nanoparticles. *J. Colloid Interface Sci.* **2008**, *318*, 88–95. [CrossRef] [PubMed]

55. Pastoriza-Santos, I.; Sanchez-Iglesias, A.; Rodriguez-Gonzales, B.; Liz-Marzan, L.M. Aerobic synthesis of Cu nanoplates with intense plasmon resonances. *Small* **2009**, *5*, 440–443. [CrossRef] [PubMed]

56. Singh, M.; Sinha, I.; Premkumar, M.; Singh, A.K.; Mandal, R.K. Structural and surface plasmon behavior of Cu nanoparticles using different stabilizers. *Colloids Surf. A* **2010**, *359*, 88–94. [CrossRef]

57. Mendez-Medrano, M.G.; Kowalska, E.; Lehoux, A.; Herissan, A.; Ohtani, B.; Bahena, D.; Briois, V.; Colbeau-Justin, C.; Rodrigues-Lopez, J.L.; Remita, H. Surface modification of TiO_2 with Ag nanoparticles and CuO nanoclusters for application in photocatalysis. *J. Phys. Chem. C* **2016**, *120*, 5143–5154. [CrossRef]
58. Janczarek, M.; Wei, Z.; Endo, M.; Ohtani, B.; Kowalska, E. Silver- and copper-modified decahedral anatase titania particles as visible light-responsive plasmonic photocatalyst. *J. Photon. Energy* **2017**, *7*, 012008. [CrossRef]
59. Wei, Z.; Endo, M.; Wang, K.; Charbit, E.; Markowska-Szczupak, A.; Ohtani, B.; Kowalska, E. Noble metal-modified octahedral anatase titania particles with enhanced activity for decomposition of chemical and microbiological pollutants. *Chem. Eng. J.* **2017**, *318*, 121–134. [CrossRef] [PubMed]
60. Muniz-Miranda, M.; Gellini, C.; Simonelli, A.; Tiberi, M.; Giammanco, F.; Giorgetti, E. Characterization of copper nanoparticles obtained by laser ablation in liquids. *Appl. Phys. A* **2013**, *110*, 829–833. [CrossRef]
61. Guo, X.; Hao, C.; Jin, G.; Zhu, H.Y.; Guo, X.Y. Copper Nanoparticles on Graphene Support: An Efficient Photocatalyst for Coupling of Nitroaromatics in Visible Light. *Angew. Chem. Int. Ed.* **2014**, *53*, 1973–1977. [CrossRef] [PubMed]
62. Gonzales-Posada, F.; Sellapan, R.; Vanpoucke, B.; Charakov, D. Oxidation of copper nanoparticles in water monitored in situ by localized surface plasmon resonance spectroscopy. *RSC Adv.* **2014**, *4*, 20659–20664. [CrossRef]
63. Rice, K.P.; Walker, E.J.; Stoykovich, M.P.; Saunders, A.E. Solvent-Dependent Surface Plasmon Response and Oxidation of Copper Nanocrystals. *J. Phys. Chem. C* **2011**, *115*, 1793–1799. [CrossRef]
64. Susman, M.D.; Feldman, Y.; Vaskevich, A.; Rubinstein, I. Chemical deposition and stabilization of plasmonic copper nanoparticle films on transparent substrates. *Chem. Mater.* **2012**, *24*, 2501–2508. [CrossRef]
65. Jimenez, J.A. Carbon as reducing agent for the precipitation of plasmonic Cu particles in glass. *J. Alloys Compd.* **2016**, *656*, 685–688. [CrossRef]
66. Fu, Q.; Wagner, T. Interaction of nanostructured metal overlayers with oxide surfaces. *Surf. Sci. Rep.* **2007**, *62*, 431–498. [CrossRef]
67. Kum, J.M.; Park, Y.J.; Kim, H.J.; Cho, S.O. Plasmon-enhanced photocatalytic hydrogen production over visible-light responsive Cu/TiO_2. *Nanotechnology* **2015**, *26*, 125402. [CrossRef] [PubMed]
68. Bessekhouad, Y.; Robert, D.; Weber, J.V. Bi_2S_3/TiO_2 and CdS/TiO_2 heterojunctions as an available configuration for photocatalytic degradation of organic pollutant. *J. Photochem. Photobiol. A Chem.* **2004**, *163*, 569–580. [CrossRef]
69. Serpone, N.; Maruthamuthu, P.; Pichat, P.; Pelizzetti, E.; Hidaka, H. Exploiting the interparticle electron transfer process in the photocatalysed oxidation of phenol, 2-chlorophenol and pentachlorophenol: Chemical evidence for electron and hole transfer between coupled semiconductors. *J. Photochem. Photobiol. A Chem.* **1995**, *85*, 247–255. [CrossRef]
70. Bessekhouad, Y.R.D.; Weber, J.-V. Photocatalytic activity of Cu_2O/TiO_2, Bi_2O_3/TiO_2 and $ZnMn_2O_4/TiO_2$ heterojunctions. *Catal. Today* **2005**, *101*, 315–321. [CrossRef]
71. Marschall, R. Semiconductor composites: Strategies for enhancing charge carrier separation to improve photocatalytic activity. *Adv. Funct. Mater.* **2014**, *24*, 2421–2440. [CrossRef]
72. Barreca, D.; Fornasiero, P.; Gasparotto, A.; Gombac, V.; Maccato, C.; Montini, T.; Tondello, E. The potential of supported Cu_2O and CuO nanosystems in photocatalytic H_2 production. *ChemSusChem* **2009**, *2*, 230–233. [CrossRef] [PubMed]
73. Hara, M.; Kondo, T.; Komoda, M.; Ikeda, S.; Shinohara, K.; Tanaka, A.; Kondo, J.N.; Domen, K. Cu_2O as a photocatalyst for overall water splitting under visible light irradiation. *Chem. Commun.* **1998**, 357–358. [CrossRef]
74. Khan, K.A. Stability of a Cu_2O photoelectrode in an electrochemical cell and the performances of the photoelectrode coated with Au and SiO thin films. *Appl. Energy* **2000**, *65*, 59–66. [CrossRef]
75. Nguyen, M.A.; Bedford, N.M.; Ren, Y.; Zahran, E.M.; Goodin, R.C.; Chagani, F.F.; Bachas, L.G.; Knecht, M.R. Direct synthetic control over the size, composition, and photocatalytic activity of octahedral copper oxide materials: Correlation between surface structure and catalytic functionality. *ACS Appl. Mater. Interfaces* **2015**, *7*, 13238–13250. [CrossRef] [PubMed]
76. Deng, X.L.; Wang, C.G.; Shao, M.H.; Xu, X.J.; Huang, J.Z. Low-temperature solution synthesis of CuO/Cu_2O nanostructures for enhanced photocatalytic activity with added H_2O_2: Synergistic effect and mechanism insight. *RSC Adv.* **2017**, *7*, 4329–4338. [CrossRef]

77. Song, K.Y.; Kwon, Y.T.; Choi, G.J.; Lee, W.I. Photocatalytic activity of Cu/TiO$_2$ with oxidation state of surface-loaded copper. *Bull. Korean Chem. Soc.* **1999**, *20*, 957–960.
78. Chiang, K.; Amal, R.; Tran, T. Photocatalytic degradation of cyanide using titanium dioxide modified with copper oxide. *Adv. Environ. Res.* **2002**, *6*, 471–485. [CrossRef]
79. Wu, S.-X.; Ma, Z.; Qin, Y.-N.; He, F.; Jia, L.-S.; Zhang, Y.-J. XPS study of copper doping TiO$_2$ photocatalyst. *Acta Phys.-Chim. Sin.* **2003**, *19*, 967–969.
80. Arana, J.; Alonso, A.P.; Rodriguez, J.M.D.; Melian, J.A.H.; Diaz, O.G.; Pena, J.P. Comparative study of MTBE photocatalytic degradation with TiO$_2$ and Cu-TiO$_2$. *Appl. Catal. B Environ.* **2008**, *78*, 355–363. [CrossRef]
81. Irie, H.; Miura, S.; Kamiya, K.; Hashimoto, K. Efficient visible light-sensitive photocatalysts: Grafting Cu(II) ions onto TiO$_2$ and WO$_3$ photocatalysts. *Chem. Phys. Lett.* **2008**, *457*, 202–205. [CrossRef]
82. Irie, H.; Kamiya, K.; Shibanuma, T.; Miura, S.; Tryk, D.A.; Yokoyama, T.; Hashimoto, K. Visible light-sensitive Cu(II)-grafted TiO$_2$ photocatalysts: Activities and X-ray absorption fine structure analyses. *J. Phys. Chem. C* **2009**, *113*, 10761–10766. [CrossRef]
83. Yu, H.; Irie, H.; Shimodaira, Y.; Hosogi, Y.; Kuroda, Y.; Miyauchi, M.; Hashimoto, K. An efficient visible-light-sensitive Fe(III)-grafted TiO$_2$ photocatalyst. *J. Phys. Chem. C* **2010**, *114*, 16481–16487. [CrossRef]
84. Arai, T.; Horiguchi, M.; Yanagida, M.; Gunji, T.; Sugihara, H.; Sayama, K. Reaction mechanism and activity of WO$_3$-catalyzed photodegradation of organic substances promoted by a CuO cocatalyst. *J. Phys. Chem. C* **2009**, *113*, 6602–6609. [CrossRef]
85. Liu, M.; Sunada, K.; Hashimoto, K.; Miyauchi, M. Visible-light sensitive Cu(II)–TiO$_2$ with sustained anti-viral activity for efficient indoor environmental remediation. *J. Mater. Chem. A* **2015**, *3*, 17312–17319. [CrossRef]
86. Li, G.; Dimitrijevic, N.M.; Chen, L.; Rajh, T.; Gray, K.A. Role of surface/interfacial Cu^{2+} sites in the photocatalytic activity of coupled CuO-TiO$_2$ nanocomposites. *J. Phys. Chem. C* **2008**, *112*, 19040–19044. [CrossRef]
87. Slamet, N.H.W.; Purnama, E.; Kosela, S.; Gunlazuardi, J. Photocatalytic reduction of CO$_2$ on copper-doped titania catalysts prepared by improved-impregnation method. *Catal. Commun.* **2005**, *6*, 313–319.
88. Kowalska, E.; Rau, S.; Ohtani, B. Plasmonic titania photocatalysts active under UV and visible-light irradiation: Influence of gold amount, size, and shape. *J. Nanotechnol.* **2012**, *2012*, 361853. [CrossRef]
89. Wang, K.; Wei, Z.; Ohtani, B.; Kowalska, E. Interparticle electron transfer in methanol dehydrogenation on platinum-loaded titania particles prepared from P25. *Catal Today* **2017**. [CrossRef]
90. Moniz, S.J.A.; Tang, J. Charge transfer and photocatalytic activity in CuO/TiO$_2$ nanoparticle heterojunctions synthesised through a rapid, one-pot, microwave solvothermal route. *Chem. Cat. Chem.* **2015**, *7*, 1659–1667.
91. Siah, W.R.; Lintang, H.O.; Shamsuddin, M.; Yoshida, H.; Tuliati, L. Masking effect of copper oxides photodeposited on titanium dioxide: Exploring UV, visible, and solar light activity. *Catal. Sci. Technol.* **2016**, *6*, 5079–5087. [CrossRef]
92. Liu, M.; Qiu, X.; Miyauchi, M.; Hashimoto, K. Cu(II) oxide amorphous nanoclusters grafted Ti^{3+} self-doped TiO$_2$: An efficient visible light photocatalyst. *Chem. Mater.* **2011**, *23*, 5282–5286. [CrossRef]
93. Siripala, W.I.A.; Jaramillo, T.F.; Baeck, S.-H.; McFarland, E.W. A Cu$_2$O/TiO$_2$ heterojunction thin film cathode for photoelectrocatalysis. *Sol. Energy Mater. Sol. C* **2003**, *77*, 229–237. [CrossRef]
94. Li, J.; Liu, L.; Yu, Y.; Tang, Y.; Li, H.; Du, F. Preparation of highly photocatalytic active nano-size TiO$_2$-Cu$_2$O particle composites with a novel electrochemical method. *Electrochem. Commun.* **2004**, *6*, 940–943. [CrossRef]
95. Huang, L.; Peng, F.; Wang, H.; Yu, H.; Li, Z. Preparation and characterization of Cu$_2$O/TiO$_2$ nano–nano heterostructure photocatalysts. *Catal. Commun.* **2009**, *10*, 1839–1843. [CrossRef]
96. Kisch, H.; Macyk, W. Visible-light photocatalysis by modified titania. *Chem. Phys. Chem.* **2002**, *3*, 399–400. [CrossRef]
97. Yan, X.; Ohno, T.; Nishijima, K.; Abe, R.; Ohtani, B. Is methylene blue an appropriate substrate for a photocatalytic activity test? A study with visible-light responsive titania. *Chem. Phys. Lett.* **2006**, *429*, 606–610. [CrossRef]
98. Zhang, J.; Zhu, H.; Zheng, S.; Pan, F.; Wang, T. TiO$_2$ Film/Cu$_2$O microgrid heterojunction with photocatalytic activity under solar light irradiation. *Appl. Mater. Interfaces* **2009**, *1*, 2111–2114. [CrossRef] [PubMed]
99. Xiong, L.B.; Yang, F.; Yan, L.L.; Yan, N.N.; Yang, X.; Qiu, M.Q.; Yu, Y. Bifunctional photocatalysis of TiO$_2$/Cu$_2$O composite under visible light: Ti^{3+} in organic pollutant degradation and water splitting. *J. Phys. Chem. Solids.* **2011**, *72*, 1104–1109. [CrossRef]

100. Wang, M.; Sun, L.; Cai, J.; Xie, K.; Lin, C. *p–n* Heterojunction photoelectrodes composed of Cu_2O-loaded TiO_2 nanotube arrays with enhanced photoelectrochemical and photoelectrocatalytic activities. *Energy Environ. Sci.* **2013**, *6*, 1211–1220. [CrossRef]

101. Chu, S.; Zheng, X.M.; Kong, F.; Wu, G.H.; Luo, L.L.; Guo, Y.; Liu, H.L.; Wang, Y.; Yu, H.X.; Zou, Z.G. Architecture of $Cu_2O@TiO_2$ core-shell heterojunction and photodegradation for 4-nitrophenol under simulated sunlight irradiation. *Mater. Chem. Phys.* **2011**, *129*, 1184–1188. [CrossRef]

102. Liu, L.M.; Yang, W.Y.; Li, Q.; Gao, S.A.; Shang, J.K. Synthesis of Cu_2O nanospheres decorated with TiO_2 nanoislands, their enhanced photoactivity and stability under visible light illumination, and their post-illumination catalytic memory. *ACS Appl. Mater. Inter.* **2014**, *6*, 5629–5639. [CrossRef] [PubMed]

103. Wei, Z.; Kowalska, E.; Verret, J.; Colbeau-Justin, C.; Remita, H.; Ohtani, B. Morphology-dependent photocatalytic activity of octahedral anatase particles prepared by ultrasonication-hydrothermal reaction of titanates. *Nanoscale* **2015**, *7*, 12392–12404. [CrossRef] [PubMed]

104. Janczarek, M.; Kowalska, E.; Ohtani, B. Decahedral-shaped anatase titania photocatalyst particles: Synthesis in a newly developed coaxial-flow gas-phase reactor. *Chem. Eng. J.* **2016**, *289*, 502–512. [CrossRef]

105. Liu, L.; Yang, W.; Sun, W.; Li, Q.; Shang, J.K. Creation of $Cu_2O@TiO_2$ composite photocatalysts with *p–n* heterojunctions formed on exposed Cu_2O facets, their energy band alignment study, and their enhanced photocatalytic activity under illumination with visible light. *ACS Appl. Mater. Inter.* **2015**, *7*, 1465–1476. [CrossRef] [PubMed]

106. Helaili, N.; Bessekhouad, Y.; Bouguelia, A.; Trari, M. Visible light degradation of Orange II using xCu_yO_z/TiO_2 heterojunctions. *J. Hazard. Mater.* **2009**, *168*, 484–492. [CrossRef] [PubMed]

107. Xing, J.; Chen, Z.P.; Xiao, F.Y.; Ma, X.Y.; Wen, C.Z.; Li, Z.; Yang, H.G. Cu-Cu_2O-TiO_2 nanojunction systems with an unusual electron-hole transportation pathway and enhanced photocatalytic properties. *Chem. Asian J.* **2013**, *8*, 1265–1270. [CrossRef] [PubMed]

108. Qiu, X.; Miyauchi, M.; Sunada, K.; Minoshima, M.; Liu, M.; Lu, Y.; Li, D.; Shimodaira, Y.; Hosogi, Y.; Kuroda, Y.; et al. Hybrid Cu_xO/TiO_2 nanocomposites as risk-reduction materials in indoor environments. *ACS Nano* **2012**, *6*, 1609–1618. [CrossRef] [PubMed]

109. Luna, A.L.; Valenzuela, M.A.; Colbeau-Justin, C.; Vazquez, P.; Rodriguez, J.; Avendano, J.R.; Alfaro, S.; Tirado, S.; Garduno, A.; De la Rosa, J.M. Photocatalytic degradation of gallic acid over CuO-TiO_2 composites under UV/Vis LEDs irradiation. *Appl. Catal. A Gen.* **2016**, *521*, 140–148. [CrossRef]

© 2017 by the authors. Licensee MDPI, Basel, Switzerland. This article is an open access article distributed under the terms and conditions of the Creative Commons Attribution (CC BY) license (http://creativecommons.org/licenses/by/4.0/).

catalysts

MDPI

Article

Photocatalytic TiO$_2$ Nanorod Spheres and Arrays Compatible with Flexible Applications

Daniela Nunes *, Ana Pimentel, Lidia Santos, Pedro Barquinha, Elvira Fortunato * and Rodrigo Martins *

i3N/CENIMAT, Department of Materials Science, Faculty of Sciences and Technology, Universidade NOVA de Lisboa and CEMOP/UNINOVA, 2829-516 Campus de Caparica, Caparica, Portugal; acgp@campus.fct.unl.pt (A.P.); ls.santos@campus.fct.unl.pt (L.S.); pmcb@fct.unl.pt (P.B.)
* Correspondence: daniela.gomes@fct.unl.pt (D.N.); emf@fct.unl.pt (E.F.); rm@uninova.pt (R.M.);
Tel.: +351-212-948-558 (D.N., E.F. & R.M.); Fax: +351-212-948-562 (D.N., E.F. & R.M.)

Academic Editors: Vladimiro Dal Santo and Alberto Naldoni
Received: 29 December 2016; Accepted: 7 February 2017; Published: 14 February 2017

Abstract: In the present study, titanium dioxide nanostructures were synthesized through microwave irradiation. In a typical microwave synthesis, nanorod spheres in the powder form were simultaneously produced with nanorod arrays grown on polyethylene terephthalate (PET) substrates. The syntheses were performed in water or ethanol with limited temperature at 80 °C and 200 °C. A simple and low-cost approach was used for the arrays growth, which involved a PET substrate with a zinc oxide seed layer deposited by spin-coating. X-ray diffraction (XRD) and Raman spectroscopy revealed that synthesis in water result in a mixture of brookite and rutile phases, while using ethanol as solvent it was only observed the rutile phase. Scanning electron microscopy (SEM) showed that the synthesized spheres were in the micrometer range appearing as aggregates of fine nanorods. The arrays maintained the sphere nanorod aggregate structures and the synthesis totally covered the flexible substrates. Transmission electron microscopy (TEM) was used to identify the brookite structure. The optical band gaps of all materials have been determined from diffuse reflectance spectroscopy. Photocatalytic activity was assessed from rhodamine B degradation with remarkable degradability performance under ultraviolet (UV) radiation. Reusability experiments were carried out for the best photocatalyst, which also revealed notable photocatalytic activity under solar radiation. The present study is an interesting and competitive alternative for the photocatalysts existing nowadays, as it simultaneously results in highly photoactive powders and flexible materials produced with low-cost synthesis routes such as microwave irradiation.

Keywords: TiO$_2$; nanorod spheres; nanorod arrays; flexible substrates; microwave irradiation; photocatalysis

---

1. Introduction

Titanium dioxide (TiO$_2$) has been extensively studied for applications ranging from sensors [1,2] to dye-solar cells [3]. However, recently, this material has been widely used as photocatalyst agents [4–10]. The great interest in TiO$_2$ as photocatalyst is related to its low-cost, non-toxicity, high stability and photoactivity, and earth-abundance [5–7,11–15]. TiO$_2$ commonly appears in the amorphous form or as three distinct crystalline phases: two tetragonal ones, anatase and rutile, and as an orthorhombic phase, brookite [16]. From these crystalline phases, rutile is the most stable, while anatase and brookite are metastable and can be converted to rutile upon heating [17]. Moreover, TiO$_2$ typically exhibits *n*-type semiconductor character, displaying optical band gaps of 3.0 and 3.2 eV for rutile and anatase, respectively [18]. The band gap values reported for brookite in the literature are varied, and range from 3.13 to 3.40 eV [17,18].

TiO$_2$ is usually used as a photocatalyst in the forms of anatase and rutile [19–22]. Recently, it has been found that the mixture of both phases displayed higher photocatalytic activity than pure phases [23]. Brookite, in contrast, is not explored for photocatalytic applications; as the other TiO$_2$ crystalline phases, however, the interest on this material has been growing lately [17,24,25]. For these materials to act as photocatalysts, redox reactions on their surface must occur [5]. In general, the photocatalytic degradation will occur when TiO$_2$ is exposed to radiation with higher energy than its band gap, creating electron–hole pairs. Then, electrons in the conduction band generate superoxides [26], while holes create hydroxyl radicals [5,27] which will decompose organic and inorganic compounds [28].

The improved photocatalytic activity of materials relies on several factors, in which the most expressive ones are crystal size, crystalline phase, specific surface area, impurities and exposed surface facets [5,6]. Several TiO$_2$ micro- and nanomaterials with numerous shapes, such as sheets [29], spheres [30], rods [31], plumes [32], and wires [33], have been reported for photocatalytic applications. Moreover, it has been previously reported that multi-scaled structures, such as nanorods in micro-spheres, are promising for improving photocatalytic activity [34], since they combine the higher surface/volume ratio of nanorods, which results in higher density of active sites for surface reactions [35,36] to the micrometer size of spheres that can be easily recycled due to the enhanced intrinsic weight sedimentation ability [34]. TiO$_2$ photocatalysts in the form of films or arrays are also frequently used, avoiding the recycling processes [5]. However, the films are often grown or deposited on rigid substrates [37], which limits their application on adaptable surfaces and increases production costs.

Several physical and chemical techniques have been reported to produce TiO$_2$ nanostructures, arrays and films over the years [38]. These techniques include sol-gel method [39], atomic layer deposition [40], thermal evaporation [41], sputtering [42], hydrothermal and solvothermal synthesis [43–46], and microwave irradiation [5]. Indeed, microwave synthesis appears as a viable and inexpensive option for the TiO$_2$ production due to its intrinsic characteristics and by the fact that it relies on the efficient heating of solvents and/or reagents [47], providing accurate temperature control, celerity, enhanced efficiency/cost balance [47–50] and uniformity [51]. Nevertheless, the type of solvent employed in microwave-assisted routes is a key parameter [47,52]. Several metal oxide-based materials have been synthesized under microwave irradiation, producing well-defined structures and morphologies, and specific phases or compositions with high purity [5,50,52–56].

The present work reports the production and photocatalytic activity of TiO$_2$ nanorod spheres and nanorod arrays grown on PET, both simultaneously synthesized under microwave irradiation with water or ethanol as solvents, and where the synthesis temperature was fixed. The aim is to challenge the present state of the art concerning the exploitation of an easy, environmentally friendly, low-cost sustainable and reliable approach to produce improved TiO$_2$ photocatalysts both in the powder form as micro-sized particles easily recycled and as uniform arrays on PET substrates that can be adaptable to different surfaces such as the ones required for water/wastewater treatment. Moreover, to the best of the author's knowledge, TiO$_2$ arrays grown at low temperatures on ZnO seeded PET substrates using inexpensive synthesis and seed deposition routes without any process to increase the PET adhesion to TiO$_2$ has never been reported before. It is also intended to offer options to commercial TiO$_2$ photocatalyst materials in the powder form, which normally presents limitations, especially in recovery due to its nanometer size (~30 nm) [20]. Structural and morphologic characterizations of the microwave synthesized materials have been carried out by X-ray diffraction (XRD), Raman spectroscopy, scanning electron microscopy (SEM) coupled with energy dispersive X-ray spectroscopy (EDS) and transmission electron microscopy (TEM). Optical properties were assessed through diffuse reflectance spectroscopy, and the photocatalytic activity has been evaluated from the evolution of rhodamine B degradation under ultraviolet (UV) and solar radiations. Rhodamine B was considered as a model-test contaminant and indicator due to its photocatalytic activity and absorption peaks in the visible range, thus its degradation can be easily monitored by optical absorption spectroscopy.

2. Results and Discussion

TiO_2 nanorod spheres and arrays were synthesized under microwave irradiation using two distinct solvents and fixing the synthesis temperature to understand the influence of these parameters on the material properties and final photocatalytic behavior. Following an easy and low-cost approach, the TiO_2 nanorod arrays were grown on seeded flexible substrates at low temperature, which opens to the possibility of employing these materials to a broad range of applications.

2.1. X-ray Diffraction and Raman Spectroscopy

Figure 1a shows powder diffractograms of the nanorod spheres (powder) synthesized using water and ethanol as solvents and with synthesis temperature of 80 °C and 200 °C. The solvent clear influences the final TiO_2 phase of the materials produced. All peaks in the experimental diffractograms could be assigned to either brookite or rutile depending on the synthesis conditions used. The materials synthesized with water showed a mixture of brookite and rutile phases, with higher amounts of brookite to the 80 °C H_2O material (53% of brookite (ICDD file No. 01-076-1937) and 47% of rutile (ICSD file No. 96-900-4143)). The 200 °C H_2O material resulted in a proportion of 26% of brookite to 74% of rutile. The fact that higher temperatures resulted in lower amounts of brookite is due to a temperature threshold where all the brookite phase is converted into rutile [17]. Lower temperatures form higher amounts of brookite, and higher temperatures form rutile-rich structures. Both materials synthesized with ethanol formed single phased materials fully assigned to rutile. The use of ethanol has been reported previously to form rutile-rich nanostructures [57,58]. No peaks associated to impurities such as $Ti(OH)_4$ were detected and all the peaks suggest that the materials are well crystallized. The broad width observed for the materials synthesized at 80 °C suggests the presence of small sized particles.

Raman measurements were also carried out (Figure 1b), as this technique allows distinguishing between TiO_2 phases [59]. Raman spectra confirmed the presence of both brookite and rutile for the 80 °C H_2O nanorod spheres. The Raman bands associated to rutile were also discernible for the 80 °C condition. Nevertheless, for the 200 °C condition, the phase mostly present is rutile. The nanorod spheres synthesized with ethanol revealed only the presence of the rutile phase for both the 80 °C and 200 °C conditions. The Raman bands associated to rutile can be assigned to B_{1g} (110 cm^{-1}), E_g (237 and 443 cm^{-1}) and A_{1g} (609 cm^{-1}), respectively, while the Raman bands of brookite can be assigned to A_{1g} (124, 150, 198 cm^{-1}), B_{1g} (214 and 278 cm^{-1}), B_{2g} (365 cm^{-1}), and B_{3g} (320 cm^{-1}) [59]. For all conditions, the characteristic Raman band of ~518 cm^{-1} [59] associated to the presence of anatase was not detected, confirming the XRD results. No additional bands could be found in both spectra.

XRD measurements of the TiO_2 arrays were not presented due to the broad and intense peaks coming from the substrate which obscured the TiO_2 signal, nevertheless from Raman spectra, the TiO_2 phases present at each material could be identified (Figure 2). The Raman spectrum from the PET substrate is presented for comparison. From the PET bands on both Raman spectra, it is possible to infer that the 80 °C EtOH nanorod arrays are thicker or denser than the ones synthesized with water (the PET bands (*) are more expressive on the 80 °C H_2O material). Raman spectrum of the 80 °C H_2O nanorod arrays confirmed the presence of both brookite and rutile for this material in accordance to the nanorod sphere results (see Figure 1a,b). The 80 °C EtOH revealed to be a single phased material fully assigned to rutile, such as the nanorod spheres (see Figure 1a,b). Some of the TiO_2 Raman bands were overlapped by the PET bands, however the ones without overlapping were in accordance to the TiO_2 Raman bands detected for the powder materials. Raman bands coming from the ZnO seed layer were not identified, nevertheless some ZnO Raman bands overlap with TiO_2 ones [52].

2.2. Electron Microscopy

Figure 3 shows the SEM images of TiO_2 nanorod spheres. All the experimental conditions resulted in micro-sized spheres composed by fine nanorods appearing radially arranged. The 80 °C

H_2O material exhibited two sphere structures: one more spherical with cracks, and the other displaying a flower-like (cauliflower) structure. Nevertheless, both sphere types displaying closely packed nanorods. The average sphere diameter was 2.4 ± 0.9 µm. Moreover, a large amount of individual nanostructured particles was also detected for this condition (see the right inset pointed by the arrow in Figure 3a or Figure S1), demonstrating that a clear mixture of structures is present, as expected from XRD and Raman results (Figure 1). For the 200 °C H_2O condition, the material evolved to homogeneous spherical structures with an average sphere diameter of 3.6 ± 0.8 µm. Other types of structures continued to surround the spheres for this synthesis condition. The 80 °C H_2O nanorods presented average widths of 11.4 ± 3.1 nm, while with the increase of synthesis temperature, the nanorods increased their size (average widths of 31.6 ± 8.3 nm) and the square shape of the nanorods could be clearly discernible (compare insets in Figure 3a,b).

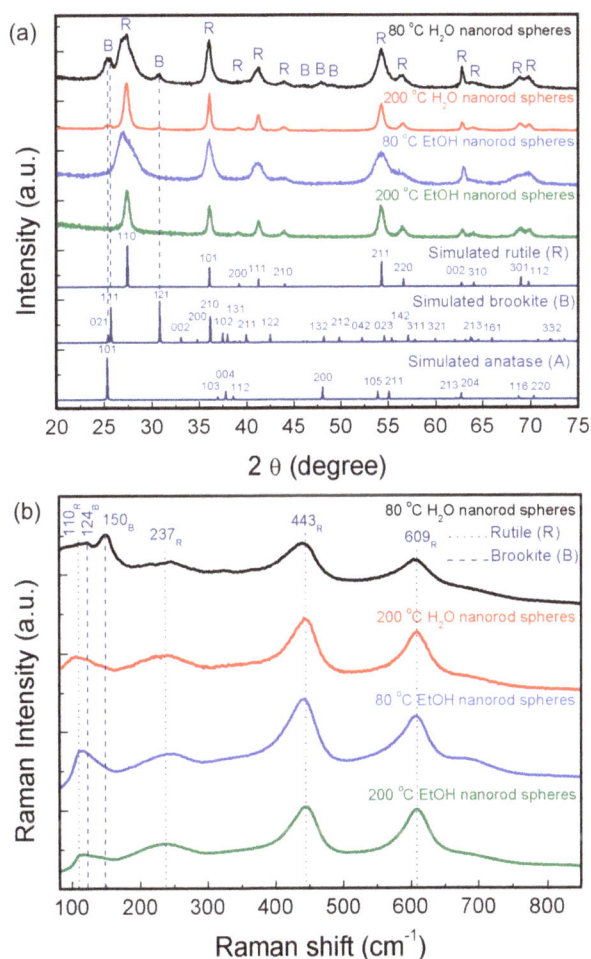

Figure 1. (a) XRD diffractograms from the TiO_2 nanorod spheres produced using water and ethanol at 80 and 200 °C. For comparison, experimental data are presented together with the simulated rutile, brookite and anatase diffractograms; **(b)** Raman spectra of the TiO_2 nanorod spheres. Dot lines indicate the rutile bands and dashed ones point out to the brookite ones.

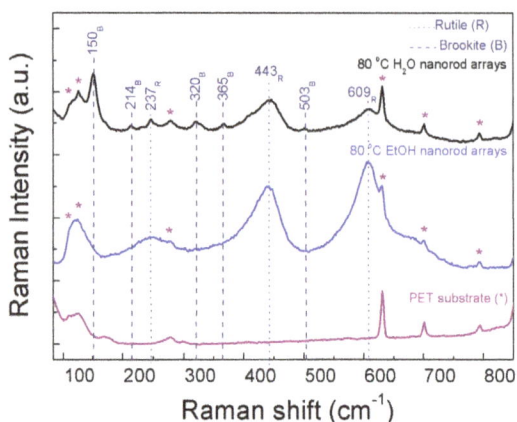

Figure 2. Raman spectra of the TiO$_2$ nanorod arrays produced with water and ethanol at 80 °C, together with the PET substrate for comparison. Dot lines indicate the rutile bands and dashed ones point out to the brookite ones.

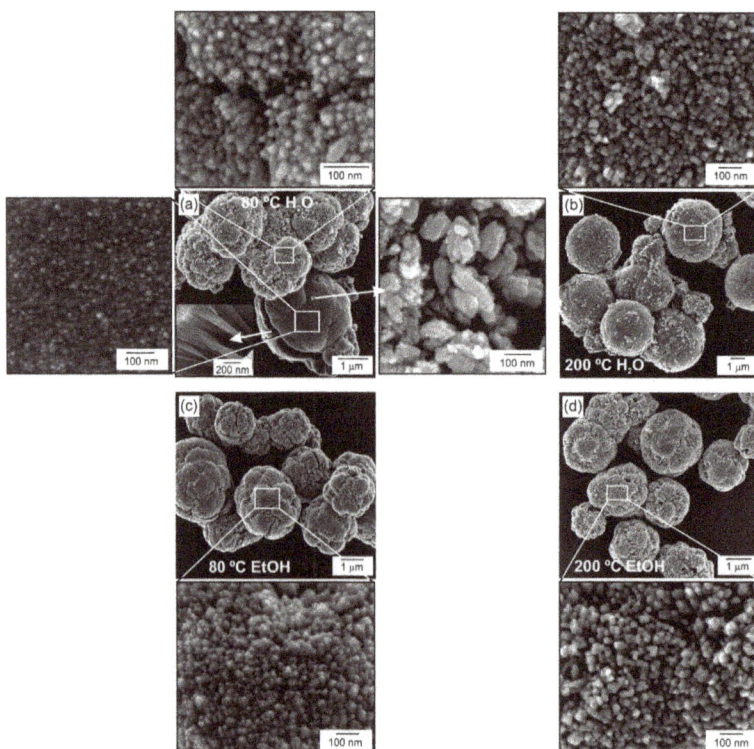

Figure 3. SEM images showing the TiO$_2$ nanorod spheres produced under microwave irradiation with water as solvent at: (**a**) 80 °C; and (**b**) 200 °C; and using ethanol at: (**c**) 80 °C; and (**d**) 200 °C. The insets magnify the nanorod structures, and the arrows in (**a**) point to the nanorod structure inside the sphere (**left** side) and to the distinct TiO$_2$ structure observed on the powder synthesized at 80 °C (**right** side) and using water as solvent.

The 80 °C EtOH nanorod spheres are highly homogeneous with quasi-spherical structures, tending to a flower-like structure with closely packed nanorods. The average sphere diameter was 2.2 ± 0.4 μm, with nanorod widths of 10.8 ± 2.8 nm. Upon the temperature increase, the 200 °C EtOH spheres maintained the uniform structure, being highly comparable to the 80 °C EtOH material. The average sphere diameter is 2.1 ± 0.3 μm with nanorods presenting widths of 27.7 ± 6.2 nm and squared shape (see inset in Figure 3d).

Comparing the micro-sized spheres regarding the solvent used, it was observed that the nanorod spheres synthesized with alcohol were smaller than the ones synthesized with water, in agreement to previous reports [60]. This behavior can be approximated to the solvent characteristics under microwave irradiation. As water is a solvent with higher boiling point than ethanol and also with lower loss tangent [52], these properties lead to lower microwave coupling efficiency [61], lower heating rate and pressure inside the microwave vessel during synthesis. Thus, the length of the nanorods can be tuned and form longer nanorods in the presence of water (larger TiO_2 nanorod spheres). Faster reaction rates result in smaller particle sizes [62], as observed for the materials synthesized with ethanol.

The 80 °C H_2O nanorod spheres were further investigated by TEM, to identify the distinct TiO_2 nanostructures observed in Figures 3a and S1. A bright-field image and respective diffraction pattern are presented in Figure 4. Small sized rod-like structures are clearly seen, and are steadily compared to the ones observed by SEM analysis (see the inset pointed by the arrow in Figure 3a). These structures displayed an average width and length of 22.4 ± 6.7 and 73.7 ± 34.8 nm, respectively. The ring diffraction pattern attested that these particles are solely from the brookite phase.

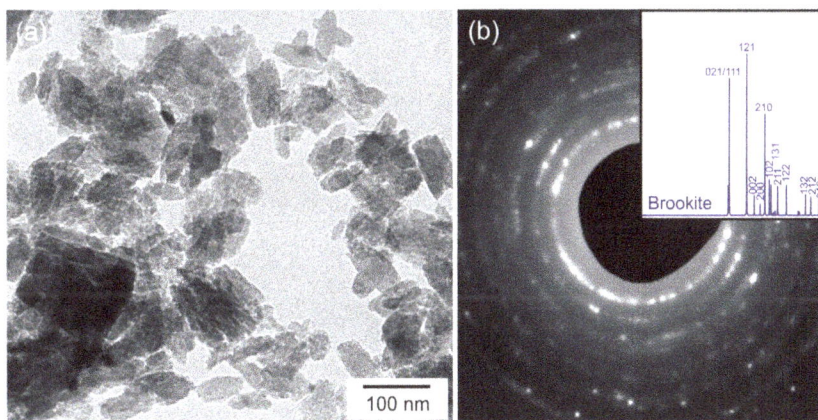

Figure 4. (a) Bright field TEM image of the TiO_2 structure detected at the 80 °C H_2O nanorod spheres; (b) Ring diffraction pattern confirming the TiO_2 brookite phase (TiO_2 brookite simulation is included).

The large TiO_2 micro-sized nanorod spheres appeared as dark structures on TEM and have not been presented. TiO_2 nanorod micro-sized spheres have been reported previously [3,34,63]. These spheres are formed during synthesis, as the nanorods produced initially in solution having the rutile phase, tend to aggregate in spheres in order to reduce the surface energy [60]. Thus, the nanorod spheres observed after microwave synthesis for all conditions studied are believed to have the rutile phase, with the individual TiO_2 nanorods having a [001] growth direction, such as previous related studies [5,63]. The nanorod spheres produced at 80 °C with water presented significant structural differences, and it can be expected that these dissimilarities are related to the nanorod sphere evolution along synthesis. The compact structure is more likely to occur as it was observed previously for synthesis using water as solvent [5]. Moreover, with the temperature increase, spherical, larger and more compact spheres were observed, supporting the assumption of evolution to compact structures.

The TiO$_2$ nanorod arrays were also successfully synthesized under microwave irradiation and using ZnO as seed layer on a PET substrate. The image of the produced material demonstrates that the substrates maintained their flexibility after microwave irradiation (see photograph in Figure 5). SEM images revealed that both TiO$_2$ nanorod arrays ensued similar final materials composed by similar nanorods agglomerates forming TiO$_2$ flower-like structures such as the ones observed for the powder spheres. The synthesis with water resulted in nanorod arrays organized as smaller individual aggregates along the substrate (Figure 5a). For the material synthesized with ethanol, it could also be observed that larger individual aggregates were also formed, but presenting a cauliflower aspect and more closed packed than the 80 °C H$_2$O material (compare Figure 5a,b). These structures were grown side-by-side along the substrate appearing as a continuous material. These results confirm the Raman spectra where the 80 °C EtOH nanorod arrays appeared to be thicker and denser due to the lower contribution of the PET substrate to the Raman spectrum (see Figure 2). The TiO$_2$ nanorods grown on the arrays presented widths comparable to the nanorods observed at the micro-sized spheres for both materials. The average aggregate sizes (thickness) were 570 ± 86.7 nm and 760 ± 114.6 nm for the 80 °C H$_2$O and 80 °C EtOH, respectively.

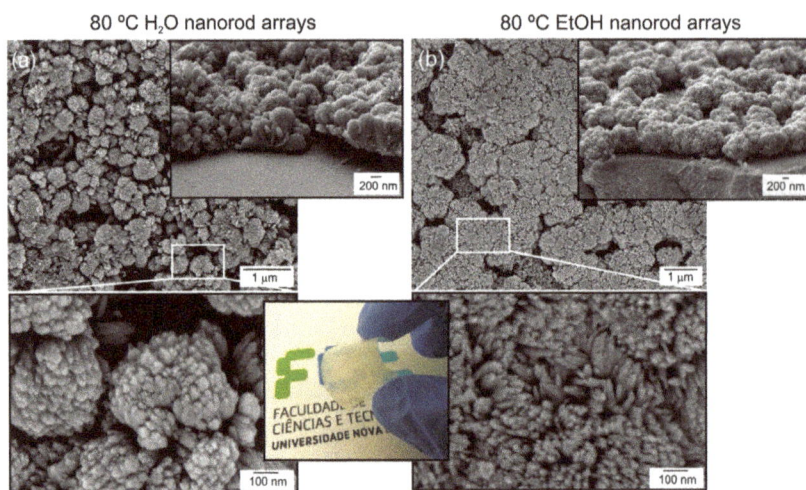

Figure 5. SEM images showing the TiO$_2$ nanorod arrays produced under microwave irradiation and using: (**a**) water; and (**b**) ethanol as solvents. Both arrays were synthesized with temperature limited to 80 °C. The insets show the magnified image of the arrays and the cross-section of the arrays produced. The array pile-up observed for the 80 °C H$_2$O condition is expected to be due to electron beam interactions with the flexible substrate during SEM analysis, causing the array detachment. A photograph of the bended PET substrate with the TiO$_2$ nanorod arrays is also shown.

EDS analyses were carried out in the material cross-sections, which revealed homogeneous distributions of Ti and O in all materials produced (see Figure S2). The presence of ZnO (Zn map) could not be confirmed through EDS due to the reduced thickness of the film [64].

2.3. Optical Characterization

Optical band gaps have been evaluated from reflectance data through the Tauc plot. The optical band gap (E_g) is related to the optical absorption coefficient and the incident photon energy as follows [65,66]:

$$\alpha h\nu = A\left(h\nu - E_g\right)^n \tag{1}$$

where α is the linear absorption coefficient of the material, $h\nu$ is the photon energy, A is a proportionality constant and n is a constant exponent which determines the type of optical transitions ($n = 1/2$ for direct allowed transition and $n = 2$ for indirect ones). Moreover, for determining the band gap, the $(\alpha h\nu)^2$ against $h\nu$ is plotted, and extracted through the intersection of the extrapolation of the linear portion with 0. The band gaps were estimated for all the produced TiO_2 nanorod spheres and arrays. The band gaps were 3.05 eV, 3.00 eV, 3.06 eV, and 3.03 eV for 80 °C H_2O, 200 °C H_2O, 80 °C EtOH and 200 °C EtOH nanorod spheres, respectively (Figure 6a). No significant band gap differences were observed between the solvents and temperatures tested. Moreover, the evaluated band gaps are within the reported values for the different TiO_2 phases [18,67,68]. The band gaps obtained for the TiO_2 nanorod arrays synthesized at 80 °C were 3.30 eV for the 80 °C H_2O and 3.04 eV for the 80 °C EtOH, respectively (Figure 6b), with both values within the values typically reported for the TiO_2 phases. A contribution from the ZnO seed layer could be expected to the values obtained (ZnO band gap has been reported to be 3.37 eV [52]), nevertheless this contribution was not evident for the 80 °C EtOH condition. An analogous study, demonstrated that the ZnO seed layer deposited by spin-coating and annealed at 200 °C, resulted in an optical band gap value of 3.26 eV [64]. Structural characteristics as array thickness and compactness can also largely influence the final band gap value [5], which may justify the differences observed. The presence of other phases with larger band gaps than rutile (brookite in the case of the 80 °C H_2O nanorod arrays) could also have played a role in the higher band gap observed; nevertheless, this effect was not confirmed for the nanorod spheres, and thus a direct relation cannot be inferred.

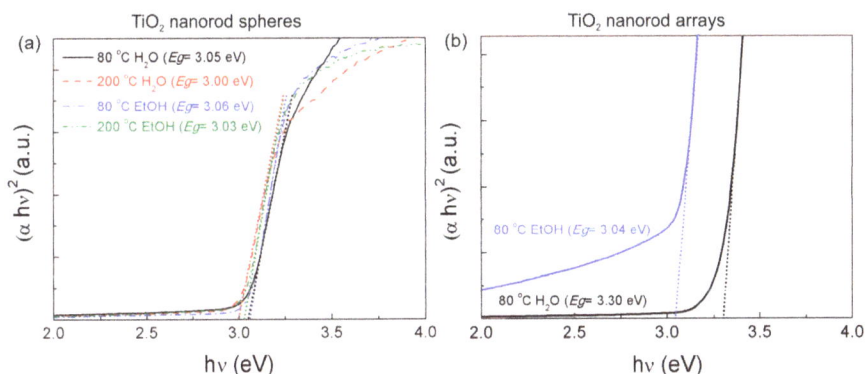

Figure 6. $(\alpha h\nu)^2$ variation versus photon energy $h\nu$ for the TiO_2 nanorod: (a) spheres; and (b) arrays.

2.4. Photocatalytic Activity

The photocatalytic activity of both the TiO_2 nanorod spheres and arrays were evaluated through the rhodamine B degradation efficiency under UV and solar radiation. For the best photocatalyst material, reusability tests were carried out. The degradation ratio (C/C_0) vs. UV exposure time is presented in Figure 7, where C is the concentration of the pollutant in the aqueous solution at each exposure time and C_0 is the initial solution concentration. The gradual rhodamine B degradation under UV radiation in all conditions could be observed in Figure 7 and Figure S3, where the 80 °C H_2O nanorod spheres showed the highest photocatalytic activity. This material reached values of 95% after 90 min, while the other powder materials required 150 min to reach closer degradation values, i.e., 87% for the 200 °C H_2O, 92% for the 80 °C EtOH, and 94% for the 200 °C EtOH. From Figure 7b, it can also be observed that the 80 °C H_2O powder photocatalyst can be reutilized despite the activity deterioration observed over the exposures [20,69], which can be related to the powder saturation of rhodamine B [69] or photocatalyst weight loss during experiments [70]. Moreover, a comparable degradation behavior was observed to the Degussa P25 catalyst (see Figure S4), however the synthesized photocatalyst has the advantage of being easily recovered and recycled due to its micrometer size.

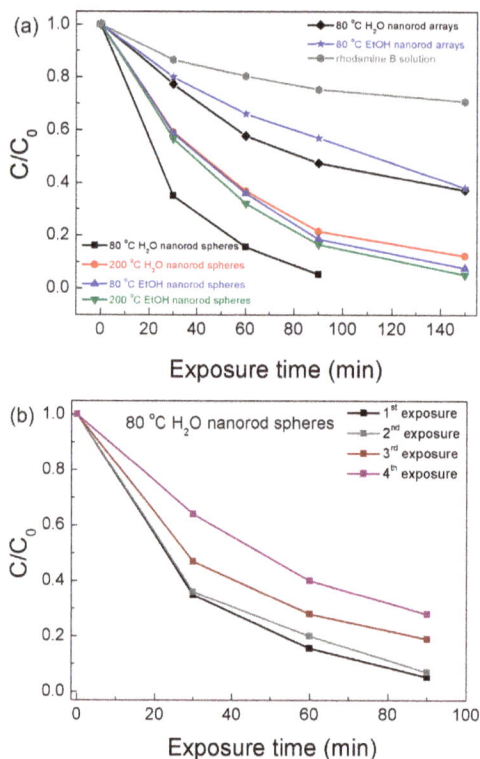

Figure 7. (**a**) Rhodamine B degradation ratio (C/C_0) vs. UV exposure time for all the materials produced. (**b**) Rhodamine B degradation ratio (C/C_0) vs. UV exposure time for the 80 °C H_2O nanorod spheres after several UV exposure experiments to attest the reusability of the material. The blank rhodamine B solution was simultaneously exposed during the degradation experiments.

The TiO_2 arrays on PET substrates with 150 min of UV exposure time reached 64% and 61% for the 80 °C H_2O and 80 °C EtOH nanorod arrays, respectively. The powder materials were more effective in rhodhamine B degradation; nevertheless, this performance is expected, as powders have a better adsorption activity and photocatalytic efficiency than films due to the larger surface area and higher amount of material [71]. A blank rhodamine B solution was also measured during the UV exposure experiments, presenting some degradation without the catalyst, however significantly lower than the degradation observed for all materials (Figure 7a).

The photocatalytic activity depends on several properties such as band gap, crystallite size, crystalline phase, specific surface area and active facets [5,13]. The TiO_2 active facets are {110} > {001} > {100} for rutile [22] and {210} for brookite [17]. In the present study, two synthesis temperatures and solvents were investigated ensuing different TiO_2 phases, as well as distinct nanorod sphere and aggregate sizes. A clear relation between the band gaps of all the materials and their photocatalytic behavior cannot be established, as no significant variations were determined (Figure 6).

Regarding the size effect, a relation with fine particles and photocatalytic activity can be stated. In the case of the materials synthesized with water, the 80 °C condition formed smaller nanorods than the 200 °C one. Thus, the spheres with finer nanorod widths resulted in the highest degradation performance (Figure 7a). The nanorod lengths must also be considered, which is deeply related to the TiO_2 sphere size as the nanorods are radially aligned forming the sphere (the 80 °C H_2O formed smaller spheres). A size effect contribution from the reduced size of the brookite particles is also expected (see Figures 3, 4 and S1), along with other property contributions such as active facets [17]. Moreover, fractions of another TiO_2 phase were identified by XRD, Raman, SEM and TEM (Figures 1, 3 and 4) resulting in a material with a mixture of TiO_2 phases (brookite and rutile), which largely increases the photocatalytic performance [72]. Brookite has been reported to display higher photocatalytic activity than anatase or rutile [59], thus an expressive contribution to the photocatalytic performance of the 80 °C H_2O material, is expected to be from the higher amount of brookite phase. The 200 °C H_2O nanorod spheres still presented a mixture of TiO_2 phases, however with lower fractions of brookite than the 80 °C H_2O one, and when compared with the pure rutile materials, the poorer rhodamine B degradation performance of the 200 °C H_2O material could be justified by the larger sphere sizes detected among all materials.

No significant differences in rhodamine B degradation under UV exposure have been detected between both spheres synthesized with ethanol. Both rutile materials resulted in similar sphere diameters despite the difference in nanorod widths. Nevertheless, the 200 °C EtOH spheres appeared to have a more open structure, which can lead to the higher specific surface area [73], justifying the slight higher rhodamine B degradation (~2% higher than the 80 °C EtOH spheres).

The TiO_2 nanorod arrays followed similar photocatalytic behavior than the spheres, where the arrays produced with water, revealed an enhanced photocatalytic activity under UV exposure than the ones synthesized with ethanol. Raman spectroscopy revealed that the 80 °C H_2O nanorod arrays is a mixture of TiO_2 phases (brookite and rutile), while the 80 °C EtOH is fully composed of the rutile phase. Additionally, a nanorod aggregate size effect can also be approached, where the smaller aggregates formed in water associated with the mixture of the crystalline phases may have resulted in an increase of the photocatalytic performance. Moreover, the 80 °C H_2O aggregates are expected to have higher surface area than the ones of the 80 °C EtOH material, in which the latter presented a closed structure with cauliflower aspect and nanorods closely packed.

All the TiO_2 nanorod spheres and arrays were then exposed to solar radiation to mimic ambient conditions (Figure 8). Once again, the 80 °C H_2O nanorod spheres revealed the greater photocatalytic activity under the solar light simulating source. This behavior was expected due to the enhanced properties of this material and to confirm these results, the absorption in function to the exposure time at a fixed wavelength of 500 nm, was measured in all powder materials (Figure S5). The 80 °C H_2O nanorod spheres revealed the higher absorption value at this wavelength suggesting that this material can be employed for pollutant degradation under visible/solar radiation. The degradation ratio (C/C_0) vs. solar light exposure time is also presented in Figure 8b. The photocatalytic activity of 80 °C H_2O material is higher under UV than under solar simulating light source (compare Figures 7a and 8b), which is expectable, however the degradation under solar radiation is expressive (55% after 150 min). The blank rhodamine B solution was not influenced under solar light so all the photocatalytic effect is due to the presence of the catalyst. The TiO_2 nanorod arrays also revealed some rhodamine B degradation under solar light; however, further investigation is required to improve the degradation rate (17.5% rhodamine B degradation after 150 min for the best array photocatalyst, i.e., the 80 °C H_2O material). Nevertheless, the TiO_2 nanorod arrays on flexible substrates show very promising results as these materials can adapt to unlike surfaces, i.e., in a photocatalytic auto-cleaning tubular line for water/wastewater treatment, moreover the approach suggested in this work is an interesting option, as it results in highly malleable materials and uses low-cost production routes.

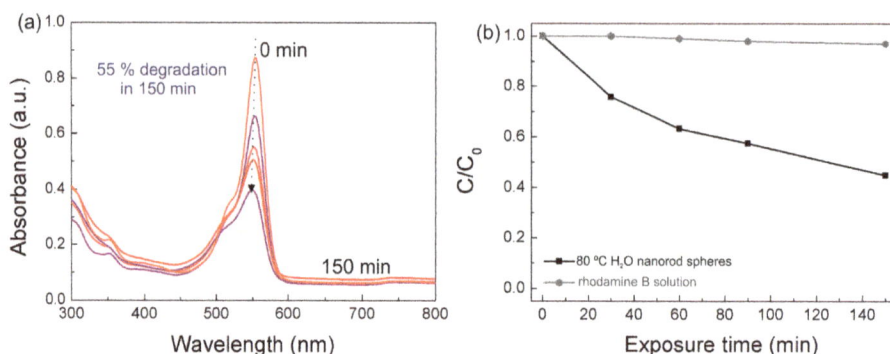

Figure 8. (**a**) Rhodamine B absorbance spectra at different solar light exposure times for the 80 °C H$_2$O nanorod spheres. (**b**) Rhodamine B degradation ratio (C/C$_0$) vs. solar light exposure time. The blank rhodamine B solution was simultaneously exposed during the degradation experiments. The absorbance spectra have not been normalized.

3. Experimental

3.1. TiO$_2$ Nanorod Sphere and Array Productions

The TiO$_2$ nanorod spheres and arrays have been synthesized simultaneously under microwave irradiation. The TiO$_2$ microwave solution has been prepared using titanium (IV) isopropoxide (Ti[OCH(CH$_3$)$_2$]$_4$, TTIP, 97%) and hydrochloric acid (HCl, 37%) both from Sigma Aldrich, St. Louis, MO, USA. Deionized water or ethanol (CH$_3$CH$_2$OH, EtOH, 99.5%) were used as solvents. In a typical synthesis, 45 mL of each solvent was mixed to 15 mL of HCl and stirred for 5 min. Afterwards, 2 mL of TTIP was added and the final mixture stirred for 10 min before microwave synthesis.

Microwave synthesis was performed using a CEM Focused Microwave Synthesis System Discover SP. Time, power and maximum pressure have been set at 75 min, 100 W and 17 bar, respectively. Solution volumes of 20 mL were transferred into capped quartz vessels of 35 mL, which were kept sealed by the constraining surrounding pressure. The synthesis temperature was fixed to 80 °C and 200 °C, and the materials produced were named: 80 °C H$_2$O, 200 °C H$_2$O, 80 °C EtOH, and 200 °C EtOH regarding the temperature and solvent used.

Each microwave reaction resulted in powder and arrays grown on the flexible substrate. In the case of the spheres, the powder composed of micro-sized spheres were collected from the bottom of the microwave vessel, washed with deionized water, centrifuged for 5 min at 4000 rpm for several times, and dried at 50 °C for 3 h. The TiO$_2$ arrays were grown on seeded substrate, i.e., polyethylene terephthalate, following a simple and efficient approach (Figure 9). For the microwave reaction, a piece of seeded PET substrate (20.0 mm × 20.0 mm) was placed at an angle against the vessel with the seed layer facing down [5]. The flexible substrates exposed to 200 °C and higher pressure became fragile after synthesis, and this condition has not been investigated. The nanorod arrays were synthesized at 80 °C and substrate kept malleable. After the synthesis, the materials were cleaned with deionized water and dried in air.

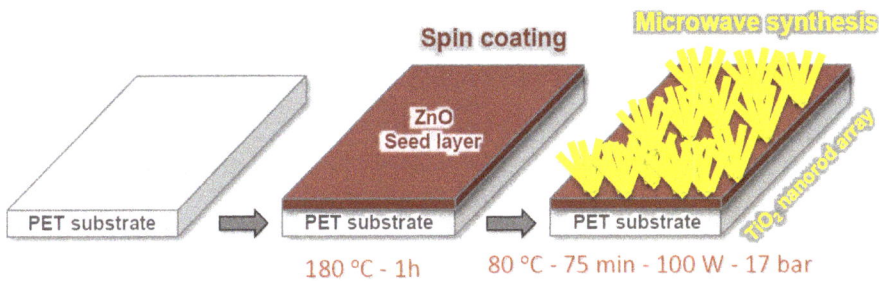

Figure 9. Scheme of the ZnO seed layer deposition and TiO$_2$ array growth. ZnO seed layer deposition on PET substrates by spin-coating and TiO$_2$ nanorod arrays growth under microwave irradiation.

Regarding the ZnO thin films to act as seed layers, their deposition was carried out using the spin-coating method. The ZnO solution was prepared from zinc acetate dihydrate (Zn(CH$_3$COO)$_2$·2H$_2$O, 98%), ethanolamine (C$_2$H$_7$NO, 99%) and 2-methoxiethanol (C$_3$H$_8$O$_2$, 99.8%). The reagents are from Sigma Aldrich and were used without further purification. In a typical reaction, the seed layer solution was prepared by dissolving zinc acetate in 2-methoxiethanol and then adding the ethanolamine. The proportion between zinc acetate and ethanolamine is 1:1, and the final mixture concentration was kept at 0.35 M [64]. The mixture was stirred for 1 h at 60 °C and filtered before depositing the ZnO thin films by the spin-coating method. Before deposition, the PET substrate was ultrasonically cleaned in isopropanol (10 min) and dried with compressed air. No treatment to increase adhesion has been performed. The film deposition was carried out at 3000 rpm for 35 s at room temperature. After each deposition, the films were dried at 180 °C for 10 min to remove the solvents, until completing a total of 6 depositions. In the final layer, the substrate with the deposited film was annealed for 1 h at 180 °C. The annealing temperature selected is between the PET glass transition temperature (~80 °C) and its melting temperature (~250 °C) [74], however an analogous study revealed that this temperature range is necessary to properly eliminate the reagents [64]. The thickness of the ZnO seed layer cannot be properly estimated as some ZnO chemical etching is expected during microwave synthesis. Nevertheless, prior to synthesis, in Figure S6, the estimated seed layer thickness was 50.7 ± 3.6 nm. Some increment to the photocatalytic behavior of the TiO$_2$ arrays can be expected from the ZnO seed layer. In fact, the use of ZnO thin films as seed layers for growth TiO$_2$ materials has been reported previously [75,76], where the combination of both oxides with high chemical stability and UV absorption, result in enhanced photocatalytic activity [75].

3.2. Structural and Optical Characterizations

X-ray diffraction experiments were performed using a PANalytical's X'Pert PRO MPD diffractometer equipped with a X'Celerator 1D detector and using CuKα radiation. The XRD data were acquired in the 20°–75° 2θ range with a step size of 0.05°. The XRD data were employed to determine the phase and composition of TiO$_2$ nanorod spheres. The PET substrate originates very broad and intense peaks on XRD, obscuring the signal of the TiO$_2$ arrays. For comparison, powder diffractograms of rutile, anatase, brookite have been simulated with PowderCell [77] using crystallographic data from Reference [78]. Raman spectroscopy experiments were performed at room temperature, using a Xplora plus Horiba spectrometer and a 532 nm laser.

Surface and cross-section SEM observations were carried out using a Carl Zeiss AURIGA CrossBeam FIB-SEM workstation equipped for EDS measurements. The dimensions of individual nanorods, spheres and arrays have been determined from SEM micrographs using the ImageJ software [79] and considering 30 distinct structures for each measurement. TEM observations were carried out with a Hitachi H8100 microscope operated at 200 kV. A drop of the sonicated dispersion was deposited onto 200-mesh copper grids covered with formvar and allowed to dry before observation.

Diffuse reflectance spectroscopy measurements of the nanorod spheres (dried powders) and arrays (TiO$_2$ nanorod arrays + ZnO seed layer + PET substrate) were performed in the 250–800 nm range with a PerkinElmer lambda 950 UV/VIS/NIR spectrophotometer equipped with a 150 mm diameter integrating sphere integrating sphere. The calibration of the system was achieved by using a standard reflector sample (reflectance, $R = 1.00$ from Spectralon disk). The optical band gap of the TiO$_2$ films was estimated from reflectance spectra using the Tauc plot method [65,80,81].

3.3. Photocatalytic Activity

The photocatalytic activity of the produced materials was evaluated at room temperature from the degradation of rhodamine B from Sigma Aldrich. The experiments considered the International standard ISO 10678. For each experiment, 25 mg of each powder containing the TiO$_2$ nanorod spheres was dispersed in 50 mL of the rhodamine B solution (5 mg/L) and then stirred for 30 min in the dark to establish absorption-desorption equilibrium. The TiO$_2$ nanorod arrays were used without any further preparation, where the flexible substrates with the arrays were placed on the bottom of the recipient and kept in the dark under stirring for 30 min prior to exposure. UV exposure was carried out using 3 lamps of 95 W aligned in parallel, from Osram, with an emission wavelength of 254 nm (ozone free). The distance between the light sources and the materials was 10 cm. Absorption spectra were recorded using a PerkinElmer lambda 950 UV/VIS/NIR spectrophotometer with intervals of 30 min for the first 90 min, and one more exposure of 60 min to complete a total exposure time of 150 min. For powder materials, after each exposure, 4 mL of the rhodamine B solution with the catalyst was collected and centrifuged for 5 min at 4000 rpm. After the absorption spectrum acquisition, the solution measured was returned to the recipient for further measurements. To perform reusability experiments, the powder was recovered by centrifugation, discarding the liquid from the previous exposure and drying the remaining material at 50 °C for 3 h. The reusability tests were carried out by the repeated UV exposure of the same sample in fresh solutions for 90 min along several weeks. The commercial TiO$_2$ (Degussa P25) was used for comparison (see Supplementary Materials). All the photocatalysts (TiO$_2$ nanorod spheres and arrays) were exposed to a solar light simulating source for equal exposure times as the UV experiments using a Xe lamp at room temperature with intensity of 100 mW/cm^2 and AM1.5 spectrum. Exposure experiments at a fixed wavelength of 500 nm over time were carried out using the PerkinElmer lambda 950 UV/VIS/NIR spectrophotometer.

4. Conclusions

Microwave irradiation proved to be an effective synthesis route to produce TiO$_2$ powders and arrays at low process temperatures, compatible with the use of flexible and low-cost substrates as PET. The approach proposed in this study appears as a valid option due to the low-priced characteristics of the synthesis and seed layer deposition routes. The material characteristics and final properties can be tuned by selecting the solvent and limiting the synthesis temperature. Water revealed to be the most adequate solvent for synthesizing materials with a mixture of phases and higher photocatalytic activities; nevertheless, ethanol formed homogeneous rutile-based materials with good photocatalytic degradation under UV radiation. Photocatalytic activity was assessed from rhodamine B degradation, with the 80 °C H$_2$O nanorod spheres showing the highest photocatalytic activity of all materials tested under UV radiation (95% after 90 min) and displaying reusability characteristics over time. All the materials were further exposed to solar radiation and the 80 °C H$_2$O powder resulted in a rhodamine B degradation of 55% after 150 min. The presence of the brookite phase in this material was determined as the main responsible for the increased photocatalytic activity. Nevertheless, a relation between the nanorod, sphere and aggregate sizes and their photocatalytic activity has been suggested for all materials tested. In the present study, both the nanorod spheres and arrays presented remarkable photocatalytic activity and, depending on the application desired, the materials can be employed as powders that are easily recycled, as they appear in the micrometer range, or as arrays grown on flexible substrates that can be adapted to several surfaces.

Supplementary Materials: The following are available online at www.mdpi.com/2073-4344/7/2/60/s1. Figure S1: SEM images of the 80 °C H₂O nanorod spheres showing the nanorods radially arranged. The inset evidences the nanorod structures inside the sphere. Figure S2: Cross-section SEM images (artificial colored) of the: (a) 80 °C H₂O; and (e) 80 °C EtOH nanorod arrays grown on ZnO seeded PET substrates, together with the corresponding X-ray maps of Ti: (b,f); O (c,g); and C (d,h). Figure S3: Rhodamine B absorbance spectra at different UV exposure times for the TiO₂ nanorod spheres produced with water at: (a) 80 °C; and (b) 200 °C; and with ethanol at: (d) 80 °C; and (e) 200 °C. The TiO₂ nanorod arrays grown on PET were also tested as photocatalystis at 80 °C for: (c) water; and (f) ethanol. The photograph illustrates the PET substrate with the TiO₂ nanorod arrays covering a tube containing the pollutant solution. The absorbance spectra have not been normalized. Figure S4: Rhodamine B degradation ratio (C/C_0) vs. UV exposure time for the 80 °C H₂O nanorod spheres and Degussa P25. Figure S5: Absorbance measurements over time at 500 nm for all the TiO₂ nanorod spheres. Figure S6: Cross-section SEM image of the ZnO seed layer prior to microwave synthesis.

Acknowledgments: The work was supported by the FCT—Portuguese Foundation for Science and Technology, through the scholarship BPD/84215/2012, as well as by the European project CEOPS with the grant agreement No.: 309984. The work was also supported by FEDER funds through the COMPETE 2020 Programme and National Funds through FCT under the project UID/CTM/50025/2013.

Author Contributions: Daniela Nunes synthesized the materials; Daniela Nunes and Ana Pimentel performed the photocatalytic experiments and structural characterization; optical characterization was performed by Lidia Santos; and the work and paper was under the supervision of Pedro Barquinha, Rodrigo Martins and Elvira Fortunato.

Conflicts of Interest: The authors declare no conflict of interest.

References

1. Bai, J.; Zhou, B. Titanium dioxide nanomaterials for sensor applications. *Chem. Rev.* **2014**, *114*, 10131–10176. [CrossRef] [PubMed]

2. Bernacka-Wojcik, I.; Senadeera, R.; Wojcik, P.J.; Silva, L.B.; Doria, G.; Baptista, P.; Aguas, H.; Fortunato, E.; Martins, R. Inkjet printed and "doctor blade" TiO₂ photodetectors for DNA biosensors. *Biosens. Bioelectron.* **2010**, *25*, 1229–1234. [CrossRef] [PubMed]

3. Lin, J.; Heo, Y.-U.; Nattestad, A.; Sun, Z.; Wang, L.; Kim, J.H.; Dou, S.X. 3D hierarchical rutile TiO₂ and metal-free organic sensitizer producing dye-sensitized solar cells 8.6% conversion efficiency. *Sci. Rep.* **2014**, *4*. [CrossRef] [PubMed]

4. Pimentel, A.; Nunes, D.; Pereira, S.; Martins, R.; Fortunato, E. Photocatalytic activity of TiO2 nanostructured arrays prepared by microwave-assisted solvothermal method. In *Semiconductor Photocatalysis—Materials, Mechanisms and Applications*; Cao, W., Ed.; InTech: Rijeka, Croatia, 2016.

5. Nunes, D.; Pimentel, A.; Pinto, J.V.; Calmeiro, T.R.; Nandy, S.; Barquinha, P.; Pereira, L.; Carvalho, P.A.; Fortunato, E.; Martins, R. Photocatalytic behavior of TiO₂ films synthesized by microwave irradiation. *Catal. Today* **2016**, *278*, 262–270. [CrossRef]

6. Nakata, K.; Fujishima, A. TiO₂ photocatalysis: Design and applications. *J. Photochem. Photobiol. C Photochem. Rev.* **2012**, *13*, 169–189. [CrossRef]

7. Schneider, J.; Matsuoka, M.; Takeuchi, M.; Zhang, J.; Horiuchi, Y.; Anpo, M.; Bahnemann, D.W. Understanding TiO₂ photocatalysis: Mechanisms and materials. *Chem. Rev.* **2014**, *114*, 9919–9986. [CrossRef] [PubMed]

8. Jin, X.; Xu, J.; Wang, X.; Xie, Z.; Liu, Z.; Liang, B.; Chen, D.; Shen, G. Flexible TiO₂/cellulose acetate hybrid film as a recyclable photocatalyst. *RSC Adv.* **2014**, *4*, 12640–12648. [CrossRef]

9. Sunada, K.; Kikuchi, Y.; Hashimoto, K.; Fujishima, A. Bactericidal and detoxification effects of TiO₂ thin film photocatalysts. *Environ. Sci. Technol.* **1998**, *32*, 726–728. [CrossRef]

10. Vukoje, I.; Kovač, T.; Džunuzović, J.; Džunuzović, E.; Lončarević, D.; Ahrenkiel, S.P.; Nedeljković, J.M. Photocatalytic ability of visible-light-responsive TiO₂ nanoparticles. *J. Phys. Chem. C* **2016**, *120*, 18560–18569. [CrossRef]

11. Kerkez, Ö.; Boz, İ. Efficient removal of methylene blue by photocatalytic degradation with TiO₂ nanorod array thin films. *React. Kinet. Mech. Catal.* **2013**, *110*, 543–557. [CrossRef]

12. Chen, D.Y.; Tsao, C.C.; Hsu, C.Y. Photocatalytic TiO₂ thin films deposited on flexible substrates by radio frequency (RF) reactive magnetron sputtering. *Curr. Appl. Phys.* **2012**, *12*, 179–183. [CrossRef]

13. Guo, Y.; Li, H.; Chen, J.; Wu, X.; Zhou, L. TiO₂ mesocrystals built of nanocrystals with exposed {001} facets: Facile synthesis and superior photocatalytic ability. *J. Mater. Chem. A* **2014**, *2*, 19589–19593. [CrossRef]

14. Kenanakis, G.; Vernardou, D.; Dalamagkas, A.; Katsarakis, N. Photocatalytic and electrooxidation properties of TiO$_2$ thin films deposited by sol–gel. *Catal. Today* **2015**, *240*, 146–152. [CrossRef]

15. Kazuhito, H.; Hiroshi, I.; Akira, F. TiO$_2$ photocatalysis: A historical overview and future prospects. *Jpn. J. Appl. Phys.* **2005**, *44*, 8269–8285.

16. Rathee, D.; Arya, S.; Kumar, M. Analysis of TiO$_2$ for microelectronic applications: Effect of deposition methods on their electrical properties. *Front. Optoelectron. China* **2011**, *4*, 349–358. [CrossRef]

17. Di Paola, A.; Bellardita, M.; Palmisano, L. Brookite, the least known TiO$_2$ photocatalyst. *Catalysts* **2013**, *3*, 36–73. [CrossRef]

18. Reyes-Coronado, D.; Rodríguez-Gattorno, G.; Espinosa-Pesqueira, M.; Cab, C.; De Coss, R.; Oskam, G. Phase-pure TiO$_2$ nanoparticles: Anatase, brookite and rutile. *Nanotechnology* **2008**, *19*, 145605. [CrossRef] [PubMed]

19. Kawahara, T.; Konishi, Y.; Tada, H.; Tohge, N.; Nishii, J.; Ito, S. A patterned TiO$_2$(anatase)/TiO$_2$ (rutile) bilayer-type photocatalyst: Effect of the anatase/rutile junction on the photocatalytic activity. *Angew. Chem.* **2002**, *114*, 2935–2937. [CrossRef]

20. Wang, R.; Cai, X.; Shen, F. Preparation of TiO$_2$ hollow microspheres by a novel vesicle template method and their enhanced photocatalytic properties. *Ceram. Int.* **2013**, *39*, 9465–9470. [CrossRef]

21. Andersson, M.; Österlund, L.; Ljungström, S.; Palmqvist, A. Preparation of nanosize anatase and rutile TiO$_2$ by hydrothermal treatment of microemulsions and their activity for photocatalytic wet oxidation of phenol. *J. Phys. Chem. B* **2002**, *106*, 10674–10679. [CrossRef]

22. Luttrell, T.; Halpegamage, S.; Tao, J.; Kramer, A.; Sutter, E.; Batzill, M. Why is anatase a better photocatalyst than rutile?—Model studies on epitaxial TiO$_2$ films. *Sci. Rep.* **2014**, *4*, 4043. [CrossRef] [PubMed]

23. Nakajima, H.; Mori, T.; Shen, Q.; Toyoda, T. Photoluminescence study of mixtures of anatase and rutile TiO$_2$ nanoparticles: Influence of charge transfer between the nanoparticles on their photoluminescence excitation bands. *Chem. Phys. Lett.* **2005**, *409*, 81–84. [CrossRef]

24. Kandiel, T.A.; Robben, L.; Alkaim, A.; Bahnemann, D. Brookite versus anatase TiO$_2$ photocatalysts: Phase transformations and photocatalytic activities. *Photochem. Photobiol. Sci.* **2013**, *12*, 602–609. [CrossRef] [PubMed]

25. Boppella, R.; Basak, P.; Manorama, S.V. Viable method for the synthesis of biphasic TiO$_2$ nanocrystals with tunable phase composition and enabled visible-light photocatalytic performance. *ACS Appl. Mater. Int.* **2012**, *4*, 1239–1246. [CrossRef] [PubMed]

26. Rochkind, M.; Pasternak, S.; Paz, Y. Using dyes for evaluating photocatalytic properties: A critical review. *Molecules* **2014**, *20*, 88–110. [CrossRef] [PubMed]

27. Lee, J.H.; Kang, M.; Choung, S.-J.; Ogino, K.; Miyata, S.; Kim, M.-S.; Park, J.-Y.; Kim, J.-B. The preparation of TiO$_2$ nanometer photocatalyst film by a hydrothermal method and its sterilization performance for giardia lamblia. *Water Res.* **2004**, *38*, 713–719. [CrossRef] [PubMed]

28. Paramasivam, I.; Jha, H.; Liu, N.; Schmuki, P. A review of photocatalysis using self-organized TiO$_2$ nanotubes and other ordered oxide nanostructures. *Small* **2012**, *8*, 3073–3103. [CrossRef] [PubMed]

29. Tan, Z.; Sato, K.; Takami, S.; Numako, C.; Umetsu, M.; Soga, K.; Nakayama, M.; Sasaki, R.; Tanaka, T.; Ogino, C.; et al. Particle size for photocatalytic activity of anatase TiO$_2$ nanosheets with highly exposed {001} facets. *RSC Adv.* **2013**, *3*, 19268–19271. [CrossRef]

30. Tang, H.; Zhang, D.; Tang, G.; Ji, X.; Li, C.; Yan, X.; Wu, Q. Low temperature synthesis and photocatalytic properties of mesoporous TiO$_2$ nanospheres. *J. Alloys Compd.* **2014**, *591*, 52–57. [CrossRef]

31. Hafez, H.S. Synthesis of highly-active single-crystalline TiO$_2$ nanorods and its application in environmental photocatalysis. *Mater. Lett.* **2009**, *63*, 1471–1474. [CrossRef]

32. Scuderi, V.; Impellizzeri, G.; Zimbone, M.; Sanz, R.; Di Mauro, A.; Buccheri, M.A.; Miritello, M.; Terrasi, A.; Rappazzo, G.; Nicotra, G.; et al. Rapid synthesis of photoactive hydrogenated TiO$_2$ nanoplumes. *Appl. Catal. B Environ.* **2016**, *183*, 328–334. [CrossRef]

33. Jitputti, J.; Suzuki, Y.; Yoshikawa, S. Synthesis of TiO$_2$ nanowires and their photocatalytic activity for hydrogen evolution. *Catal. Commun.* **2008**, *9*, 1265–1271. [CrossRef]

34. Bai, H.; Liu, Z.; Liu, L.; Sun, D.D. Large-scale production of hierarchical TiO$_2$ nanorod spheres for photocatalytic elimination of contaminants and killing bacteria. *Chem. Eur. J.* **2013**, *19*, 3061–3070. [CrossRef] [PubMed]

35. Wang, W.-Y.; Chen, B.-R. Characterization and photocatalytic activity of TiO$_2$ nanotube films prepared by anodization. *Int. J. Photoenergy* **2013**, *2013*. [CrossRef]

36. Xingtao, J.; Wen, H.; Xudong, Z.; Hongshi, Z.; Zhengmao, L.; Yingjun, F. Microwave-assisted synthesis of anatase TiO₂ nanorods with mesopores. *Nanotechnology* **2007**, *18*, 075602.

37. Gregori, D.; Benchenaa, I.; Chaput, F.; Therias, S.; Gardette, J.L.; Leonard, D.; Guillard, C.; Parola, S. Mechanically stable and photocatalytically active TiO₂/SiO₂ hybrid films on flexible organic substrates. *J. Mater. Chem. A* **2014**, *2*, 20096–20104. [CrossRef]

38. Chen, X.; Mao, S.S. Titanium dioxide nanomaterials: Synthesis, properties, modifications, and applications. *Chem. Rev.* **2007**, *107*, 2891–2959. [CrossRef] [PubMed]

39. Antonelli, D.M.; Ying, J.Y. Synthesis of hexagonally packed mesoporous TiO₂ by a modified sol–gel method. *Angew. Chem. Int. Ed. Engl.* **1995**, *34*, 2014–2017. [CrossRef]

40. Scuderi, V.; Impellizzeri, G.; Romano, L.; Scuderi, M.; Nicotra, G.; Bergum, K.; Irrera, A.; Svensson, B.G.; Privitera, V. TiO₂-coated nanostructures for dye photo-degradation in water. *Nanoscale Res. Lett.* **2014**, *9*, 458. [CrossRef] [PubMed]

41. Wu, J.-M.; Shih, H.C.; Wu, W.-T.; Tseng, Y.-K.; Chen, I.C. Thermal evaporation growth and the luminescence property of TiO₂ nanowires. *J. Cryst. Growth* **2005**, *281*, 384–390. [CrossRef]

42. Boyadzhiev, S.; Georgieva, V.; Rassovska, M. Characterization of reactive sputtered TiO₂ thin films for gas sensor applications. *J. Phys.* **2010**, *253*, 012040.

43. Lee, D.; Rho, Y.; Allen, F.I.; Minor, A.M.; Ko, S.H.; Grigoropoulos, C.P. Synthesis of hierarchical TiO₂ nanowires with densely-packed and omnidirectional branches. *Nanoscale* **2013**, *5*, 11147–11152. [CrossRef] [PubMed]

44. Yu, J.; Wang, Y.; Xiao, W. Enhanced photoelectrocatalytic performance of SnO₂/TiO₂ rutile composite films. *J. Mater. Chem. A* **2013**, *1*, 10727–10735. [CrossRef]

45. Zhao, X.; Liu, M.; Zhu, Y. Fabrication of porous TiO₂ film via hydrothermal method and its photocatalytic performances. *Thin Solid Films* **2007**, *515*, 7127–7134. [CrossRef]

46. Chen, Q.; Qian, Y.; Chen, Z.; Wu, W.; Chen, Z.; Zhou, G.; Zhang, Y. Hydrothermal epitaxy of highly oriented TiO₂ thin films on silicon. *Appl. Phys. Lett.* **1995**, *66*, 1608–1610. [CrossRef]

47. Bilecka, I.; Niederberger, M. Microwave chemistry for inorganic nanomaterials synthesis. *Nanoscale* **2010**, *2*, 1358–1374. [CrossRef] [PubMed]

48. Zhao, Y.; Zhu, J.-J.; Hong, J.-M.; Bian, N.; Chen, H.-Y. Microwave-induced polyol-process synthesis of copper and copper oxide nanocrystals with controllable morphology. *Eur. J. Inorg. Chem.* **2004**, *2004*, 4072–4080. [CrossRef]

49. Lidström, P.; Tierney, J.; Wathey, B.; Westman, J. Microwave assisted organic synthesis—A review. *Tetrahedron* **2001**, *57*, 9225–9283. [CrossRef]

50. Nunes, D.; Pimentel, A.; Barquinha, P.; Carvalho, P.A.; Fortunato, E.; Martins, R. Cu₂O polyhedral nanowires produced by microwave irradiation. *J. Mater. Chem. C* **2014**, *2*, 6097–6103. [CrossRef]

51. Yang, M.; Ding, B.; Lee, S.; Lee, J.-K. Carrier transport in dye-sensitized solar cells using single crystalline TiO₂ nanorods grown by a microwave-assisted hydrothermal reaction. *J. Phys. Chem. C* **2011**, *115*, 14534–14541. [CrossRef]

52. Pimentel, A.; Rodrigues, J.; Duarte, P.; Nunes, D.; Costa, F.M.; Monteiro, T.; Martins, R.; Fortunato, E. Effect of solvents on ZnO nanostructures synthesized by solvothermal method assisted by microwave radiation: A photocatalytic study. *J. Mater. Sci.* **2015**, *50*, 5777–5787. [CrossRef]

53. Pimentel, A.; Nunes, D.; Duarte, P.; Rodrigues, J.; Costa, F.M.; Monteiro, T.; Martins, R.; Fortunato, E. Synthesis of long ZnO nanorods under microwave irradiation or conventional heating. *J. Phys. Chem. C* **2014**, *118*, 14629–14639. [CrossRef]

54. Marques, A.C.; Santos, L.; Costa, M.N.; Dantas, J.M.; Duarte, P.; Gonçalves, A.; Martins, R.; Salgueiro, C.A.; Fortunato, E. Office paper platform for bioelectrochromic detection of electrochemically active bacteria using tungsten trioxide nanoprobes. *Sci. Rep.* **2015**, *5*, 9910. [CrossRef] [PubMed]

55. Gonçalves, A.; Resende, J.; Marques, A.C.; Pinto, J.V.; Nunes, D.; Marie, A.; Goncalves, R.; Pereira, L.; Martins, R.; Fortunato, E. Smart optically active VO₂ nanostructured layers applied in roof-type ceramic tiles for energy efficiency. *Sol. Energy Mater. Sol. C* **2016**, *150*, 1–9. [CrossRef]

56. Herring, N.P.; Panda, A.B.; AbouZeid, K.; Almahoudi, S.H.; Olson, C.R.; Patel, A.; El-Shall, M.S. Microwave synthesis of metal oxide nanoparticles. In *Metal Oxide Nanomaterials for Chemical Sensors*; Carpenter, A.M., Mathur, S., Kolmakov, A., Eds.; Springer: New York, NY, USA, 2013.

57. Wang, Y.; Zhang, L.; Deng, K.; Chen, X.; Zou, Z. Low temperature synthesis and photocatalytic activity of rutile TiO$_2$ nanorod superstructures. *J. Phys. Chem. C* **2007**, *111*, 2709–2714. [CrossRef]

58. Yan, J.; Feng, S.; Lu, H.; Wang, J.; Zheng, J.; Zhao, J.; Li, L.; Zhu, Z. Alcohol induced liquid-phase synthesis of rutile titania nanotubes. *Mater. Sci. Eng. B* **2010**, *172*, 114–120. [CrossRef]

59. Wang, Y.; Li, L.; Huang, X.; Li, Q.; Li, G. New insights into fluorinated TiO$_2$ (brookite, anatase and rutile) nanoparticles as efficient photocatalytic redox catalysts. *RSC Adv.* **2015**, *5*, 34302–34313. [CrossRef]

60. Wang, M.; Li, Q.; Yu, H.; Hur, S.H.; Kim, E.J. Phase-controlled preparation of TiO$_2$ films and micro(nano)spheres by low-temperature chemical bath deposition. *J. Alloys Compd.* **2013**, *578*, 419–424. [CrossRef]

61. Hayes, B.L. *Microwave Synthesis: Chemistry at the Speed of Light*; CEM Pub.: Matthews, NC, USA, 2002.

62. Cao, G.; Wang, Y. *Nanostructures and Nanomaterials: Synthesis, Properties, and Applications*; World Scientific: Singapore, 2011.

63. Zhou, W.; Liu, X.; Cui, J.; Liu, D.; Li, J.; Jiang, H.; Wang, J.; Liu, H. Control synthesis of rutile TiO$_2$ microspheres, nanoflowers, nanotrees and nanobelts via acid-hydrothermal method and their optical properties. *CrystEngComm* **2011**, *13*, 4557–4563. [CrossRef]

64. Pimentel, A.; Ferreira, S.; Nunes, D.; Calmeiro, T.; Martins, R.; Fortunato, E. Microwave synthesized ZnO nanorod arrays for UV sensors: A seed layer annealing temperature study. *Material* **2016**, *9*, 299. [CrossRef]

65. Du, Y.; Zhang, M.S.; Wu, J.; Kang, L.; Yang, S.; Wu, P.; Yin, Z. Optical properties of SrTiO$_3$ thin films by pulsed laser deposition. *Appl. Phys. A* **2003**, *76*, 1105–1108. [CrossRef]

66. Aydın, C.; Benhaliliba, M.; Al-Ghamdi, A.; Gafer, Z.; El-Tantawy, F.; Yakuphanoglu, F. Determination of optical band gap of ZnO:ZnAl$_2$O$_4$ composite semiconductor nanopowder materials by optical reflectance method. *J. Electroceram.* **2013**, *31*, 265–270. [CrossRef]

67. Yu, J.-G.; Yu, H.-G.; Cheng, B.; Zhao, X.-J.; Yu, J.C.; Ho, W.-K. The effect of calcination temperature on the surface microstructure and photocatalytic activity of TiO$_2$ thin films prepared by liquid phase deposition. *J. Phys. Chem. B* **2003**, *107*, 13871–13879. [CrossRef]

68. Wu, J.-M.; Shih, H.C.; Wu, W.-T. Formation and photoluminescence of single-crystalline rutile TiO$_2$ nanowires synthesized by thermal evaporation. *Nanotechnology* **2006**, *17*, 105. [CrossRef]

69. Nagaveni, K.; Sivalingam, G.; Hegde, M.S.; Madras, G. Solar photocatalytic degradation of dyes: High activity of combustion synthesized nano TiO$_2$. *Appl. Catal. B Environ.* **2004**, *48*, 83–93. [CrossRef]

70. Chen, L.; Yang, S.; Mader, E.; Ma, P.-C. Controlled synthesis of hierarchical TiO$_2$ nanoparticles on glass fibres and their photocatalytic performance. *Dalton Trans.* **2014**, *43*, 12743–12753. [CrossRef] [PubMed]

71. Chen, Y.-H.; Tu, K.-J. Thickness dependent on photocatalytic activity of hematite thin films. *Int. J. Photoenergy* **2011**, *2012*. [CrossRef]

72. Wang, H.; Gao, X.; Duan, G.; Yang, X.; Liu, X. Facile preparation of anatase–brookite–rutile mixed-phase N-doped TiO$_2$ with high visible-light photocatalytic activity. *J. Environ. Chem. Eng.* **2015**, *3*, 603–608. [CrossRef]

73. Qiu, S.; Ben, T. *Porous Polymers: Design, Synthesis and Applications*; Royal Society of Chemistry: Cambridge, UK, 2015.

74. Fann, D.M.; Huang, S.K.; Lee, J.Y. DSC studies on the crystallization characteristics of poly(ethylene terephthalate) for blow molding applications. *Polym. Eng. Sci.* **1998**, *38*, 265–273. [CrossRef]

75. Kim, W.-Y.; Kim, S.-W.; Yoo, D.-H.; Kim, E.J.; Hahn, S.H. Annealing effect of ZnO seed layer on enhancing photocatalytic activity of ZnO/TiO$_2$ nanostructure. *Int. J. Photoenergy* **2013**, *2013*. [CrossRef]

76. Chou, H.-T.; Hsu, H.-C. The effect of annealing temperatures to prepare ZnO seeds layer on ZnO nanorods array/TiO$_2$ nanoparticles photoanode. *Solid-State Electr.* **2016**, *116*, 15–21. [CrossRef]

77. Kraus, W.; Nolze, G. Powder cell—A program for the representation and manipulation of crystal structures and calculation of the resulting X-Ray powder patterns. *J. Appl. Crystallogr.* **1996**, *29*, 301–303. [CrossRef]

78. Pearson, W.B.; Villars, P.; Calvert, L.D. *Pearson's Handbook of Crystall Ographic Data for Intermetallic Phases*; American Society for Metals: Cleveland, OH, USA, 1985.

79. Schneider, C.A.; Rasband, W.S.; Eliceiri, K.W. NIH image to image J: 25 years of image analysis. *Nat. Methods* **2012**, *9*, 671–675. [CrossRef]

80. Santos, L.; Nunes, D.; Calmeiro, T.; Branquinho, R.; Salgueiro, D.; Barquinha, P.; Pereira, L.; Martins, R.; Fortunato, E. Solvothermal synthesis of gallium-indium-zinc-oxide nanoparticles for electrolyte-gated transistors. *ACS Appl. Mater. Interface* **2015**, *7*, 638–646. [CrossRef] [PubMed]

81. Tauc, J. Optical properties and electronic structure of amorphous Ge and Si. *Mater. Res. Bull.* **1968**, *3*, 37–46. [CrossRef]

© 2017 by the authors. Licensee MDPI, Basel, Switzerland. This article is an open access article distributed under the terms and conditions of the Creative Commons Attribution (CC BY) license (http://creativecommons.org/licenses/by/4.0/).

catalysts

MDPI

Article

Fast and Large-Scale Anodizing Synthesis of Pine-Cone TiO$_2$ for Solar-Driven Photocatalysis

Yan Liu [1], Yanzong Zhang [1,*], Lilin Wang [1], Gang Yang [1], Fei Shen [1], Shihuai Deng [1], Xiaohong Zhang [1], Yan He [1], Yaodong Hu [2] and Xiaobo Chen [3,*]

[1] College of Environment, Sichuan Agricultural University, Chengdu 611130, China; liuly6262@hotmail.com (Y.L.); shadowwll@126.com (L.W.); yg8813@163.com (G.Y.); fishensjtu@gmail.com (F.S.); shdeng8888@163.com (S.D.); zxh19701102@126.com (X.Z.); hy2005127220@126.com (Y.H.)

[2] Institute of Animal Genetics and Breeding, College of Animal Science and Technology, Sichuan Agricultural University, Chengdu 611130, China; tianxiaojohn007@163.com

[3] Department of Chemistry, University of Missouri-Kansas City, Kansas City, MO 64110, USA

* Correspondence: yzzhang@sicau.edu.cn (Y.Z.); chenxiaobo@umkc.edu (X.C.); Tel.: +86-28-86291132 (Y.Z.); +1-816-235-6420 (X.C.)

Received: 6 July 2017; Accepted: 27 July 2017; Published: 1 August 2017

Abstract: Anodization has been widely used to synthesize nanostructured TiO$_2$ films with promising photocatalytic performance for solar hydrogen production and pollution removal. However, it usually takes a few hours to obtain the right nanostructures even on a small scale (e.g., 10 mm × 20 mm). In order to attract interest for industrial applications, fast and large-scale fabrication is highly desirable. Herein, we demonstrate a fast and large-scale (e.g., 300 mm × 360 mm) synthesis of pine-cone TiO$_2$ nanostructures within two min. The formation mechanism of pine-cone TiO$_2$ is proposed. The pine-cone TiO$_2$ possesses a strong solar absorption, and exhibits high photocatalytic activities in photo-oxidizing organic pollutants in wastewater and producing hydrogen from water under natural sunlight. Thus, this study demonstrates a promising method for fabricating TiO$_2$ films towards practical photocatalytic applications.

Keywords: pine-cone TiO$_2$ nanoclusters; formation mechanism; lattice defects; optical absorption; large-sized films; printing and dyeing wastewater

1. Introduction

Nano-structured photoactive TiO$_2$ materials are believed to have a great promise for many photocatalytic applications such as pollution degradation [1], watersplitting [2], and dye-sensitized solar cells [3]. Many methods have been created to fabricate TiO$_2$, such as sol-gel [4,5], hydrothermal treatment [6,7], assisted-template method [8–10], laser ablation [11–14], and electrochemical anodic oxidation [15,16]. However, these methods have a common shortcoming that it needs a long reaction time to obtain the photocatalysts. For example, it usually takes several days for the sol-gel and hydrothermal methods to obtain TiO$_2$ [17–21]. Konishi et al. synthesized monolithic TiO$_2$ by sol-gel method from the starting solution containing titanium n-propoxide, HCl, formamide, and H$_2$O, followed by aging for 24 h, drying for seven days, and heat-treatment for three hours in air [22]. Lu's group prepared TiO$_2$ nanosheets by a hydrothermal method with titanium butoxide and hydrofluoric acid solution mixed in a Telfon-lined autoclave at 200 °C for 24 h, and subsequent drying at 50 °C overnight [23]. Although it is faster to prepare TiO$_2$ with the anodic oxidation method, it still takes a few hours [24–27]. For example, well-aligned TiO$_2$ nanotube arrays were obtained after first anodizing for two hours in an electrolyte containing NH$_4$F, ethylene glycol, and H$_2$O at 60 V, and again anodizing for 20 min at 60 V, and finally annealing for three hours at 450 °C [28]. On the other hand, in real applications, TiO_2 nanotube films obtained by the anodization method have many advantages, such as high photocatalytic activity, simple installation and easy recycling. However, so far, it is rarely reported on the large-scale production of anodized films, leaving their practical promise in industrial application unknown. For practical industrial applications, it is highly desirable to develop a fast and large-scale fabrication method.

In the present study, we demonstrate a fast and large-scale anodizing synthesis of TiO_2 within two minutes. The synthesized TiO_2 has a pine-cone-like structure (pine-cone TiO_2: PCT). This structure is formed as a result of continuous deposition of TiO_2-coated graphene layers and subsequent removal of graphene during the annealing process. It has a strong visible-light absorption and exhibits impressive photocatalytic performances for photocatalytic oxidation of organic pollutants in wastewater and photocatalytic generation of hydrogen from water under natural sunlight irradiation.

2. Results and Discussion

2.1. Morphology

The morphologies of the TiO_2 films were investigated with scanning electron microscope (SEM). Figure 1A and Figure S1 showed that three-dimensional pine-cone-like nanoclusters were grown on the surface of the Ti substrate when the Ti foil was anodized with grapheme in the electrolyte. When the Ti foil was anodized without graphene in the electrolyte, only a rough oxide layer was formed on the Ti foil without the formation of any nanoclusters, as shown in Figure S2. These indicated that the graphene played a crucial role for the formation of the pine-cone nanostructure. The pine-cone nanocluster was made of many layered structures (Figure 1B and Figure S3); there were layers of pores between adjacent layers within the pine-cone structure, and each layer was comprised of small TiO_2 nanoparticles (Figure 1C and Figure S4). In addition, the areas between the pine-cones were comprised of nanotubes structures (Figure 1D, Figures S5 and S6). These unique pine-core structures provided a large specific surface area of the film.

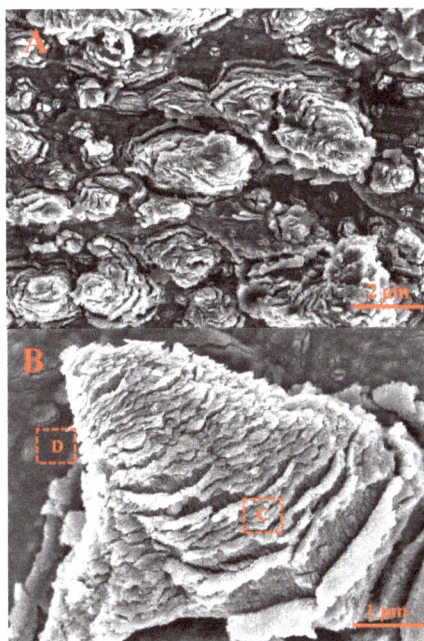

Figure 1. *Cont.*

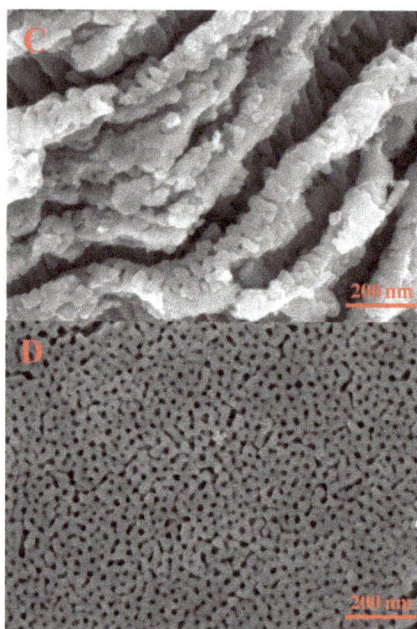

Figure 1. (**A**) Field-emission scanning electron microscopy (FE-SEM) image of the pine-cone TiO$_2$ (PCT) film anodized in the electrolyte of graphene (5.0 mg) and magnesium nitrate solution (100.0 mg L^{-1}, 50.0 mL) at 60 V for 2 min, (**B**) enlarged scale of the PCT nanoclusters, (**C**) enlarged scale of the marked region in B. (**D**) enlarged scale of the marked region in B.

2.2. Proposed Formation Mechanism

To understand the formation process of these unique pine-cone structures, we examined the morphology changes of the films formed after various anodization periods. Figure 2 showed a series of TiO$_2$ films obtained at different anodization times from 1 to 3 min which apparently displayed the evolution of the pine-cone structures. Figure 2A showed that, after being anodized for 1 min, an uneven oxide layer with many small holes was formed, along with many bumps which had some small cracks in the center. After being anodized for 1.5 min, the size of the small holes increased, along with the expansion of the bumps and the cracks in the center (Figure 2B). After 2 min anodization, pine-cone structures were formed in the center of the cracks (Figure 2C). Further anodization led to the partial dissolution or collapse of the pine-cone structure (Figure 2D).

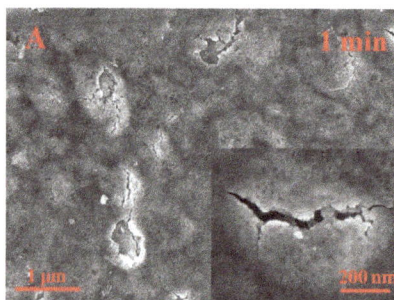

Figure 2. *Cont.*

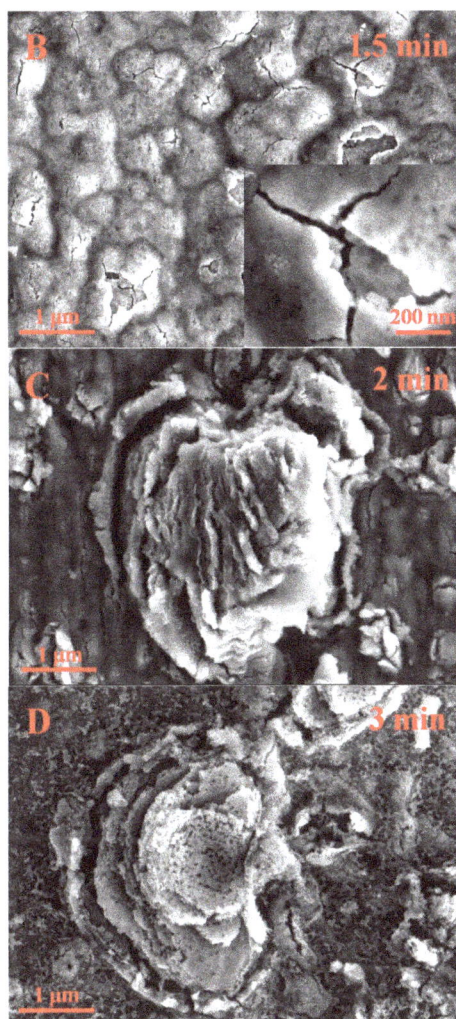

Figure 2. Growth process of PCT film on Ti foil prepared at 60 V for different times. FE-SEM images of film anodized for (**A**) 1 min, (**B**) 1.5 min, (**C**) 2 min, (**D**) 3 min.

Based on the observation above, a possible mechanism as shown in Figure 3 was proposed to explain the formation of the PCT film. As illustrated in Figure 3A, at the beginning, a compact oxide layer with small holes was formed. Mg^{2+} and NO_3^- ions in the solution moved to the cathode and anode under the electrical field, respectively. The NO_3^- ions reacted with the oxide layer and soluble species such as $[TiO_{2-x}(NO_3)_x]^{m-n}$ ($m > n$, $0 < x < 2$) were formed, leading to the local dissolution/thinning of the oxide layer and forming many small pores rapidly. Meanwhile, as shown in Figure 3B, as NO_3^- and OH^- reacted with the oxide layer/Ti interface to form soluble $[Ti(NO_3)_n]^{m-n}$ ($m = 3, 4$; $n > 4$) and $Ti(OH)_4$ species [29] the pores continued to grow. However, the formation of oxide layers led to lattice expansion and generated stress at the metal/oxide interface. Accumulation of the stress led to the deformation of the film and formation of bumps and eventually cracks (Figure S7) [30]. Some of the TiO_2 fragments were washed to the electrolyte by the O_2 gas (Figure S8). In this experiment, the pH of the electrolyte was 6.8. As the point of zero charge (pH_{pzc}) of TiO_2 was approximately

6 to 6.5 [31–33], the surface of TiO$_2$ was negatively charged in the electrolyte. In addition, it was found that most of the graphene sheets were quickly adsorbed on the cathode. So, the graphene sheets were likely positively charged on the surface. Therefore, the negatively charged TiO$_2$ fragments could be easily adsorbed on the surface of the positively charged graphene sheets to form layered structures. As negatively charged TiO$_2$ accumulated on the graphene sheets, this eventually turned the graphene-TiO$_2$ complex negatively charged on the surface to prevent further accumulation of TiO$_2$ on the surface. As shown in Figure 3C, the complex was attracted to the anode and reversed on charge on the surface under the electrical field. The cracks on the expansion parts were excellent landing positions where the electric field intensity was the strongest due to the shortest distance between the two electrodes. As this process proceeded, a layer-by-layer stacking structure was obtained. At the end, the graphene was removed in the annealing process at 723 K in the air, and the pine-cone was formed as shown in Figure 3D.

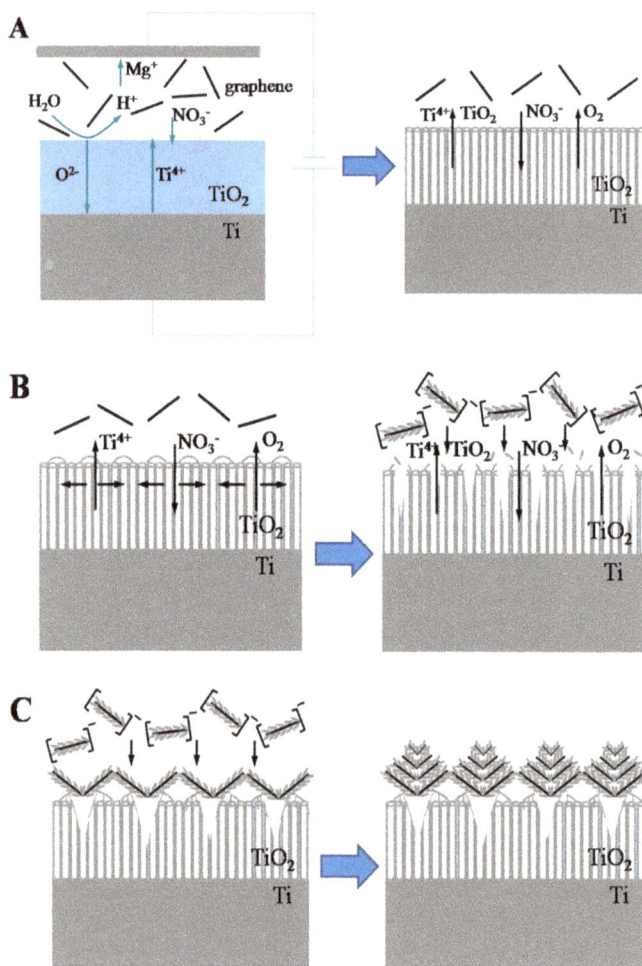

Figure 3. *Cont.*

D

Figure 3. Schematic illustration of the formation mechanism of the PCT film: (**A**) formation of the anodic oxide layer and pore growth; (**B**) formation of the cracks and TiO_2-graphene layers; (**C**) deposition of the TiO_2-graphene layers on the cracks; (**D**) removal of the graphene and formation of the pine-cone structure.

2.3. Crystal Phase and Surface Chemical Composition

Both XRD and Raman measurements suggested the PCT film annealed at 737 K had an anatase/rutile mixed phase. The grain sizes estimated with the Scherrer equation from the XRD pattern were around 18 and 20 nm for the anatase and rutile phases, respectively (Figure S9 and Table S1). The mixed crystal parameters obtained were consistent with their theoretical values [34]. The Raman peaks (Figure S10) at 147, 198, 396, 515, and 636 cm^{-1} were from the $E_{g(1)}$, $E_{g(2)}$, B_{1g}, A_{1g} or B_{1g}, and $E_{g(3)}$ modes of the anatase phase, respectively. The peak located at 448 cm^{-1} was due to the E_g mode of the rutile phase. XPS measurements (Figure S11) revealed that the PCT film was found to contain Ti, O, C, and N, while TNTs was comprised of only Ti, O, and C. The C element existing at the surface of films was derived from carbon contaminants or residual organic carbons [35]. The concentration of N calculated from the N 1s XPS spectrum of the PCT film was ~2.93 at %, which included interstitial N (Ti-O-N; 399.6 eV; 81.3%) [36], substitutional N (Ti-N; 397.1 eV; 8.7%) [37], and molecularly chemisorbed γ-N_2 (401.3 eV; 10%) [38], as shown in Figure S12. N atoms were introduced by the anodizing technique, likely from the nitric irons attacking the oxide layer and some adsorbed complexes, such as $[TiO_{2-x}(NO_3)_x]^{m-n}$ ($m > n$, $0 < x < 2$) and $[Ti(NO_3)_n]^{m-n}$ ($m = 3, 4; n > 4$), during the formation of TiO_2 [29,39]. These complex anions $[Ti(NO_3)_n]^{m-n}$ were thermodynamically unstable and decomposed into thermodynamically stable TiO_2 with N doping [29]. In addition, the NO species from the decomposition of the nitrate radical might be chemisorbed on the surface or incorporated into the defects of the resultant TiO_2 [40]. The binding energies of the Ti 2p3/2 (458.4 eV) and lattice O^{2-} (529.5 eV) states of the TNTs shifted 0.2 eV and 0.5 eV toward lower binding energies of 458.6 eV and 530.0 eV, as shown in Figures S13 and S14, respectively. These suggested the reduction of the valence state of titanium from Ti^{4+} to Ti^{3+} [41], and the presence of N species and oxygen vacancies (Vo) [42].

2.4. Lattice Defects and Optical Absorption

ESR results (Figure 4A) showed that the PCT film exhibited a strong signal at g = 2.004, while TNTs did not show observable signals. This signal was attributed to the electron traps associated with Vo [43,44], possibly associated with N doping, Ti^{3+} and/or disordered structures. N doping could lead to thermal instability and the formation of Vo [45,46]. The interface between the Ti substrate and the oxide layer could undergo the reaction Ti + $TiO_2 \rightarrow TiO_x$ ($x = 0$–2) to form Vo [47]. Moreover, the fast cooling process employed was a key component for maintaining stable Vo defects and forming a disordered structure. As shown in Figure S15, the fast cooling sample possessed more Vo defects than a slow cooling sample. The metastable defective phase on the surface, such as Vo, can subsequently be oxidized when a sample annealed at high temperature is rapidly exposed to oxygen-rich air [48], but the Vo defects are maintained at a high concentration in the internal lattice.

Figure 4B showed that the valence band maximum (VBM) of the PCT film was located at 0.72 eV, and the band tail region was shifted toward the vacuum level, ending at approximately −0.34 eV, while TNTs displayed the typical valence band value of TiO$_2$ with the VBM located at ~1.25 eV below the Fermi energy [49]. The substantial shift of the VBM (1.59 eV) for the PCT film was likely caused by the substitutional/interstitial N atoms and the disorder of the anatase/rutile interfaces [50].

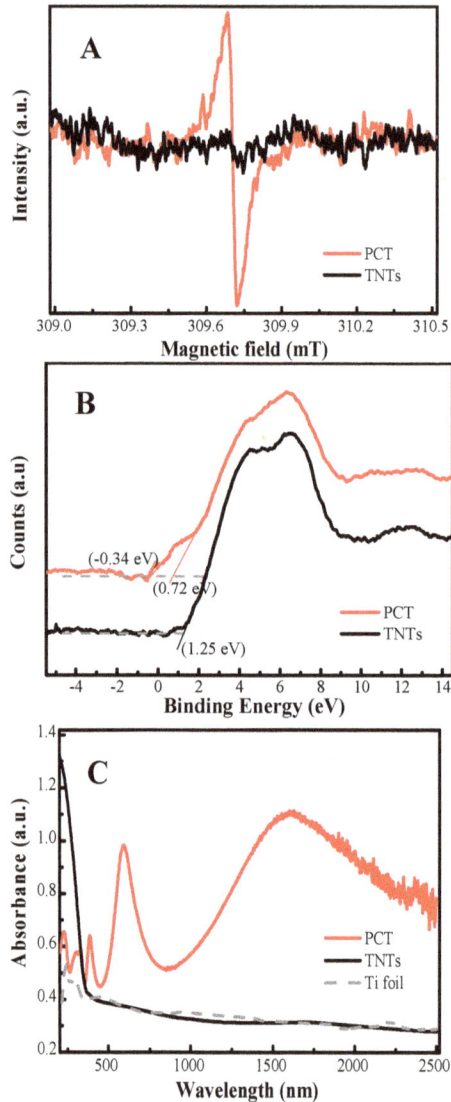

Figure 4. *Cont.*

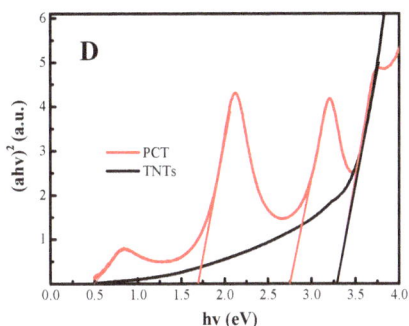

Figure 4. (**A**) Electron spin resonance (ESR) spectra of the PCT film and typical TiO$_2$ nanotube arrays (TNTs) in air at 77 K. (**B**) Valence band edge spectra of the PCT film and TNTs. (**C**) Ultraviolet-visible-near-infrared (UV-vis-NIR) absorption spectra of the PCT film and TNTs. (**D**) Bandgap evaluation from the plots of $(\alpha h \nu)^2$ vs. the energy (h) of the absorbed light.

The UV-vis-NIR absorption spectra in Figure 4C showed that both PCT film and TNTs had a substantial absorption from 200 to 400 nm due to electron excitation from VB to CB. The PCT film had absorption above 500 nm and the absorption edge remarkably extended to the infrared region (760–2500 nm). This enhanced visible-infrared absorption might be attributed to the N doping and/or Vo and/or the nanoporous pine-cone structure. Nanoporous-layer-covered on the TiO$_2$ nanotube arrays were reported as photonic crystals with a substantial optical absorption even in the infrared region [51–53]. Some oscillations in absorbance might also be due to optical interference. The plots of $(\alpha h \nu)^2$ versus the energy of the absorbed light ($h\nu$) [37] shown in Figure 4D hinted that the PCT film might have several distinct electronic transitions, while the TNTs had one of 3.27 eV.

2.5. Photoelectrochemical Activity

The photoelectrochemical activity of the PCT film under natural sunlight was evaluated with its transient photocurrent response, linear sweeps voltammogram and calculated its photoconversion efficiency. As shown in Figure 5A, the PCT film had a strong photocurrent density of 0.28 mA cm^{-2}, 14-fold of the TNTs (0.02 mA cm^{-2}). Figure 5B displayed that the photocurrent density of the PCT film increased rapidly with increasing bias. Figure 5C showed that a maximum photoconversion efficiency of 0.21% was obtained at −0.34 V vs. Ag/AgCl (0.66 V vs. a reversible hydrogen electrode) for the PCT film, approximately 13 times higher than that of TNTs.

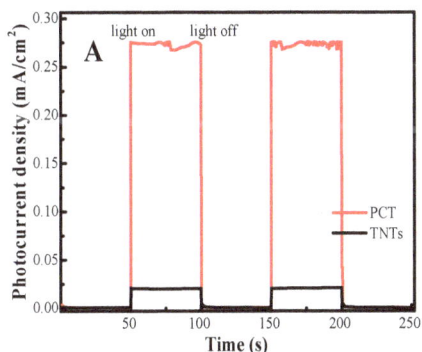

Figure 5. *Cont.*

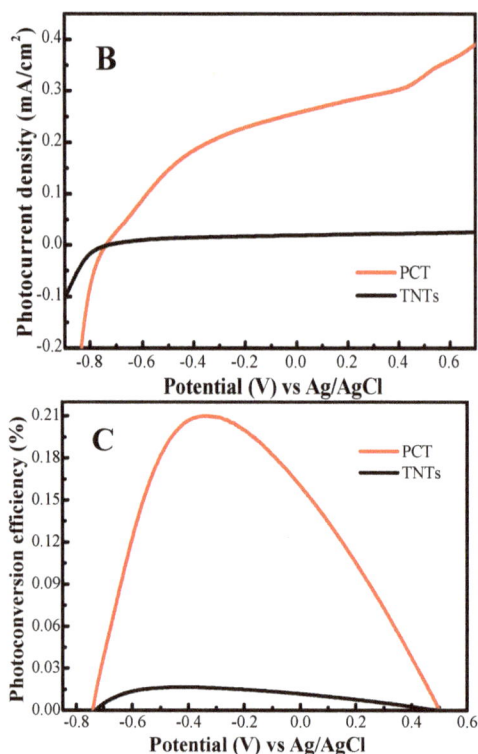

Figure 5. Photoelectrochemical activities of the PCT film and TNTs: (**A**) Transient photocurrent responses; (**B**) Linear sweep voltammograms in a potential range of −0.9 to 0.7 V vs. Ag/AgCl; (**C**) Calculated photoconversion efficiencies under natural sunlight.

2.6. Photoelectrocatalytic Activity

The solar-driven photoelectrocatalytic activity of the PCT film (30 mm × 40 mm) was further evaluated with hydrogen generation and MO degradation under natural sunlight. As shown in Figure 6A, the PCT film exhibited stable activity during a 10-day testing period, and maintained a hydrogen yield of up to 0.8 mL h^{-1} cm^{-2}. The PCT film also displayed a high photoactivity in MO degradation (Figure 6B). The MO (20 mg L^{-1}, 40.0 mL) degradation rate was 97% and 82% after 2 h on a sunny and cloudy day, respectively (the illumination intensity of sunlight is summarized in Table S2). Measurements showed that the self-decomposition of MO under the light and the adsorption by the photocatalyst were negligible (Figure 6C). The PCT film maintained a stable degradation rate over 10 cycles (Figure S16). Thus, the PCT film is a durable, stable, and efficient photocatalyst.

The activity of a large-sized PCT film (300 mm × 360 mm, Figure 6D and Figure S17) was also evaluated with the decomposition of 2.0 L of MO (20 mg L^{-1}) under three-types of light sources: UV-light, natural sunlight and fluorescent lamp. It had a good photocatalytic activity under UV light and sunlight (Table 1), with 94% and 85% of MO removed within 20 min and 2 h, respectively.

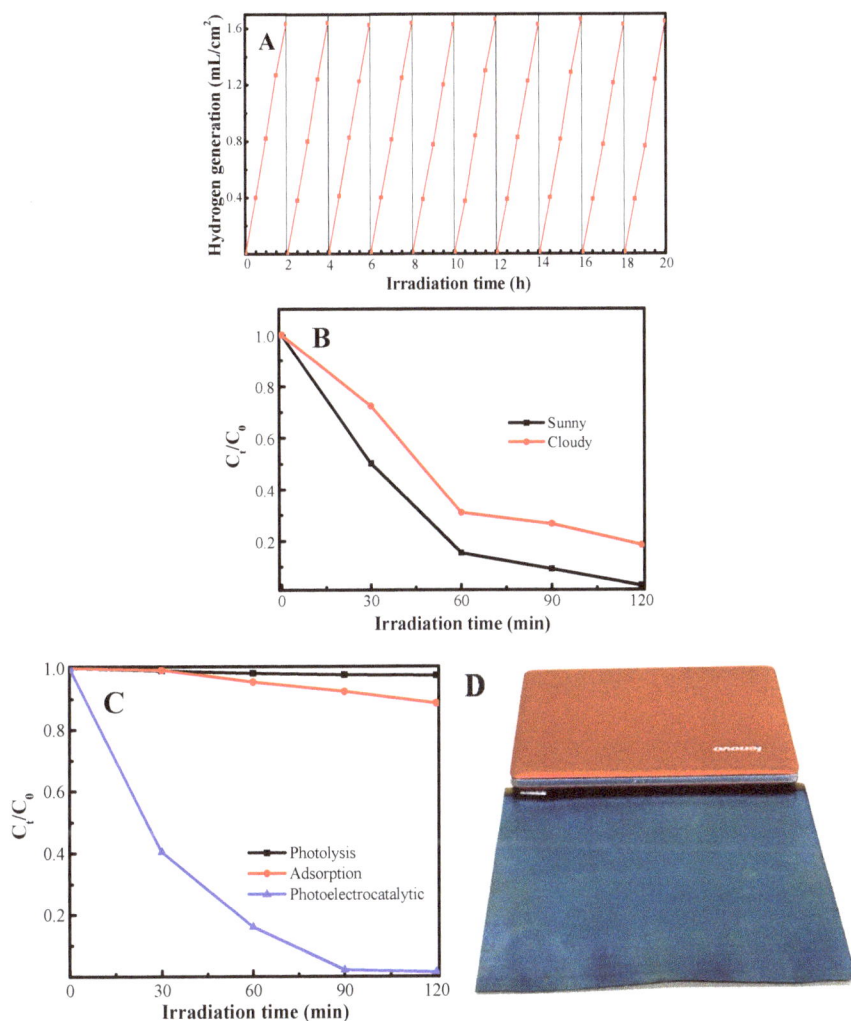

Figure 6. (**A**) Cycling measurements of hydrogen gas generation through photoelectrocatalytic water splitting with the PCT film under natural sunlight. Experiments were conducted over a 10-day period, with 20 h of overall irradiation time. (**B**) Comparison of the solar-driven photoelectrocatalytic activity of PCT film in sunny day and cloudy day. (**C**) Degradation rates of methyl orange (MO) solution with PCT film in three different processes. Photolysis means the MO degradation under light in the absence of PCT film, adsorption means the MO degradation in the darkness with PCT film, and photoelectrocatalytic is the MO degradation under light and bias with PCT film. (**D**) Picture of the large-size PCT film (300 mm × 360 mm) compared with the laptop.

Table 1. Degradation of MO solution under three different light sources with a large-scale PCT film.

Light Sources	Irradiation Time (min)	Degradation Rate (%)
UV	20	94
sunlight	120	85
fluorescent lamp	180	16

2.7. Photoelectrocatalytic Degradation of Wastewater

The practical performance of the PCT film (30 mm × 40 mm) was further tested with the printing and dyeing wastewater. Figure 7 showed the treatment process. After flocculation and precipitation, the upper wastewater (50.0 mL) was taken out for the photoelectrochemical degradation under UV light. The UV-visible absorbance spectra of waste water (Figure S18) showed that the peaks around 230 and 280 nm obviously decreased after 1 h irradiation, and completely disappeared after 2 h. The wastewater after the test almost had no color. The removal rates of COD and TOC reached 90% and 72.8%, respectively (Table 2). These results indicated that the PCT film had a good photocatalytic activity for purifying wastewater from printing and dyeing industries. According to the previous MO degradation tests, we can calculate that 2.5 L upper wastewater can be purified in 2 h with one large-scale PCT film (300 mm × 360 mm). That means, one 300 mm × 360 mm PCT film can purify 5 L of printing and dyeing wastewater within 2 h. Accordingly, around 556 L wastewater can be purified everyday with 1 m^2 of our PCT film.

| Waste water | After flocculation and precipitation | After photocatalytic degradation of 2 h |

Figure 7. A schematic diagram of the treatment process for printing and dyeing wastewater.

Table 2. Chemical oxygen demand (COD) and total organic carbon (TOC) removal rates after photoelectrocatalytic degradation.

	Initial Concentration (mg L^{-1})	Final Concentration (mg L^{-1})	Removal Rate (%)
COD	2620	253	90
TOC	1030	280	72.8

3. Experimental Section

3.1. Fast Anodization Synthesis of the TiO$_2$ Nanostructured Film

Ti foils (purity 99.9%, 0.3 mm × 30 mm × 40 mm) are polished with abrasive papers, cleaned with acetone, anhydrous ethanol, and distilled water in an ultrasonic bath, each for 10 min. The foils are then washed in a solution of $HF/HNO_3/H_2O$ (1:4:5 $v/v/v$) for 30 s, and rinsed with deionized water several times. A mixture of graphene (5.0 mg) (transmission electron microscopy (TEM) and high resolution transmission electron microscopy (HRTEM) images are shown in Figures S19 and S20) and magnesium nitrate solution (100.0 mg L^{-1}, 50.0 mL) is sonicated for 30 min to form a uniform aqueous electrolyte. In the anodizing process, two Ti foils are used as the anode and cathode, and are subjected to a constant voltage of 60 V for 2 min at 298 K. The sample is then rinsed with distilled water, and calcinated at 723 K for 2 h in a muffle oven. This is followed by a fast cooling step, where the sample is immediately taken out from the muffle oven after calcination, and cooled to room temperature. For comparison, the film anodized without graphene in the electrolyte (100.0 mg L^{-1}, 50.0 mL of $Mg(NO_3)_2$) is obtained in the same condition, and typical TiO$_2$ nanotube arrays (TNTs) are prepared as well at 60 V in 0.6 wt % NH$_4$F-10 vol % H$_2$O-EG electrolyte for 2 h. In the anodizing process, a Ti foil and a graphite rod are used as the anode and the cathode, respectively.

3.2. Fast Anodization Synthesis of Large-Scale TiO$_2$ Film

Large Ti foils (purity 99.9%, 0.3 mm × 300 mm × 360 mm) are cleaned the same way as the small Ti foils. They are first polished using abrasive papers, then cleaned with acetone, anhydrous ethanol, and distilled water in order. The pre-treatment, composition of the electrolyte, anodization conditions, and calcinations are the same as those applied to the small foils in Section 3.1.

3.3. Characterization

The micromorphologies are analyzed with a Hitachi S-4800 field-emission scanning electron microscopy (FE-SEM) instrument and Tecnai G2 F20 TEM instrument. The crystalline structures are obtained with glancing angle X'pert Pro X-ray diffraction (XRD) with Cu-Kα radiation (0.15418 nm). Ultraviolet-visible-near-infrared (UV-vis-NIR) absorption, Fourier Transform Infrared (FIIR) and Raman spectra are recorded on a Hitachi U-4100 UV-vis-NIR spectrometer, a Nicolet-5700 FTIR spectrophotometer and a LeiCA DMLM micro-spectrometer with excitation wavelength of 514.5 nm, respectively. The chemical states of samples are examined on a Perkin-Elmer X-ray photoelectron spectroscopy (XPS) instrument with an Mg K anode. The binding energy values are calibrated with C 1s = 284.4 eV. Surface defect is studied with a JES FA-200 electron spin resonance (ESR) spectrometer in air at 77 K.

3.4. Photoelectrochemical Performance

The photoelectrochemical measurements are conducted under sunlight in a three-electrode cell using a CHI 830C electrochemical workstation. A 1.0 M NaOH solution, the TiO$_2$ film (1.0 cm^2), Ag/AgCl and Pt foil, are used as the electrolyte, the working electrode, the reference and the counter electrode, respectively. The linear sweep voltammetry and photocurrent density are recorded. The efficiency of photoconversion is calculated as follows [54].

$$\begin{aligned} \eta(\%) &= [(total\ power\ output - electrical\ power\ input)/[light\ power\ input] \times 100\% \\ &= j_p \left[(E_{rev}^0 - |E_{app}|)/I_0 \right] \times 100\% \end{aligned} \tag{1}$$

here, j_p refers to the photocurrent density (mA cm^{-2}), E_{rev}^0 = 1.23 V the standard potential for H$_2$ evolution, I_0 the incident-light power density, and the applied potential $E_{app} = E_{means} - E_{aoc}$, where E_{means} refers to the working electrode potential at j_p, E_{aoc} the open circuit working electrode potential under equivalent conditions, and E_{aoc} is the voltage where the photocurrent becomes zero.

3.5. Photoelectrocatalytic Activity in Hydrogen Generation and Removal of Model Organic Pollutants

The photoelectrocatalytic activity is assessed with photocatalytic hydrogen generation and pollutants degradation under sunlight. Hydrogen generation is conducted in a quartz cell with a 1.0 M NaOH and 10% vol of methanol. The TiO$_2$ film (30 mm × 40 mm) and Pt foil are used as the photoanode and the cathode, respectively. A bias potential of 1.0 V is applied to the two electrodes separated by a Nafion membrane. Prior to irradiation, this system is purged to remove the air completely with argon. The hydrogen gas produced at the Pt electrode is collected in an inverted burette.

Methylene blue (MB) discoloration test is one of the most popular methods for assessing photocatalytic activity of films and powders [55–57]. Here we use methyl orange (MO) as the model organic pollutant as an example. The MO degradation experiments are performed in 40.0 mL of MO solution (20.0 mg L^{-1}, pH = 6.8) with 0.50 M NaCl as the supporting electrolyte. The TiO$_2$ film (30 mm × 40 mm) and a cleanly Ti foil are used as the working electrode and the counter electrode, respectively. A bias potential of 2.0 V is applied. Cycling tests are conducted for 2.5 h at the same time of day (12:00 to 14:30) under direct sunlight for 10 days. 1.0 mL of MO solution is taken out every 30 min, and the absorption spectrum of the MO solution is measured with a UV-vis spectrophotometer. The concentration of MO is determined with the absorbance value at 464 nm. The efficiency of MO decolorization (*D*%) is calculated as follows:

$$D\% = \frac{C_0 - C_t}{C_0} \times 100\% \tag{2}$$

here, C_0 and C_t are the initial absorbance and the absorbance after reaction time *t* of MO.

3.6. Removal of Organic Pollutants in Wastewater

The performance of the PCT film for practical pollution removal is assessed with the degradation of wastewater from printing and dyeing industries (Sichuan Mianyang Jialian Printing and Dyeing Company, Mianyang, China) under UV light. The wastewater is dark brown and its composition is very complex (pH = 13, Chemical oxygen demand (COD) is 2620 mg/L. Total organic carbon (TOC) is 1030 mg/L). Then, 100 mL wastewater is taken out to test, after pre-treatment of flocculation and precipitation, 50 mL upper solutions is then taken for the photoelectrocatalytic degradation. The large TiO_2 film (30 mm × 40 mm) and a cleanly Ti foil are used as the working and counter electrodes, respectively. A bias potential of 2.0 V is applied to the two electrodes. 0.50 M NaCl is added as the supporting electrolyte. Then, 1 mL MO solution is taken out every 30 min to measure its absorption spectrum. After 2 h irradiation, its COD is measured by Potassium dichromate oxidation method, and its TOC is determined using a Shimadzu TOC analyzer. COD and TOC removal are calculated using the following equations, respectively.

$$COD\,removal\,(\%) = (COD_0 - COD_t)/COD_0 \times 100 \tag{3}$$

$$TOC\,removal\,(\%) = (TOC_0 - TOC_t)/TOC_0 \times 100 \tag{4}$$

where COD_0 (mg L^{-1}) and COD_t (mg L^{-1}) are the initial concentration of chemical oxygen demand and its remaining concentration of chemical oxygen demand after reaction, respectively. TOC_0 (mg L^{-1}) and TOC_t (mg L^{-1}) are the initial concentration of total organic carbon and its remaining concentration of total organic carbon after reaction, respectively.

4. Conclusions

In this study, we have demonstrated a fast anodizing method to synthesize large-scale (e.g., 300 mm × 360 mm) pine-cone nanostructured TiO_2 film. The pine-cone TiO_2 possesses a strong solar absorption, and exhibits high photocatalytic activities in photo-oxidizing organic pollutants in wastewater and producing hydrogen from water under natural sunlight. This work has showed a promising future for practical utilization of anodized TiO_2 films in renewable energy and clean environment applications.

Supplementary Materials: The following are available online at www.mdpi.com/2073-4344/7/8/229/s1, Figure S1: FE-SEM image of overall PCT film, Figure S2: FE-SEM images of the TiO_2 film anodized in the electrolyte of magnesium nitrate solution (100 mg L^{-1}, 50 mL) at 60 V for 2 min, Figure S3: FE-SEM image of top view of the PCT film, Figure S4: FE-SEM image of layer structure of the PCT film, Figure S5: FE-SEM image of cracks on the oxide layer, Figure S6: FE-SEM image of the nanoporous oxide layer, Figure S7: FE-SEM image of small cracks in the Figure 1A, Figure S8: FE-SEM image of TiO^2 fragment washed up by O_2 gas, Figure S9: XRD patterns of the PCT films. Specific diffraction peaks for anatase (JCPDS#21−1272) and rutile (JCPDS#21−1276) are labeled according to A(hkl) and R(hkl), respectively, Figure S10: Raman scattering patterns of the PCT films, where the anatase and rutile vibration modes are labeled according to mode (a) and mode (r), respectively, Figure S11: Full XPS spectra of PCT film and TNTs, Figure S12: N 1s XPS spectrum of the PCT film, Figure S13: Ti 2p XPS spectra of the PCT film and TNTs, Figure S14: O 1s XPS spectra of the PCT film and TNTs, Figure S15: ESR spectra of PCT film with different cooling styles after calcination at 723 K for 2 h, Figure S16: Stability test of methyl orange degradation for ten cycles under natural sunlight (CMO = 20 mg L^{-1}, VMO = 40 mL), Figure S17: Pictures for the large-scale PCT films prepared by different anodization conditions. A laptop was used as a reference, Figure S18:

UV-Vis absorbance of the printing and dyeing wastewater during the photoelectrochemical degradation process. ☆ represents the absorbance of the waste water after flocculation and precipitation, Figure S19: TEM images of prepared graphene. The as-prepared graphene was almost completely transparent and had an average length and width of about 500 and 300 nm, Figure S20: HRTEM images of prepared graphene. The graphene was no more than 7 layers thick, Table S1: Lattice parameters and particle sizes of the PCT film, Table S2: Average illumination intensity of sunlight and outside temperature during the MO photodegradation.

Acknowledgments: This work was supported by Program for Changjiang Scholars and Innovative Research Team in University (IRT13083) from Ministry of Education of The People's Republic of China. The authors thank the support from China Scholarship Council and the College of Arts and Science, University of Missouri-Kansas City, and University of Missouri Research Board.

Author Contributions: As for authors, Yan Liu and Yanzong Zhang conceived and designed the experiments; Yan Liu performed all experiments and finished the manuscript with assistance from Lilin Wang, Gang Yang, Fei Shen, Shihuai Deng, and Xiaohong Zhang. He Yan and Yaodong Hu helped revised the manuscript. Xiaobo Chen provided the analysis tools and helped analyzed the mechanism.

Conflicts of Interest: The authors declare no conflict of interest.

References

1. Dong, J.; Han, J.; Liu, Y.; Nakajima, A.; Matsushita, S.; Wei, S.; Gao, W. Defective Black TiO_2 Synthesized via Anodization for Visible-Light Photocatalysis. *ACS Appl. Mater. Interfaces* **2014**, *6*, 1385–1388. [CrossRef] [PubMed]

2. Cui, H.; Zhao, W.; Yang, C.; Yin, H.; Lin, T.; Shan, Y.; Xie, Y.; Gu, H.; Huang, F. Black TiO_2 nanotube arrays for high-efficiency photoelectrochemical water-splitting. *J. Mater. Chem. A* **2014**, *2*, 8612–8616. [CrossRef]

3. Su, T.; Yang, Y.; Na, Y.; Fan, R.; Li, L.; Wei, L.; Yang, B.; Cao, W. An Insight into the Role of Oxygen Vacancy in Hydrogenated TiO_2 Nanocrystals in the Performance of Dye-Sensitized Solar Cells. *ACS Appl. Mater. Interfaces* **2015**, *7*, 3754–3763. [CrossRef] [PubMed]

4. Lee, J.; Hwang, S.H.; Yun, J.; Jang, J. Fabrication of SiO_2/TiO_2 Double-Shelled Hollow Nanospheres with Controllable Size via Sol-Gel Reaction and Sonication-Mediated Etching. *ACS Appl. Mater. Interfaces* **2014**, *6*, 15420–15426. [CrossRef] [PubMed]

5. Sarkar, A.; Jeon, N.J.; Noh, J.H.; Seok, S.I. Well-Organized Mesoporous TiO_2 Photoelectrodes by Block Copolymer-Induced Sol-Gel Assembly for Inorganic-Organic Hybrid Perovskite Solar Cells. *J. Phys. Chem. C* **2014**, *118*, 16688–16693. [CrossRef]

6. Lee, J.S.; You, K.H.; Park, C.B. Highly photoactive, low bandgap TiO_2 nanoparticles wrapped by graphene. *Adv. Mater.* **2012**, *24*, 1084–1088. [CrossRef] [PubMed]

7. Perera, S.D.; Mariano, R.G.; Vu, K.; Nour, N.; Seitz, O.; Chabal, Y.; Balkus, K.J. Hydrothermal Synthesis of Graphene-TiO_2 Nanotube Composites with Enhanced Photocatalytic Activity. *ACS Catal.* **2012**, *2*, 949–956. [CrossRef]

8. Chen, M.; Shen, X.; Wu, Q.; Li, W.; Diao, G. Template-assisted synthesis of core-shell α-Fe_2O_3@TiO_2 nanorods and their photocatalytic property. *J. Mater. Sci.* **2015**, *50*, 4083–4094. [CrossRef]

9. Zhang, X.; Huang, Y.; Huang, X.; Huang, C.; Li, H. Synthesis and characterization of polypyrrole using TiO_2 nanotube@poly(sodium styrene sulfonate) as dopant and template. *Polym. Compos.* **2016**, *37*, 462–467. [CrossRef]

10. Dutta, S.; Patra, A.K.; De, S.; Bhaumik, A.; Saha, B. Self-Assembled TiO_2 Nanospheres By Using a Biopolymer as a Template and Its Optoelectronic Application. *ACS Appl. Mater. Interfaces* **2012**, *4*, 1560–1564. [CrossRef] [PubMed]

11. Zimbone, M.; Cacciato, G.; Boutinguiza, M.; Privitera, V.; Grimaldi, M.G. Laser irradiation in water for the novel, scalable synthesis of black TiOx photocatalyst for environmental remediation. *Beilstein J. Nanotechnol.* **2017**, *8*, 196–202. [CrossRef] [PubMed]

12. Zimbone, M.; Cacciato, G.; Sanz, R.; Carles, R.; Gulino, A.; Privitera, V.; Grimaldi, M.G. Black TiO^x photocatalyst obtained by laser irradiation in water. *Catal. Commun.* **2016**, *84*, 11–15. [CrossRef]

13. Zimbone, M.; Cacciato, G.; Buccheri, M.A.; Sanz, R.; Piluso, N.; Reitano, R.; Via, F.L.; Grimaldi, M.G.; Privitera, V. Photocatalytical activity of amorphous hydrogenated TiO_2 obtained by pulsed laser ablation in liquid. *Mater. Sci. Semicond. Process.* **2016**, *42*, 28–31. [CrossRef]

14. Zimbone, M.; Buccheri, M.A.; Cacciato, G.; Sanz, R.; Rappazzo, G.; Boninelli, S.; Reitano, R.; Romano, L.; Privitera, V.; Grimaldi, M.G. Photocatalytical and antibacterial activity of TiO_2 nanoparticles obtained by laser ablation in water. *Appl. Catal. B Environ.* **2015**, *165*, 487–494. [CrossRef]

15. Du, Y.; Cai, H.; Wen, H.; Wu, Y.; Huang, L.; Ni, J.; Li, J.; Zhang, J. Novel Combination of Efficient Perovskite Solar Cells with Low Temperature Processed Compact TiO_2 Layer via Anodic Oxidation. *ACS Appl. Mater. Interfaces* **2016**, *8*, 12836–12842. [CrossRef] [PubMed]

16. Li, H.; Cheng, J.-W.; Shu, S.; Zhang, J.; Zheng, L.; Tsang, C.K.; Cheng, H.; Liang, F.; Lee, S.-T.; Li, Y.Y. Selective Removal of the Outer Shells of Anodic TiO_2 Nanotubes. *Small* **2013**, *9*, 37–44. [CrossRef] [PubMed]

17. Liu, G.; Sun, C.; Yang, H.G.; Smith, S.C.; Wang, L.; Lu, G.Q.; Cheng, H.-M. Nanosized anatase TiO_2 single crystals for enhanced photocatalytic activity. *Chem. Commun.* **2010**, *46*, 755–757. [CrossRef] [PubMed]

18. Jiao, W.; Wang, L.; Liu, G.; Lu, G.Q.; Cheng, H.-M. Hollow Anatase TiO_2 Single Crystals and Mesocrystals with Dominant {101} Facets for Improved Photocatalysis Activity and Tuned Reaction Preference. *ACS Catal.* **2012**, *2*, 1854–1859. [CrossRef]

19. Wu, X.; Lu, G.Q.; Wang, L. Shell-in-shell TiO_2 hollow spheres synthesized by one-pot hydrothermal method for dye-sensitized solar cell application. *Energy Environ. Sci.* **2011**, *4*, 3565–3572. [CrossRef]

20. Joo, J.B.; Dahl, M.; Li, N.; Zaera, F.; Yin, Y. Tailored synthesis of mesoporous TiO_2 hollow nanostructures for catalytic applications. *Energy Environ. Sci.* **2013**, *6*, 2082–2092. [CrossRef]

21. Joo, J.B.; Zhang, Q.; Dahl, M.; Lee, I.; Goebl, J.; Zaera, F.; Yin, Y. Control of the nanoscale crystallinity in mesoporous TiO_2 shells for enhanced photocatalytic activity. *Energy Environ. Sci.* **2012**, *5*, 6321–6327. [CrossRef]

22. Konishi, J.; Fujita, K.; Nakanishi, K.; Hirao, K. Monolithic TiO_2 with Controlled Multiscale Porosity via a Template-Free Sol-Gel Process Accompanied by Phase Separation. *Chem. Mater.* **2006**, *18*, 6069–6074. [CrossRef]

23. Wu, X.; Chen, Z.; Lu, G.Q.; Wang, L. Nanosized Anatase TiO_2 Single Crystals with Tunable Exposed (001) Facets for Enhanced Energy Conversion Efficiency of Dye-Sensitized Solar Cells. *Adv. Funct. Mater.* **2011**, *21*, 4167–4172. [CrossRef]

24. Sánchez-Tovar, R.; Lee, K.; García-Antón, J.; Schmuki, P. Formation of anodic TiO_2 nanotube or nanosponge morphology determined by the electrolyte hydrodynamic conditions. *Electrochem. Commun.* **2013**, *26*, 1–4. [CrossRef]

25. Yoo, J.E.; Lee, K.; Altomare, M.; Selli, E.; Schmuki, P. Self-Organized Arrays of Single-Metal Catalyst Particles in TiO_2 Cavities: A Highly Efficient Photocatalytic System. *Angew. Chem. Int. Ed.* **2013**, *52*, 7514–7517. [CrossRef] [PubMed]

26. Yoo, J.; Altomare, M.; Mokhtar, M.; Alshehri, A.; Al-Thabaiti, S.A.; Mazare, A.; Schmuki, P. Photocatalytic H_2 Generation Using Dewetted Pt-Decorated TiO_2 Nanotubes: Optimized Dewetting and Oxide Crystallization by a Multiple Annealing Process. *J. Phys. Chem. C* **2016**, *120*, 15884–15892. [CrossRef]

27. Liang, F.; Luo, L.-B.; Tsang, C.-K.; Zheng, L.; Cheng, H.; Li, Y.Y. TiO_2 nanotube-based field effect transistors and their application as humidity sensors. *Mater. Res. Bull.* **2012**, *47*, 54–58. [CrossRef]

28. Zheng, L.; Han, S.; Liu, H.; Yu, P.; Fang, X. Hierarchical MoS2 Nanosheet@ TiO_2 Nanotube Array Composites with Enhanced Photocatalytic and Photocurrent Performances. *Small* **2016**, *12*, 1527–1536. [CrossRef] [PubMed]

29. Chu, S.; Inoue, S.; Wada, K.; Hishita, S.; Kurashima, K. A New Electrochemical Lithography Fabrication of Self-Organized Titania Nanostructures on Glass by Combined Anodization. *J. Electrochem. Soc.* **2005**, *152*, B116–B124. [CrossRef]

30. Ghicov, A.; Schmuki, P. Self-ordering electrochemistry: A review on growth and functionality of TiO_2 nanotubes and other self-aligned MO_x structures. *Chem. Commun.* **2009**, 2791–2808. [CrossRef] [PubMed]

31. Giammar, D.E.; Maus, C.J.; Xie, L. Effects of particle size and crystalline phase on lead adsorption to titanium dioxide nanoparticles. *Environ. Eng. Sci.* **2007**, *24*, 85–95. [CrossRef]

32. Keller, A.A.; Wang, H.; Zhou, D.; Lenihan, H.S.; Cherr, G.; Cardinale, B.J.; Miller, R.; Ji, Z. Stability and aggregation of metal oxide nanoparticles in natural aqueous matrices. *Environ. Sci. Technol.* **2010**, *44*, 1962–1967. [CrossRef] [PubMed]

33. Shih, Y.-H.; Liu, W.-S.; Su, Y.-F. Aggregation of stabilized TiO_2 nanoparticle suspensions in the presence of inorganic ions. *Environ. Toxicol. Chem.* **2012**, *31*, 1693–1698. [CrossRef] [PubMed]

34. Deskins, N.A.; Kerisit, S.; Rosso, K.M.; Dupuis, M. Molecular Dynamics Characterization of Rutile-Anatase Interfaces. *J. Phys. Chem. C* **2007**, *111*, 9290–9298. [CrossRef]

35. Liang, H.C.; Li, X.Z. Effects of structure of anodic TiO_2 nanotube arrays on photocatalytic activity for the degradation of 2,3-dichlorophenol in aqueous solution. *J. Hazard. Mater.* **2009**, *162*, 1415–1422. [CrossRef] [PubMed]

36. Hasegawa, G.; Sato, T.; Kanamori, K.; Nakano, K.; Yajima, T.; Kobayashi, Y.; Kageyama, H.; Abe, T.; Nakanishi, K. Hierarchically Porous Monoliths Based on N-Doped Reduced Titanium Oxides and Their Electric and Electrochemical Properties. *Chem. Mater.* **2013**, *25*, 3504–3512. [CrossRef]

37. Wang, D.-H.; Jia, L.; Wu, X.-L.; Lu, L.-Q.; Xu, A.-W. One-step hydrothermal synthesis of N-doped TiO_2/C nanocomposites with high visible light photocatalytic activity. *Nanoscale* **2012**, *4*, 576–584. [CrossRef] [PubMed]

38. Asahi, R.; Morikawa, T.; Ohwaki, T.; Aoki, K.; Taga, Y. Visible-Light Photocatalysis in Nitrogen-Doped Titanium Oxides. *Science* **2001**, *293*, 269–271. [CrossRef] [PubMed]

39. Liu, Y.; Mu, K.; Zhong, J.; Chen, K.; Zhang, Y.; Yang, G.; Wang, L.; Deng, S.; Shen, F.; Zhang, X. Design of a solar-driven TiO_2 nanofilm on Ti foil by self-structure modifications. *RSC Adv.* **2015**, *5*, 41437–41444. [CrossRef]

40. Hadjiivanov, K.; Knözinger, H. Species formed after NO adsorption and NO + O_2 co-adsorption on TiO_2: An FTIR spectroscopic study. *Phys. Chem. Chem. Phys.* **2000**, *2*, 2803–2806. [CrossRef]

41. Yang, J.; Bai, H.; Tan, X.; Lian, J. IR and XPS investigation of visible-light photocatalysis-Nitrogen-carbon-doped TiO_2 film. *Appl. Surf. Sci.* **2006**, *253*, 1988–1994. [CrossRef]

42. Wang, J.; Tafen, D.N.; Lewis, J.P.; Hong, Z.; Manivannan, A.; Zhi, M.; Li, M.; Wu, N. Origin of photocatalytic activity of nitrogen-doped TiO_2 nanobelts. *J. Am. Chem. Soc.* **2009**, *131*, 12290–12297. [CrossRef] [PubMed]

43. Long, M.; Qin, Y.; Chen, C.; Guo, X.; Tan, B.; Cai, W. Origin of Visible Light Photoactivity of Reduced Graphene Oxide/TiO_2 by in Situ Hydrothermal Growth of Undergrown TiO_2 with Graphene Oxide. *J. Phys. Chem. C* **2013**, *117*, 16734–16741. [CrossRef]

44. Zhuang, J.; Weng, S.; Dai, W.; Liu, P.; Liu, Q. Effects of Interface Defects on Charge Transfer and Photoinduced Properties of TiO_2 Bilayer Films. *J. Phys. Chem. C* **2012**, *116*, 25354–25361. [CrossRef]

45. Batzill, M.; Morales, E.H.; Diebold, U. Influence of nitrogen doping on the defect formation and surface properties of TiO_2 rutile and anatase. *Phys. Rev. Lett.* **2006**, *96*, 026103. [CrossRef] [PubMed]

46. Di Valentin, C.; Pacchioni, G.; Selloni, A.; Livraghi, S.; Giamello, E. Characterization of paramagnetic species in N-doped TiO_2 powders by EPR spectroscopy and DFT calculations. *J. Phys. Chem. B* **2005**, *109*, 11414–11419. [CrossRef] [PubMed]

47. Fang, D.; Luo, Z.; Huang, K.; Lagoudas, D.C. Effect of heat treatment on morphology, crystalline structure and photocatalysis properties of TiO_2 nanotubes on Ti substrate and freestanding membrane. *Appl. Surf. Sci.* **2011**, *257*, 6451–6461. [CrossRef]

48. Naldoni, A.; Allieta, M.; Santangelo, S.; Marelli, M.; Fabbri, F.; Cappelli, S.; Bianchi, C.L.; Psaro, R.; Dal Santo, V. Effect of Nature and Location of Defects on Bandgap Narrowing in Black TiO_2 Nanoparticles. *J. Am. Chem. Soc.* **2012**, *134*, 7600–7603. [CrossRef] [PubMed]

49. Chen, X.; Liu, L.; Yu, P.Y.; Mao, S.S. Increasing Solar Absorption for Photocatalysis with Black Hydrogenated Titanium Dioxide Nanocrystals. *Science* **2011**, *331*, 746–750. [CrossRef] [PubMed]

50. Xia, T.; Li, N.; Zhang, Y.; Kruger, M.B.; Murowchick, J.; Selloni, A.; Chen, X. Directional Heat Dissipation across the Interface in Anatase-Rutile Nanocomposites. *ACS Appl. Mater. Interfaces* **2013**, *5*, 9883–9890. [CrossRef] [PubMed]

51. Zhang, Z.; Zhang, L.; Hedhili, M.N.; Zhang, H.; Wang, P. Plasmonic Gold Nanocrystals Coupled with Photonic Crystal Seamlessly on TiO_2 Nanotube Photoelectrodes for Efficient Visible Light Photoelectrochemical Water Splitting. *Nano Lett.* **2013**, *13*, 14–20. [CrossRef] [PubMed]

52. Zhang, Z.; Wang, P. Optimization of photoelectrochemical water splitting performance on hierarchical TiO_2 nanotube arrays. *Energy Environ. Sci.* **2012**, *5*, 6506–6512. [CrossRef]

53. Xiao, F.-X.; Hung, S.-F.; Miao, J.; Wang, H.-Y.; Yang, H.; Liu, B. Metal-Cluster-Decorated TiO_2 Nanotube Arrays: A Composite Heterostructure toward Versatile Photocatalytic and Photoelectrochemical Applications. *Small* **2015**, *11*, 554–567. [CrossRef] [PubMed]

54. Sun, Y.; Yan, K.; Wang, G.; Guo, W.; Ma, T. Effect of Annealing Temperature on the Hydrogen Production of TiO_2 Nanotube Arrays in a Two-Compartment Photoelectrochemical Cell. *J. Phys. Chem. C* **2011**, *115*, 12844–12849. [CrossRef]

55. Ryu, J.; Choi, W. Substrate-Specific Photocatalytic Activities of TiO_2 and Multiactivity Test for Water Treatment Application. *Environ. Sci. Technol.* **2008**, *42*, 294–300. [CrossRef] [PubMed]

56. Mills, A.; Mcfarlane, M. Current and possible future methods of assessing the activities of photocatalyst films. *Catal. Today* **2007**, *129*, 22–28. [CrossRef]

57. Mills, A.; Hill, C.; Robertson, P.K.J. Overview of the current ISO tests for photocatalytic materials. *J. Photochem. Photobiol. A Chem.* **2012**, *237*, 7–23. [CrossRef]

© 2017 by the authors. Licensee MDPI, Basel, Switzerland. This article is an open access article distributed under the terms and conditions of the Creative Commons Attribution (CC BY) license (http://creativecommons.org/licenses/by/4.0/).

catalysts

MDPI

Article

TiO$_2$ Nanotubes on Transparent Substrates: Control of Film Microstructure and Photoelectrochemical Water Splitting Performance

Matus Zelny [1], Stepan Kment [1,*], Radim Ctvrtlik [1], Sarka Pausova [2], Hana Kmentova [1], Jan Tomastik [1], Zdenek Hubicka [3], Yalavarthi Rambabu [1], Josef Krysa [2], Alberto Naldoni [1], Patrik Schmuki [1,4] and Radek Zboril [1,*]

[1] Regional Centre of Advanced Technologies and Materials, Faculty of Science, Palacky University, 17. Listopadu 1192/12, 771 46 Olomouc, Czech Republic; matus.zelny01@upol.cz (M.Z.); radim.ctvrtlik@upol.cz (R.C.); hana.kmentova@upol.cz (H.K.); jan.tomastik@upol.cz (J.T.); rambabu.yalavarthi@upol.cz (Y.R.); alberto.naldoni@upol.cz (A.N.); schmuki@ww.uni-erlangen.de (P.S.)
[2] Department of Inorganic Technology, University of Chemical Technology Prague, Technicka 5, 166 28 Prague, Czech Republic; Sarka.Pausova@vscht.cz (S.P.); josef.krysa@vscht.cz (J.K.)
[3] Institute of Physics, Academy of Sciences of the Czech Republic, Na Slovance 2, 14800 Prague, Czech Republic; hubicka@fzu.cz
[4] Department of Materials Science and Engineering, University of Erlangen-Nuremberg, Martensstrasse 7, D-91058 Erlangen, Germany
[*] Correspondence: stepan.kment@upol.cz (S.K.); radek.zboril@upol.cz (R.Z.); Tel.: +42-058-563-4365 (S.K.)

Received: 13 December 2017; Accepted: 11 January 2018; Published: 15 January 2018

Abstract: Transfer of semiconductor thin films on transparent and or flexible substrates is a highly desirable process to enable photonic, catalytic, and sensing technologies. A promising approach to fabricate nanostructured TiO$_2$ films on transparent substrates is self-ordering by anodizing of thin metal films on fluorine-doped tin oxide (FTO). Here, we report pulsed direct current (DC) magnetron sputtering for the deposition of titanium thin films on conductive glass substrates at temperatures ranging from room temperature to 450 °C. We describe in detail the influence that deposition temperature has on mechanical, adhesion and microstructural properties of titanium film, as well as on the corresponding TiO$_2$ nanotube array obtained after anodization and annealing. Finally, we measure the photoelectrochemical water splitting activity of different TiO$_2$ nanotube samples showing that the film deposited at 150 °C has much higher activity correlating well with the lower crystallite size and the higher degree of self-organization observed in comparison with the nanotubes obtained at different temperatures. Importantly, the film showing higher water splitting activity does not have the best adhesion on glass substrate, highlighting an important trade-off for future optimization.

Keywords: titanium; anodization; TiO$_2$ nanotubes; hardness; adhesion; photoelectrochemistry

1. Introduction

Titanium dioxide (TiO$_2$) is one of the most widely studied semiconductor photocatalysts owing to its ability to catalyze numerous redox reactions [1], high stability, nontoxicity and low cost [2,3]. Nanocrystalline TiO$_2$ has highly promising optical, photocatalytic and photoelectrochemical (PEC) performance [4], which have been extensively used for a broad range of applications including environmental purification, organic oxidations [5], solar cells, PEC water splitting, and hydrogen peroxide production [1,2,4,6–12].

The photocatalytic performance of TiO$_2$ is strongly affected by its morphology and structure. With the decrease of the material's dimensions to nanoscale, surface-to-volume ratio and specific surface area increase as well as electronic properties that may also deviate from ideal behavior [13]. Crystal phase and orientation in nanocrystals and thin films have shown a dramatic role in enhancing charge separation, band gap, and surface catalytic properties of TiO$_2$ nanomaterials [6,11–13].

Furthermore, TiO$_2$ nanomaterials have shown shape-dependent photocatalytic performance [14]. For thin films, various TiO$_2$ one-dimensional (1D) nanostructures involving self-organized nanotubes, highly-ordered nanorod arrays, and nanowires, have attracted much attention due to the combination of highly functional features and controllable nanoscale geometry with the possibility of adjusting length, diameter, and spacing [13,15–17]. In particular, 1D TiO$_2$ nanotubes (TNT) show better PEC performance than thin compact layers due to higher surface area, favorable charge transfer along the nanotube *y*-axis perpendicular to charge collecting bottom layer, and enhanced light harvesting efficiency [18].

Self-organized oxide tube arrays can be fabricated by electrochemical anodization of a suitable metal foil under specific anodic conditions and proper electrolytes [18–23]. For instance, highly ordered TiO$_2$ nanotube arrays are typically grown by electrochemical anodization of titanium foils [23]. However, fully transparent photoelectrodes or photonic devices are highly desirable in many applications including photovoltaics and PEC cells [18,24]. For instance, TNT reaching transparency higher than 65% and producing stable photocurrent can be employed in PEC tandem cells, in which the illumination from the support—semiconductor interface is a required condition, enabling the efficient excitation of both photoanode and photocathode materials [24].

Recently, the anodization of titanium thin films deposited on glass substrates via different Physical Vapor Deposition methods [18,25], especially magnetron sputtering, have proven to be a promising route to achieve devices with high stability and performance [16,24–29].

In particular, the properties of TNT are highly influenced by the properties of the starting magnetron sputtered metal films, which in turn have shown strong tenability depending on energy of impinging ions, degree of ionization of the sputtered-particles [30], deposition temperature [31], and concentration of oxygen or nitrogen [32]. Nevertheless, a study exclusively focused on the influence of temperature of a glass substrate during the pulsed DC magnetron sputtering on the properties of the deposited titanium films and consequently on the properties and functionality of final TiO$_2$ nanotubes is still missing. In this work, we report a detailed investigation of mechanical and adhesion properties of Ti films sputtered at different temperatures, as well as how these different sputtering conditions influence crystallographic and photoelectrochemical water spitting activity of the self-organized TiO$_2$ nanotubes grown from these Ti films.

2. Results and Discussion

2.1. Structure and Morphology of Sputtered Ti Films

Titanium films were sputtered onto fluorine-doped tin oxide (FTO) substrate at different temperatures such as room temperature (RT), 150 °C, 300 °C, and 450 °C. For the sake of clarity, despite the fact that the RT condition did not include intentional heating, the substrate's temperature slightly rose up to 80 °C due to the bombardment of the FTO surface by plasma discharge ions. The deposition rate follows a volcano-shaped trend with increasing the deposition temperature and reaches the maximum of 47 nm/min for 150 °C (Figure 1). The grain size follows an opposite trend and the minimum value of ~80 nm is reached for the Ti sample deposited at 150 °C. This dependence can be ascribed to the processes occurring during the sputtering deposition such as extent of shadowing effect at low deposition temperatures, and surface and volume diffusion of condensing atoms at higher temperatures in accordance with the structure zone models [33].

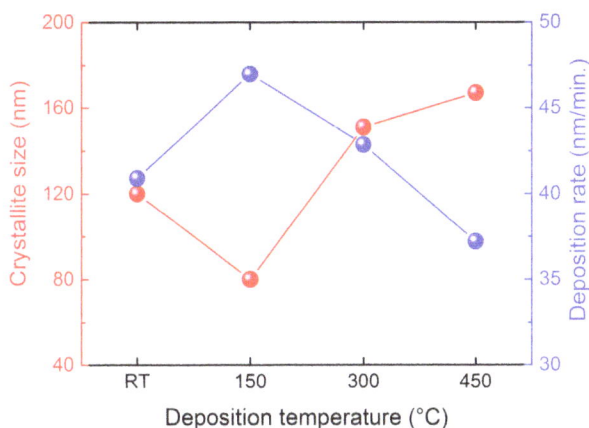

Figure 1. Crystallite size and deposition rate as a function of deposition temperature for the direct current magnetron sputtered Ti films. Deposition rate shows a volcano shape dependence on temperature, while crystallite size has an opposite trend. RT: room temperature.

The as-deposited titanium samples were well adherent to the FTO substrate regardless of the temperature used without any visually observable defects. The morphological images of all deposited films are displayed in the scanning electron microscopy (SEM) micrographs reported in Figure 2a–d. With increasing temperature, the shape of grains and their size distribution varied considerably: Ti films deposited at RT showed globular morphology, while fully developed hexagonal platelets with sharp edges were clearly seen for the film deposited at 150 °C. The platelets were randomly distributed and placed on top of each other. The change of morphology is associated with a decrease of the crystallite size from 120 nm (RT) to 80 nm (150 °C) as summarized in Table 1 and Figure 1. Further increase of the deposition temperature led to a higher surface and bulk diffusivity of sputtered atoms promoting formation of angular grains, especially for film deposited at 450 °C, thus leading to a substantial increase of crystallite size (see Table 1).

Table 1. Thickness and roughness from profilometry of Ti films; crystallite size of the deposited Ti films and anodized TiO_2 nanotubes. RT: room temperature.

Temperature, °C	Thickness, µm	Roughness, µm	Crystallite Size Ti, nm	Crystallite Size TiO_2, nm
RT	1.35	0.080	120	39
150	1.61	0.125	80	14
300	1.50	0.118	151	25
450	1.48	0.028	167	87

X-ray diffraction was used to determine the crystalline structures of deposited Ti films. Irrespective of the deposition temperature, the as-prepared films exhibited a variation of peak intensities corresponding to hexagonal polycrystalline structure of titanium with preferential orientation along the (002) crystalline plane that is generally the densest plane for hexagonal close packed structures. It is worth noting the evolution of (100) and (110) peaks with the increase of the deposition temperature in agreement with previous work [34]. The development of these crystalline planes may be attributed to the presence of a compressive stress induced in the films' microstructure [34,35]. Crystallite size was estimated by using the Scherrer's equation and retrieved values are shown in Table 1. Figure 2f shows the energy dispersive X-ray spectroscopy (EDS) spectrum of the titanium film deposited at 150 °C on FTO glass. The SEM-EDS analysis revealed traces of Sn in all samples (not shown here) due to the presence of Sn ions into the FTO (i.e., F-doped SnO_2) underneath substrate. The surface roughness (R_a) of deposited Ti layers was tested by contact profilometry and the obtained values of standard R_a parameters are provided in Table 1. Surprisingly, the smoothest surface was identified for the films composed of the biggest crystalline size. It was also visually observable as a very flat mirror-like surface. Film thickness obtained from profilometry measurements ranges from 1.35 to 1.60 μm, thus being comparable for all Ti films.

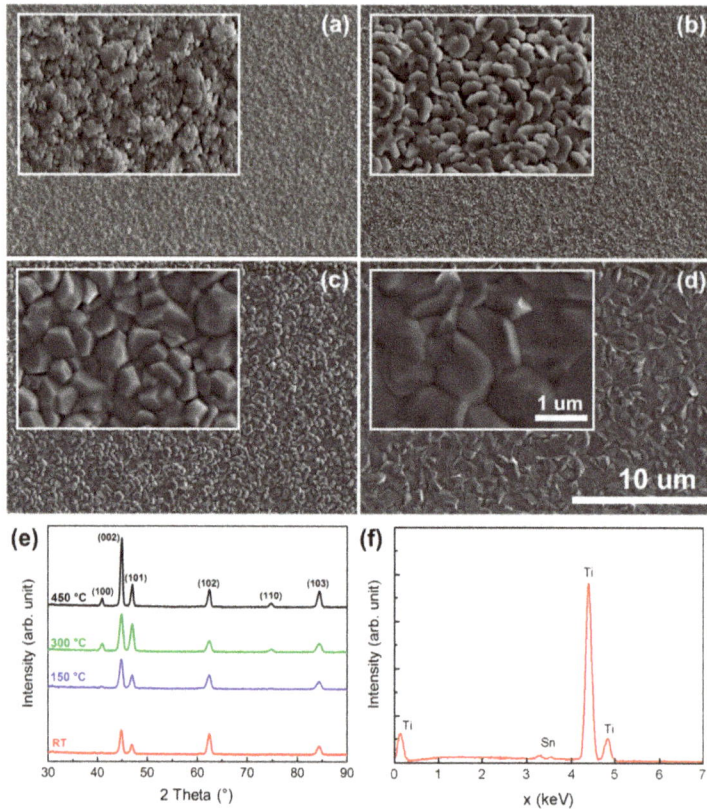

Figure 2. Scanning electron microscope (SEM) images of titanium samples deposited without substrate heating at (**a**) RT, (**b**) 150 °C, (**c**) 300 °C and (**d**) 450 °C. (**e**) X-ray diffraction (XRD) spectra of all deposited Ti films; (**f**) energy dispersive X-ray spectroscopy spectrum of titanium film deposited at 150 °C on fluorine-doped tin oxide (FTO) substrate.

2.2. Mechanical Properties of Sputtered Titanium Films

Analysis of nanoindentation data (Figure 3) evidenced significant differences in mechanical properties between the deposited Ti films. Substrate heating led to an increase in reduced modulus from 109 GPa for RT sample (and similar values observed for 150 and 300 °C) to 130 GPa for the film deposited at 450 °C. Similarly, hardness values are almost the same regardless of the deposition temperature up to 300 °C (~3.75 GPa), while a small increase can be observed for the film deposited at 450 °C. Nanoindentation hardness of the films is only slightly higher in comparison to the coarse grained Ti bulk sheet (~2 GPa) measured at the same experimental setup, especially for the films deposited up to 300 °C. This fact correlates well with the fine-grained structure, where dislocation activity for crystallite with size around 100 nm is suppressed as explained by the well know Hall–Petch effect [36]. Nevertheless, the relative proximity of hardness values of the Ti films and pure Ti metal bulk reflects the high purity of the films, since oxygen or nitrogen impurities strongly affect the mechanical properties and increase both the hardness and elastic modulus values [32]. No traces of oxygen were detected using EDS. Hardness increase up to 4 GPa was reported for Ti sputtered films under similar conditions but under Ar/O_2 gas mixture [18]. It should be noted that either oxides (TiO_2) or nitrides (TiN_x) can reach much higher hardness. In case of magnetron sputtered TiO_2 films hardness values of pure anatase is in the range of 6–11 GPa, whereas rutile can reach around 20 GPa [37].

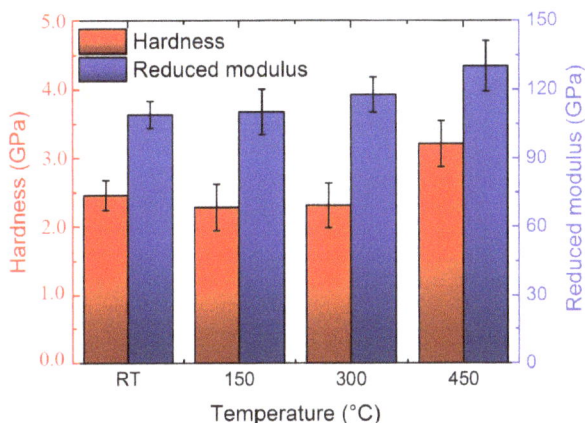

Figure 3. Hardness and reduced modulus of the Ti films deposited at different temperatures.

Progressive load scratch tests revealed increasing endurance of Ti films with the increase of deposition temperature, as demonstrated from the residual grooves tracks shown in Figure 4. Sample deposited without applying external heating (RT) exhibits full coating delamination starting from 1/3 of the scratch track. Large spalled areas uncovering the bare substrate, far beyond the residual groove, show that coating-to-substrate adhesion as well as cohesion strength are weak and become a main reason for the system failure. Substrate heating up to 150 °C led to better adhesion, as no delamination is observed and coating is scratched through after approximately half of the scratch track. Deposition temperature of 300 °C and especially 450 °C has a significant impact on scratch resistance as coatings were not scratched through. The residual scratch tracks' surface morphologies are almost smooth and dominated by plastic deformation (see beginnings of the scratches). This is in accordance with the high level of plasticity index of approx. 83%, defined as the ratio of the plastic work to total indentation work. With increasing load, formation of faint pile up around the wear track occurs. Only slightly worn particle packing is observed at the sides of the wear track. It should be noted that findings of scratch test performed at lower maximum force of 100 mN coincide with those reported in [18].

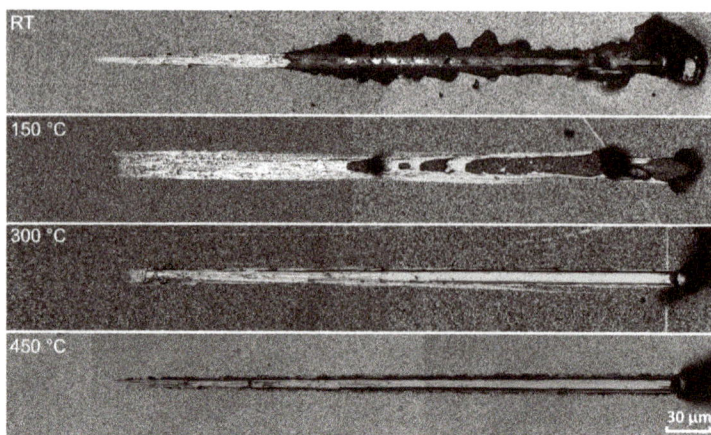

Figure 4. Typical scratch tracks for Ti films deposited at temperatures ranging from room temperature (RT) to 450 °C.

2.3. Electrochemical Anodization to Grow TiO₂ Nanotubes

Figure 5a shows the current density plot in time during the anodic oxidation of the sputtered Ti layers sputtered on FTO glass at different temperatures (RT, 150, 300 and 450 °C). All the current transient curves can be divided into three typical stages, already described in detail elsewhere [16,38]. Briefly, the very sharp current density peak observed during the first seconds of the anodization process is associated with the formation of an initial compact TiO₂ layer. A relatively steady state region followed, denoting the self-organizing electrochemical reaction underlying the formation of TiO₂ nanotubes. Finally, a sudden increase of current density marks the end point of the reaction, which may be accompanied with the creation of random cracks within the TNT and/or their partial delamination [39].

Notably, two trends can be observed when the current density curves are compared. The titanium films deposited at higher temperature required higher current density to be anodically oxidized, while electrochemical reaction lasted for a much shorter time than the titanium films prepared under lower temperature. Due to the elevated temperatures, the crystallite size and density of layers are significantly increased, along with a change in crystallite preferential orientation towards the (002) planes. The shorter the anodization duration the larger the crystallite sizes; at the same time the denser the films the higher current density is required for TNT formation. Interestingly, the TNT peeled off in the center of the anodized area only for the 450 °C sample (Figure 5a). The photographs were captured after the thermal annealing of the as-grown amorphous TiO₂ nanotubes in air at 500 °C for 1 h to obtain the crystalline structure. The delamination already occurred during the anodization process despite thermal annealing. It should be noted that the dominant (002) plane is the one with the highest thermally induced strain energy per unit volume [40]. Taking into account the crystallographic planes observed in Figure 2e, the strain energy decreases in the following order (002), (103), (102), (101), (100) and (110), where the last two are equal [40]. Hence, the problematic anodization of the film deposited at 450 °C stems from the combination of high current density (high thermal load) during anodization and the highest thermally induced strains of the film's dominant (002) plane.

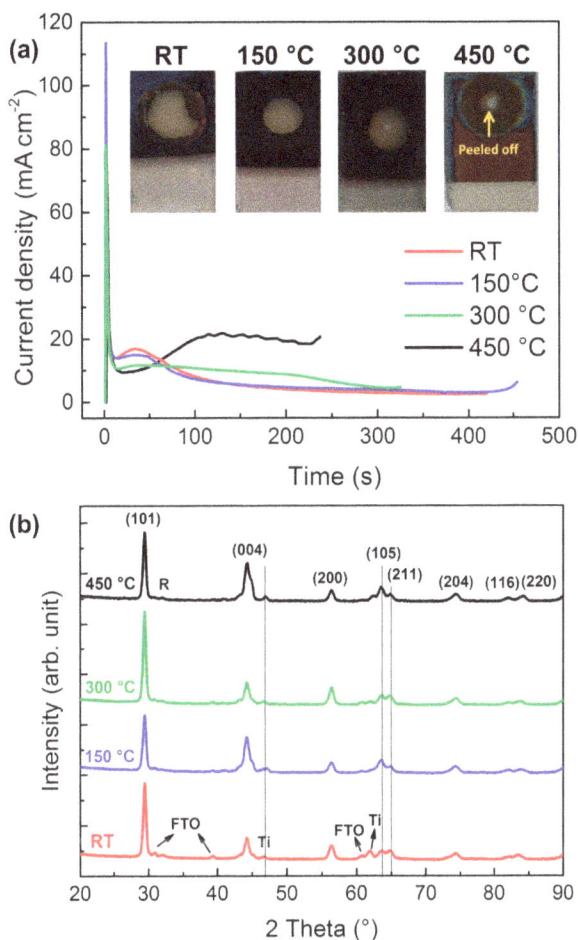

Figure 5. (a) Current density versus time plot recorder during the electrochemical anodization of titanium films with the pictures of grown and annealed TiO$_2$ nanotubes; (b) XRD spectra of TiO$_2$ nanotubes after thermal annealing at 450 °C for 3 h in air.

XRD spectra of the thermally annealed TiO$_2$ nanotubes show only characteristic peaks related to the polycrystalline anatase phase (Figure 5b). The crystallite size, obtained through the Sherrer equation, followed the same trend as observed for titanium films, i.e., the smallest grain size of 14 nm was revealed for the TNT grown from the titanium films deposited at 150 °C, while with the increase of the deposition temperature the grain sizes increased up to 87 nm for 450 °C. In the diffractograms, the signals related to metallic titanium as well as the cassiterite (SnO$_2$) of the FTO substrate were detected. The source of Ti signal is probably due to the side unanodized parts of the samples rather than residual Ti impurities in the TNTs.

The cross-sectional and surface SEM morphology images of the prepared nanotubes are shown in Figure 6. The anodization formed self-organized arrays of highly transparent nanotubes grown vertically on the FTO substrate. All the prepared nanotubes had a similar thickness of 3 μm, which corresponds to a volume expansion factor of ~2 due to the anodization procedure. This value is in agreement with findings from Albu and Schmuki, describing how key parameters such as content of water in the electrolyte and overall anodization potential influence the expansion factor of TNTs [41].

The surface SEM images (see inserts in Figure 6) are very similar for RT, 150 °C, and 300 °C deposited Ti films. The similarity is due to a nanoporous thin initiation layer which is always present at the top of nanotubes due to the TiO$_2$ layer formed at the first stage of the anodization process (see description above). By comparing these three types of nanotube arrays a widening of the tube diameters can be indicated. The surface morphology is slightly different for the film deposited at 450 °C. In this case, pores with higher diameters are formed at edges of the very large titanium grains (see Figures 2d and 6d) [39]. The highest quality of nanotubes in terms of homogeneity, degree of organization, smoothness, and compactness was observed for the TNT grown from Ti films deposited at 150 °C. A much higher number of defects was observed for the TNT grown from RT and 300 °C titanium films. For comparison, the parts of TNT made from Ti films deposited at 450 °C, which were not delaminated from the FTO substrate are also shown. The very low quality of these nanotubes is mainly due to numerous cracks and is evident (Figure 6d).

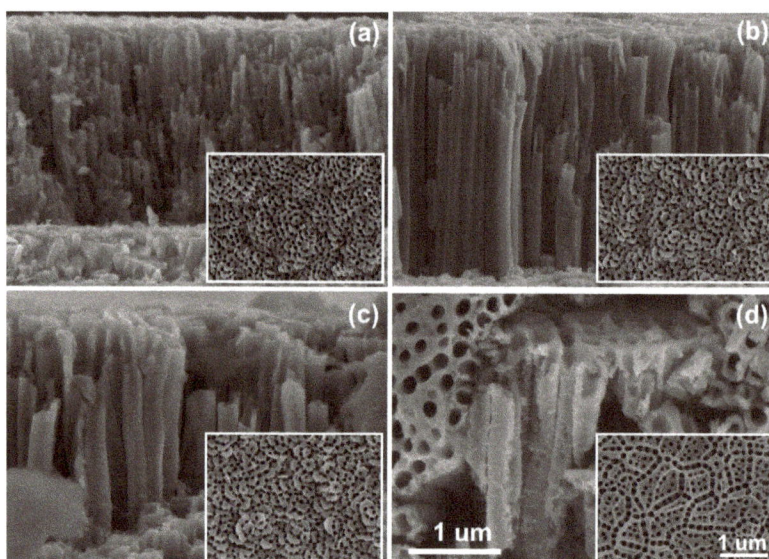

Figure 6. Cross-sectional and corresponding surface (inserts) SEM images of the TiO$_2$ nanotubes (TNT) arrays grown from titanium films deposited at RT (**a**), 150 °C (**b**), 300 °C (**c**), and 450 °C (**d**).

2.4. Photoelectrochemical Measurements

Photoelectrochemical properties were investigated based on linear sweep voltammetry and electrochemical impedance spectroscopy (EIS) measurements. The polarization curves showing the dependence of current density on applied potential are presented in Figure 7a. The performance of TNT obtained from Ti film sputtered at 450 °C was not measured due to the collapse of the structure upon air annealing. The experiments were carried out in a conventional three-electrode configuration in 1 M NaOH and under standard AM1.5G (intensity 100 mW/cm^2) illumination. All three measured photoanodes showed a similar voltammetry profile characterized by an onset potential at ~0.74 V vs. Ag/AgCl and reaching a photocurrent plateau. From the polarization curves (see Figure 7a), the photocurrent density at 0.5 V (at the end point of the steady-state plateau-like photocurrents and before the electrochemical oxygen evolution onset potential) are as follows: 175 µA cm^{-2}, 125 µA cm^{-2}, and 116 µA cm^{-2} for 150 °C, RT, and 300 °C TNT photoanodes, respectively. The highest photocurrent value obtained for the sample TNT-150 can be ascribed to the smallest crystallite size, the defect-free morphology of high quality nanotubes for the 150 °C sample.

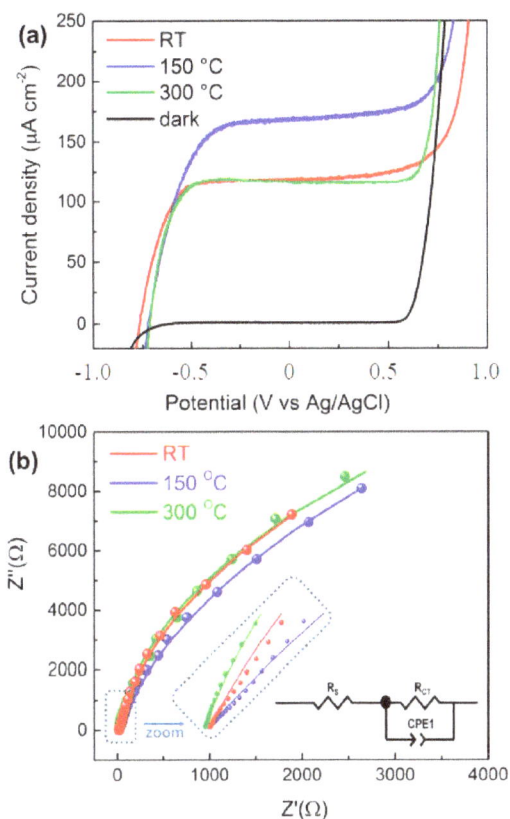

Figure 7. Linear sweep voltammetry curves of the TiO_2 nanotubes (**a**). Nyquist plots for TNT-RT, TNT-150, and TNT-300 measured at an applied bias potential of 0 V vs. Ag/AgCl under illumination of AM1.5G with the intensity of 100 mW cm^{-2} (**b**). Symbols are experimentally measured impedance data; solid curves are fitted to the equivalent circuit shown (inset). R_s is solution resistance, R_{CT} is charge transfer resistance, Constant phase element (CPE1) is capacitance element. Photoelectrochemical (PEC) measurements were carried out in 1 M NaOH electrolyte under simulated solar light irradiation (air mass-AM1.5G, 100 mW/cm^2).

To investigate the reasons underlying the photocurrent trend, we carried out EIS measurements under AM1.5G illumination at 0 V vs. Ag/AgCl (Figure 7b). The semicircular arch diameter indicates the charge transfer ability of the examined photoelectrode. To extract the charge transfer parameters associated with EIS curves, we fitted the curves with an equivalent Randle's circuit (see inset of Figure 7b), where R_s is the series/solution resistance, R_{CT} is the charge transfer resistance, and CPE1 is the constant phase element (capacitance) of semiconductor/electrolyte interface. The fitted parameters for each sample are shown in Table 2. R_s slightly decreases with increasing temperature of Ti sputtered films and shows comparable values. The double layer capacitance CPE1 is higher for TNT-RT due to higher accumulation of charge at the electrode/electrolyte interface; it may be because of grain size differences or accumulation of more charges at the grain boundaries of the sample. The charge transfer resistance R_{CT} for TNT-150 sample is 33.7 kΩ, lower compared to TNT-RT (39.7 kΩ) and TNT-300 (36.4 kΩ) suggesting a higher charge transfer rate at the semiconductor/electrolyte interface that, thus, underlies the observed enhancement in the photocurrent.

Table 2. Electrochemical Impedance spectroscopy data from measurement taken at 0 V vs. Ag/AgCl under 1 sun illumination.

Sample	R_S, Ω	R_{CT}, Ω	CPE1, μF
RT	16.71	39,799 $\pm$ 1149	201 $\pm$ 0.57
150	15.70	33,698 $\pm$ 279	172 $\pm$ 0.33
300	13.37	36,430 $\pm$ 841	165 $\pm$ 0.44

3. Experimental

3.1. Deposition of Titanium Films by Magnetron Sputtering

Titanium thin films, with thickness of ~1.5 μm, were deposited by pulsed DC magnetron sputtering of the titanium target (size of 4″ and purity of 99.995%) on a commercially available FTO coated glasses substrates (Solaronix, Aubonne, Switzerland) with dimensions of 25 × 15 × 2 mm. A standardized three-step cleaning protocol was used before film deposition. rinsing the substrates in an ultrasound bath in acetone, ethanol and distilled water, each step lasting 5 min. Subsequent drying at RT was applied to remove residual water from the glass samples. The chamber was evacuated to the base pressure of 1×10^{-4} Pa. The depositions were performed at a pressure of 0.2 Pa for 240 min on unheated as well as heated substrates at various temperatures of 150, 300 and 450 °C. Although no intentional heating was applied the substrate temperature increased up to 90 °C during deposition as a result of its interaction with plasma. The DC power of 700 W (power density of 8.6 W/cm^2) was applied in a pulsed mode at pulse frequency of 50 kHz and duty cycle of 50%. Prior the deposition a substrate pre-treatment was employed. First the substrate surfaces were cleaned using radio frequency (RF) (13.56 MHz) plasma etching in argon and then activated in the RF discharge in the mixture 50:50 of Ar and forming gas (10% of H_2 and 90% of N_2). The RF power of 130 W was typically used.

3.2. Mechanical and Tribological Properties

Mechanical and tribological characteristics were explored using a fully calibrated NanoTest instrument (MicroMaterials, Wrexham, UK) in a load-controlled mode. Nanoindentation at a peak force of 3 mN with a diamond pyramidal Berkovich indenter was employed for hardness and elastic modulus measurement [42,43]. The indentation curves were analyzed using the standard method [44].

Sphero-conical Rockwell indenter with a nominal radius of 10 μm was used for scratch test to assess the adhesion-cohesion properties of the films. During the standard scratch procedure, the initially constant topographic load of 0.02 mN was applied over the first 50 μm and then ramped to 500 mN at constant loading rate of 13 mN/s to initiate films failure and reveal their cohesive and/or adhesive limits. Evaluation of the scratch test was performed on the basis of the indenter on-load depth record and analysis of the residual scratch tracks. Laser scanning confocal microscope LEXT OLS 3100 (Olympus, Tokyo, Japan) was used for high-resolution imaging.

3.3. Electrochemical Anodization to Grow Self-Organized TiO$_2$ Nanotubes

The titanium films on FTO glass were washed with ethanol. TiO$_2$ nanotubes were then grown at 60 V using a power source (STATRON 3253.3, Statron AG, Mägenwil, Switzerland) in a two-electrode configuration with a counter electrode made of platinum (cathode) and the working electrode was the titanium film (anode). The electrolyte contained 0.2 mol dm^{-3} NH$_4$F and 4 mol dm^{-3} H$_2$O in ethylene glycol. After the anodization process, the samples were washed in ethanol and then dried in a nitrogen stream. The as-prepared amorphous TNT were annealed at 500 °C for 1h in air using cylindrical furnace (Clasic CLARE 4.0, CLASIC, Revnice, Czech Republic) with temperature increase 5 °C min^{-1} to obtain the crystalline phase. The nanotubes grown from titanium films deposited at different temperatures such as RT, 150 °C, 300 °C, and 450 °C, are in the text coded as TNT-RT, TNT-150, TNT-300, and TNT-450, respectively.

3.4. Characterization of the Titanium Films and TiO₂ Nanotubes

Structure of the Ti films was determined using the Empyrean (PANalytical, Almelo, The Netherlands) diffractometer equipped with Co. radiation source, focusing mirror, and Pixcell detector via grazing angle regime with incident angle 2°. Mean crystallite size was determined using the Scherrer equation. Surface of the films and its cross-sections were observed using Scanning Electron Microscope Hitachi SU6600 (Hitachi, Tokyo, Japan).

3.5. Photoelectrochemistry

The photoelectrochemical data were collected using a standard three-electrode electrochemical cell with a Gamry Series G 300 Potentiostat (Warminster, PA, USA). The TiO₂ nanotubes served as working electrode (photoanode), the Ag/AgCl (3 M KCl) as the reference electrode and the Pt wire was used as the counter electrode. A 150 W Xenon lamp coupled with an AM1.5G filter was used as a light source. The power intensity was kept at 1 sun (100 mW/cm²) which was calibrated though a silicon reference solar cell (Newport Corporation, Irvine, CA, USA). The photoelectrochemical behavior of prepared electrodes was investigated by means of linear sweep voltammetry measurements in 1 M NaOH electrolyte (pH 13.5). The electrochemical impedance spectroscopy (EIS) data were recorded using a Gamry instrument (ESA 410, Gamry, Warminster, PA, USA) in the frequency range from 0.1 Hz to 100 kHz under 1 sun illumination at a bias of 0 V vs. Ag/AgCl. At least three electrodes of each type were fabricated and tested. All electrodes showed similar J-V curves, and representative data are reported.

4. Conclusions

In this study, we have reported a detailed investigation of mechanical and adhesion properties of Ti films sputtered at different temperatures, showing that temperatures as high as 450 °C produce Ti films with well-defined platelet texture and with best mechanical and adhesion properties. However, we have found that these different sputtering conditions strongly influence crystallographic and photoelectrochemical water spitting activity of self-organized TiO₂ nanotubes grown from Ti films. The more active TiO₂ nanotube sample towards photoelectrochemical water splitting was obtained from Ti substrate sputtered at 150 °C showing the lowest crystallite size, best degree of self-organization, and enhanced charge transfer at the semiconductor/liquid interface. This work remarks the challenge behind achieving highly active and durable materials for photonics applications and shows that advanced magnetron sputtering may enable good control over microstructural properties and, thus, performance of semiconductor thin films.

Acknowledgments: The authors gratefully acknowledge the support by the Operational Programme Research, Development and Education—European Regional Development Fund, project no. CZ.02.1.01/0.0/0.0/15_003/0000416 and the project 8E15B009 of the Ministry of Education, Youth and Sports of the Czech Republic. The authors also acknowledge the financial support from Grant Agency of Czech Republic (project number 15-19705S and 17-20008S) and the Internal Grant of Palacky University (IGA_PrF_2017_005).

Author Contributions: S.K. and Z.H. conceived and designed the experiments; M.Z., R.C., H.K., S.P., J.T., Y.R., A.N., J.K. performed the experiments, analyzed the data and contributed reagents/materials/analysis tools; M.Z., R.C., S.K., A.N. wrote the paper, P.S. and R.Z. supervised the project.

Conflicts of Interest: The authors declare no conflicts of interest.

References

1. Papoutsi, D.; Lianos, P.; Yianoulis, P.; Koutsoukos, P. Sol-gel derived TiO₂ microemulsion gels and coatings. *Langmuir* **1994**, *10*, 1684–1689. [CrossRef]
2. Kment, S.; Kmentova, H.; Kluson, P.; Krysa, J.; Hubicka, Z.; Cirkva, V.; Gregora, I.; Solcova, O.; Jastrabik, L. Notes on the photo-induced characteristics of transition metal-doped and undoped titanium dioxide thin films. *J. Colloid Interface Sci.* **2010**, *348*, 198–205. [CrossRef] [PubMed]

3. Krysa, J.; Zlamal, M.; Kment, S.; Brunclikova, M.; Hubicka, Z. TiO_2 and Fe_2O_3 films for photoelectrochemical water splitting. *Molecules* **2015**, *20*, 1046. [CrossRef] [PubMed]
4. Kment, S.; Kluson, P.; Stranak, V.; Virostko, P.; Krysa, J.; Cada, M.; Pracharova, J.; Kohout, M.; Morozova, M.; Adamek, P.; et al. Photo-induced electrochemical functionality of the TiO_2 nanoscale films. *Electrochim. Acta* **2009**, *54*, 3352–3359. [CrossRef]
5. Naldoni, A.; Riboni, F.; Marelli, M.; Bossola, F.; Ulisse, G.; Di Carlo, A.; Pis, I.; Nappini, S.; Malvestuto, M.; Dozzi, M.V.; et al. Influence of TiO_2 electronic structure and strong metal-support interaction on plasmonic au photocatalytic oxidations. *Catal. Sci. Technol.* **2016**, *6*, 3220–3229. [CrossRef]
6. Chen, X.; Mao, S.S. Titanium dioxide nanomaterials: Synthesis, properties, modifications, and applications. *Chem. Rev.* **2007**, *107*, 2891–2959. [CrossRef] [PubMed]
7. Kavan, L.; Grätzel, M.; Rathouský, J.; Zukalb, A. Nanocrystalline TiO_2 (anatase) electrodes: Surface morphology, adsorption, and electrochemical properties. *J. Electrochem. Soc.* **1996**, *143*, 394–400. [CrossRef]
8. Naldoni, A.; Montini, T.; Malara, F.; Mróz, M.M.; Beltram, A.; Virgili, T.; Boldrini, C.L.; Marelli, M.; Romero-Ocaña, I.; Delgado, J.J.; et al. Hot electron collection on brookite nanorods lateral facets for plasmon-enhanced water oxidation. *ACS Catal.* **2017**, *7*, 1270–1278. [CrossRef]
9. Ren, L.; Li, Y.; Hou, J.; Zhao, X.; Pan, C. Preparation and enhanced photocatalytic activity of TiO_2 nanocrystals with internal pores. *ACS Appl. Mater. Interfaces* **2014**, *6*, 1608–1615. [CrossRef] [PubMed]
10. Chen, X.; Shen, S.; Guo, L.; Mao, S.S. Semiconductor-based photocatalytic hydrogen generation. *Chem. Rev.* **2010**, *110*, 6503–6570. [CrossRef] [PubMed]
11. Ma, Y.; Wang, X.; Jia, Y.; Chen, X.; Han, H.; Li, C. Titanium dioxide-based nanomaterials for photocatalytic fuel generations. *Chem. Rev.* **2014**, *114*, 9987–10043. [CrossRef] [PubMed]
12. Straňák, V.; Čada, M.; Quaas, M.; Block, S.; Bogdanowicz, R.; Kment, S.; Wulff, H.; Hubička, Z.; Helm, C.A.; Tichý, M.; et al. Physical properties of homogeneous TiO_2 films prepared by high power impulse magnetron sputtering as a function of crystallographic phase and nanostructure. *J. Phys. D: Appl. Phys.* **2009**, *42*, 105204. [CrossRef]
13. Roy, P.; Berger, S.; Schmuki, P. TiO_2 nanotubes: Synthesis and applications. *Angew. Chem. Int. Ed.* **2011**, *50*, 2904–2939. [CrossRef] [PubMed]
14. Gordon, T.R.; Cargnello, M.; Paik, T.; Mangolini, F.; Weber, R.T.; Fornasiero, P.; Murray, C.B. Nonaqueous synthesis of TiO_2 nanocrystals using TiF_4 to engineer morphology, oxygen vacancy concentration, and photocatalytic activity. *J. Am. Chem. Soc.* **2012**, *134*, 6751–6761. [CrossRef] [PubMed]
15. Dong, F.; Zhao, W.; Wu, Z. Characterization and photocatalytic activities of C, N and S co-doped TiO_2 with 1D nanostructure prepared by the nano-confinement effect. *Nanotechnology* **2008**, *19*, 365607. [CrossRef] [PubMed]
16. Kment, S.; Riboni, F.; Pausova, S.; Wang, L.; Wang, L.; Han, H.; Hubicka, Z.; Krysa, J.; Schmuki, P.; Zboril, R. Photoanodes based on TiO_2 and α-Fe_2O_3 for solar water splitting—Superior role of 1D nanoarchitectures and of combined heterostructures. *Chem. Soc. Rev.* **2017**, *46*, 3716–3769. [CrossRef] [PubMed]
17. Paramasivam, I.; Jha, H.; Liu, N.; Schmuki, P. A review of photocatalysis using self-organized TiO_2 nanotubes and other ordered oxide nanostructures. *Small* **2012**, *8*, 3073–3103. [CrossRef] [PubMed]
18. Krysa, J.; Lee, K.; Pausova, S.; Kment, S.; Hubicka, Z.; Ctvrtlik, R.; Schmuki, P. Self-organized transparent 1D TiO_2 nanotubular photoelectrodes grown by anodization of sputtered and evaporated ti layers: A comparative photoelectrochemical study. *Chem. Eng. J.* **2017**, *308*, 745–753. [CrossRef]
19. Fahim, N.F.; Sekino, T.; Morks, M.F.; Kusunose, T. Electrochemical growth of vertically-oriented high aspect ratio titania nanotubes by rabid anodization in fluoride-free media. *J. Nanosci. Nanotechnol.* **2009**, *9*, 1803–1818. [CrossRef] [PubMed]
20. Kmentova, H.; Kment, S.; Wang, L.; Pausova, S.; Vaclavu, T.; Kuzel, R.; Han, H.; Hubicka, Z.; Zlamal, M.; Olejnicek, J.; et al. Photoelectrochemical and structural properties of TiO_2 nanotubes and nanorods grown on FTO substrate: Comparative study between electrochemical anodization and hydrothermal method used for the nanostructures fabrication. *Catal. Today* **2017**, *287*, 130–136. [CrossRef]
21. Lee, K.; Kim, D.; Berger, S.; Kirchgeorg, R.; Schmuki, P. Anodically formed transparent mesoporous TiO_2 electrodes for high electrochromic contrast. *J. Mater. Chem.* **2012**, *22*, 9821–9825. [CrossRef]

22. Macak, J.M.; Schmuki, P. Anodic growth of self-organized anodic TiO$_2$ nanotubes in viscous electrolytes. *Electrochim. Acta* **2006**, *52*, 1258–1264. [CrossRef]

23. Zwilling, V.; Aucouturier, M.; Darque-Ceretti, E. Anodic oxidation of titanium and TA6V alloy in chromic media. An electrochemical approach. *Electrochim. Acta* **1999**, *45*, 921–929. [CrossRef]

24. Paušová, Š.; Kment, Š.; Zlámal, M.; Baudys, M.; Hubička, Z.; Krýsa, J. Transparent nanotubular TiO$_2$ photoanodes grown directly on fto substrates. *Molecules* **2017**, *22*, 775. [CrossRef] [PubMed]

25. Berger, S.; Ghicov, A.; Nah, Y.C.; Schmuki, P. Transparent TiO$_2$ nanotube electrodes via thin layer anodization: Fabrication and use in electrochromic devices. *Langmuir* **2009**, *25*, 4841–4844. [CrossRef] [PubMed]

26. Krýsa, J.; Zlámal, M.; Paušová, Š.; Kotrla, T.; Kment, Š.; Hubička, Z. Hematite photoanodes for solar water splitting: Directly sputtered vs. Anodically oxidized sputtered Fe. *Catal. Today* **2017**, *287*, 99–105. [CrossRef]

27. Sadek, A.Z.; Zheng, H.; Latham, K.; Wlodarski, W.; Kalantar-zadeh, K. Anodization of Ti thin film deposited on ito. *Langmuir* **2009**, *25*, 509–514. [CrossRef] [PubMed]

28. Tang, Y.; Tao, J.; Zhang, Y.; Wu, T.; Tao, H.; Bao, Z. Preparation and characterization of TiO$_2$ nanotube arrays via anodization of titanium films deposited on fto conducting glass at room temperature. *Acta Physico-Chim. Sin.* **2008**, *24*, 2191–2197. [CrossRef]

29. Wang, J.; Wang, H.; Li, H.; Wu, J. Synthesis and characterization of TiO$_2$ nanotube film on fluorine-doped tin oxide glass. *Thin Solid Films* **2013**, *544*, 276–280. [CrossRef]

30. Olejníček, J.; Hubička, Z.; Kment, Š.; Čada, M.; Kšírová, P.; Adámek, P.; Gregora, I. Investigation of reactive HiPIMS + MF sputtering of TiO$_2$ crystalline thin films. *Surf. Coat. Technol.* **2013**, *232*, 376–383. [CrossRef]

31. Bukauskas, V.; Kaciulis, S.; Mezzi, A.; Mironas, A.; Niaura, G.; Rudzikas, M.; Šimkienė, I.; Šetkus, A. Effect of substrate temperature on the arrangement of ultra-thin TiO$_2$ films grown by a dc-magnetron sputtering deposition. *Thin Solid Films* **2015**, *585*, 5–12. [CrossRef]

32. Firstov, S.; Kulikovsky, V.; Rogul, T.; Ctvrtlik, R. Effect of small concentrations of oxygen and nitrogen on the structure and mechanical properties of sputtered titanium films. *Surf. Coat. Technol.* **2012**, *206*, 3580–3585. [CrossRef]

33. Petrov, I.; Barna, P.B.; Hultman, L.; Greene, J.E. Microstructural evolution during film growth. *J. Vacuum Sci. Technol. A* **2003**, *21*, S117–S128. [CrossRef]

34. Chawla, V.; Jayaganthan, R.; Chawla, A.K.; Chandra, R. Microstructural characterizations of magnetron sputtered Ti films on glass substrate. *J. Mater. Process. Technol.* **2009**, *209*, 3444–3451. [CrossRef]

35. Savaloni, H.; Taherizadeh, A.; Zendehnam, A. Residual stress and structural characteristics in Ti and Cu sputtered films on glass substrates at different substrate temperatures and film thickness. *Phys. B Condens. Matter* **2004**, *349*, 44–55. [CrossRef]

36. Arzt, E. Size effects in materials due to microstructural and dimensional constraints: A comparative review. *Acta Mater.* **1998**, *46*, 5611–5626. [CrossRef]

37. Kulikovsky, V.; Ctvrtlik, R.; Vorlicek, V.; Filip, J.; Bohac, P.; Jastrabik, L. Mechanical properties and structure of TiO$_2$ films deposited on quartz and silicon substrates. *Thin Solid Films* **2013**, *542*, 91–99. [CrossRef]

38. Lee, K.; Mazare, A.; Schmuki, P. One-dimensional titanium dioxide nanomaterials: Nanotubes. *Chem. Rev.* **2014**, *114*, 9385–9454. [CrossRef] [PubMed]

39. Kuzmych, O. Defect Minimization and Morphology Optimization in TiO$_2$ Nanotube Thin Films, Grown on Transparent Conducting Substrate, for Dye Synthesized Solar Cell Application. *Thin Solid Films* **2012**, *522*, 71–78. [CrossRef]

40. Witt, F.; Vook, R.W. Thermally induced strains in diamond cubic, tetragonal, orthorhombic, and hexagonal films. *J. Appl. Phys.* **1969**, *40*, 709–719. [CrossRef]

41. Albu, S.P.; Schmuki, P. Influence of anodization parameters on the expansion factor of TiO$_2$ nanotubes. *Electrochim. Acta* **2013**, *91*, 90–95. [CrossRef]

42. Ctvrtlik, R.; Al-Haik, M.; Kulikovsky, V. Mechanical properties of amorphous silicon carbonitride thin films at elevated temperatures. *J. Mater. Sci.* **2015**, *50*, 1553–1564. [CrossRef]

43. Ctvrtlik, R.; Kulikovsky, V.; Vorlicek, V.; Tomastik, J.; Drahokoupil, J.; Jastrabik, L. Mechanical properties and microstructural characterization of amorphous SiC$_x$N$_y$ thin films after annealing beyond 1100 °C. *J. Am. Ceram. Soc.* **2016**, *99*, 996–1005. [CrossRef]

44. Oliver, W.C.; Pharr, G.M. An improved technique for determining hardness and elastic modulus using load and displacement sensing indentation experiments. *J. Mater. Res.* **1992**, *7*, 1564–1583. [CrossRef]

© 2018 by the authors. Licensee MDPI, Basel, Switzerland. This article is an open access article distributed under the terms and conditions of the Creative Commons Attribution (CC BY) license (http://creativecommons.org/licenses/by/4.0/).

catalysts

MDPI

Article

Novel Synthesis of Plasmonic Ag/AgCl@TiO$_2$ Continues Fibers with Enhanced Broadband Photocatalytic Performance

Nan Bao [1,*], Xinhan Miao [1], Xinde Hu [1], Qingzhe Zhang [2], Xiuyan Jie [1] and Xiyue Zheng [1]

[1] Shandong Key Laboratory of Water Pollution Control and Source Reuse,
School of Environmental Science and Engineering, Shandong University, Jinan 250100, China;
miaoxh1991@outlook.com (X.M.); hxd7887@163.com (X.H.); jxiuyan0413@163.com (X.J.);
zhengxiyue1995@163.com (X.Z.)

[2] Institut National de la Recherche Scientifique (INRS), Centre Énergie Materiaux et Télécommunications,
Université du Québec, 1650 Boulevard Lionel-Boulet, Varennes, Québec, QC J3X 1S2, Canada;
zhangqingzhe17@163.com

* Correspondence: baonan@sdu.edu.cn; Tel.: +86-531-8836-4724

Academic Editors: Vladimiro Dal Santo and Alberto Naldoni
Received: 28 February 2017; Accepted: 13 April 2017; Published: 17 April 2017

Abstract: The plasmonic Ag/AgCl@TiO$_2$ fiber (S-CTF) photocatalyst was synthesized by a two-step approach, including the sol-gel and force spinning method for the preparation of TiO$_2$ fibers (TF), and the impregnation-precipitation-photoreduction strategy for the deposition of Ag/AgCl onto the fibers. NaOH aqueous solution was utilized to hydrolyze TiCl$_4$, to synthesize TF and remove the byproduct HCl, and the produced NaCl was recycled for the formation and deposition of Ag/AgCl. The surface morphology, specific surface area, textural properties, crystal structure, elemental compositions and optical absorption of S-CTF were characterized by a series of instruments. These results revealed that the AgCl and Ag^0 species were deposited onto TF successfully, and the obtained S-CTF showed improved visible light absorption due to the surface plasmon resonance of Ag^0. In the photocatalytic degradation of X-3B, S-CTF exhibited significantly enhanced activities under separate visible or UV light irradiation, in comparison to TF.

Keywords: Ag/AgCl@TiO$_2$ fibers; plasmonic photocatalyst; photodegradation

---

1. Introduction

As a potential technology, photocatalytic oxidation is widely applied in water treatment [1,2] and energy conversion area [3,4]. TiO$_2$ is considered to be one of the most promising semiconductor photocatalysts due to its low cost, chemical stability and non-toxicity [5]. However, its large band gap (3.2 eV) and high recombination rate of photo-generation carrier [6] limit its applications as an efficient and visible-light driven photocatalyst. Numerous strategies have been explored to enhance its activity, especially doping with nonmetal elements (N and S, etc.) [7,8], coupling with other semiconductors (GO and BiVO$_4$, etc.) [9,10], and depositing with noble metals (Au and Ag, etc.) [11–14]. Ag^0 nanoparticles-modified TiO$_2$ exhibits excellent activity under visible light resulting from the surface plasmon resonance (SPR) [15,16], which is induced by the nano-size and morphology of Ag^0 [17,18].

Kakuta et al. [19] observed that Ag^0 species were formed on the surface of AgBr under ultraviolet (UV) irradiation. Afterward, Huang et al. [20] found that the modification of TiO$_2$ with Ag/AgCl obviously improved its photocatalytic activity. However, previous research has mainly focused on the fabrication of nano-sized Ag/AgCl/TiO$_2$ catalysts, such as nanoparticles [21,22], nanospheres [23], nanoarrays [20,24] and nanofibers [25]. These nano-sized catalysts are difficult to separate directly from post-slurry for recycling. Therefore, it is crucial to develop an easily recycled catalyst with a macroscopic shape and self-supporting structure. TiO_2 continuous fibers, fabricated be our group using sol-gel and force spinning method, is a promising strategy.

In addition, for most studies on Ag/AgCl-deposited TiO_2, HCl [26] or cetyltrimethylammonium chloride (CTAC) [27] were employed as chlorine sources, which were costly, toxic and harmful to environment. Hence, it is necessary to explore low-cost and non-toxic alternative sources. The cheap and extensive $TiCl_4$ can be employed as a Ti source for TiO_2 fabrication. However, its hydrolytic process is accompanied with the formation of hazardous HCl [28,29]. For this purpose, Xu et al. [30] reported a method in which NaOH was used to neutralize the produced HCl and form NaCl during the hydrolysis of $TiCl_4$. This significant improvement allows us to recycle the produced NaCl, and also invites the investigation of the feasibility of using the NaCl as Cl source for Ag/AgCl deposition.

In this study, TiO_2 fibers (TF) with a macroscopic shape were fabricated by a sol-gel method according to the previously developed TiO_2 continuous fibers [31], and Ag/AgCl nanoparticles were deposited on TF by an impregnation-precipitation-photoreduction method [20]. $TiCl_4$ served as the Ti source for TF, and NaOH aqueous solution was utilized to hydrolyze $TiCl_4$, neutralize the produced HCl and form NaCl. The recycled NaCl was employed as a chlorine source for Ag/AgCl deposition. The morphology, textural properties, chemical component and optical property of Ag/AgCl@TiO_2 fibers (S-CTF) were analyzed by scanning electron microscopy (SEM), N_2 absorption and desorption isotherms, X-ray diffractometer (XRD), X-ray photoelectron spectroscopy (XPS) and UV-vis spectrophotometer (UV-vis DRS). The photocatalytic activity and stability were evaluated by brilliant red X-3B degradation under visible and UV light.

2. Results and Discussion

2.1. Morphology and Textural Properties

2.1.1. SEM

Figure 1 presents typical SEM images of TF, AgCl@TiO_2 fibers (CTF) and S-CTF. The diameters of the fibers are about 30 μm. In Figure 1a,c,e, the surface of TF is homogeneous and smooth, while the surface of CTF and S-CTF are rough. Local enlarged images of TF, CTF and S-CTF are shown in Figure 1b,d,f, in which the porous structure that formed by the agglomeration of TiO_2 nanoparticles can be observed. AgCl particles with diameters less than 300 nm (Figure 1d) were successfully deposited on TF. However, there is no significant difference between the SEM images of CTF and S-CTF. The existence of Ag^0 cannot be observed through SEM, but will be confirmed by XRD, XPS and UV-vis DRS.

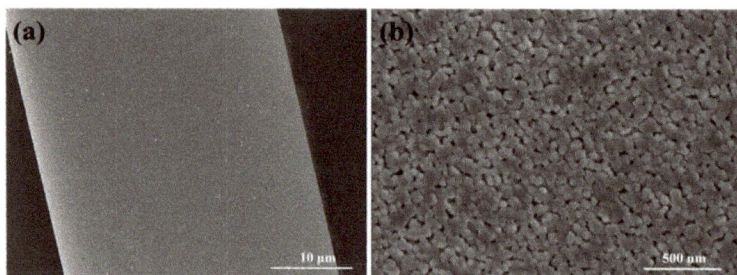

Figure 1. *Cont.*

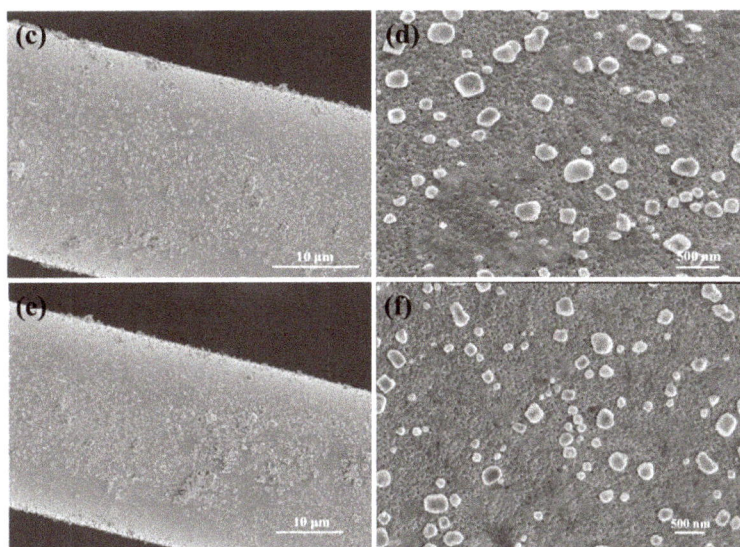

Figure 1. SEM images of TF (**a**,**b**), CTF (**c**,**d**) and S-CTF (**e**,**f**).

2.1.2. Textural Properties

The textural parameters of TF and S-CTF are presented in Table 1. After heat treatment, the Brunauer-Emmett-Teller (BET) surface area of TF increases from 16.0 of precursor fibers (PF) to 27.9 $m^2 g^{-1}$ (TF), due to the removal of organic matters and the formation of porous structure. With Ag/AgCl particles deposited on TF, the surface area of S-CTF increases to 30.6 $m^2 g^{-1}$, which is due to its rough surface [32], as shown in Figure 1.

Table 1. Textural Parameters of TF (TiO_2 fibers) and S-CTF (Ag/AgCl@TiO_2 fibers).

Samples	BET Surface Area ($m^2 g^{-1}$)	Pore Volume ($cm^3 g^{-1}$)	Pore Diameter (nm)
TF	27.9	0.111	16.7
S-CTF	30.6	0.074	10.8

N_2 absorption and desorption isotherms of TF and S-CTF are shown in Figure 2a. Isotherms of TF and S-CTF belong to *typical IV* adsorption behavior and *Type H1* hysteresis loop, according to International Union of Pure and Applied Chemistry (IUPAC) classification [33], which is consistent with pores with homogeneous and spindle cylindrical shapes. This reveals the formation of a mesoporous structure by the heat treatment. According to the pore-size distribution isotherms (Figure 2b), S-CTF shows narrower pore-size distribution (3–22 nm) and smaller average pore size (10.8 nm) than TF (5–40 nm and 16.7 nm). Hence, S-CTF (0.111 $cm^3 g^{-1}$) possesses smaller pore volume compared to TF (0.074 $cm^3 g^{-1}$). The decrease of average pore size and volume results from the deposition of nano-sized Ag/AgCl particles into macro-sized pores. In addition, S-CTF shows bimodal porous structure with two pore-size distribution peaks at 3.80 and 10.8 nm, which has also resulted from the pore size narrowing by AgCl deposition. The enlarged surface area and narrowed bimodal pore-size distribution obtained by S-CTF are both benefited to dye adsorption [34,35].

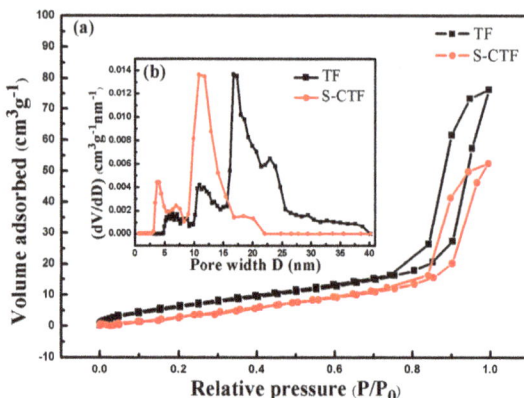

Figure 2. N_2 adsorption-desorption isotherms (a) and pore-size distribution (b) of TF and S-CTF.

2.2. XRD

Figure 3a shows XRD patterns of NaCl produced in the neutral reaction of NaOH and HCl. According to Joint Committee on Powder Diffraction Standards (JCPDS) standards, its crystal form is confirmed to be halite NaCl by diffraction peaks at 27.4°, 31.7°, 45.4°, 53.7°, 56.4°, 66.2° and 57.3°. XRD patterns of AgCl, Ag/AgCl, TF, CTF and S-CTF are shown in Figure 3b. Diffraction peaks of TF, Ag@TiO₂(CTF) and S-CTF at about 25.3°, 37.8°, 48.1°, 53.9°, 55.1°, 62.7°, 68.7°, 70.3° and 75.1° belong to (101), (004), (200), (105), (211), (204), (116), (220) and (215) reflections [36] of anatase TiO_2 (JCPDS No. 21-1272), while diffraction peaks of AgCl, Ag/AgCl, CTF and S-CTF at about 27.9°, 32.3°, 46.3°, 54.9°, 57.5°, 67.5° and 76.7° are in accordance with the (111), (200), (220), (311), (222), (400) and (420) planes [37] of cubic phase AgCl (JCPDS No. 31-1238). A weak diffraction peak at $2\theta = 38.3°$ of cubic phase Ag (JCPDS No. 65-2871) in Ag/AgCl confirmed the successful photoreduction of AgCl and the formation of Ag^0. However, no obvious diffraction peak of Ag^0 occurred in S-CTF, which is due to the peak overlap with anatase (at about $2\theta = 37.8°$) and its high dispersion. Otherwise, anatase peaks of CTF and S-CTF show about 0.45 and 0.2 shifts compared with TF (inset in Figure 3b), which is mainly caused by the combination of AgCl and the formation of Ag^0. The shift of anatase peaks towards higher angles indicate a lattice expansion [38] due to the AgCl and Ag^0 nanoparticles in the interlayer space of anatase gains agglomerated fibers.

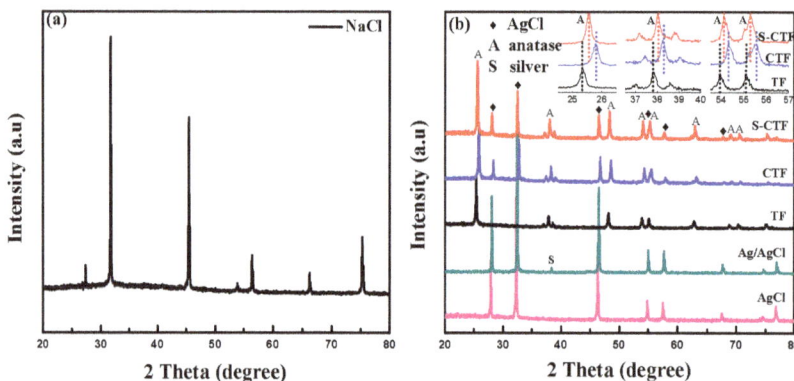

Figure 3. XRD patterns of NaCl separated from reaction solution (a), AgCl, Ag/AgCl, TF, CTF and S-CTF (b).

2.3. XPS

The XPS survey spectrum of S-CTF (Figure 4a) shows all peaks of Ti, O, Ag, Cl and C elements. The peak for C 1s at 284.6 eV is mainly from the adventitious hydrocarbon, due to the XPS instrument itself. The result of the survey of the XPS spectrum is consistent with the XRD results.

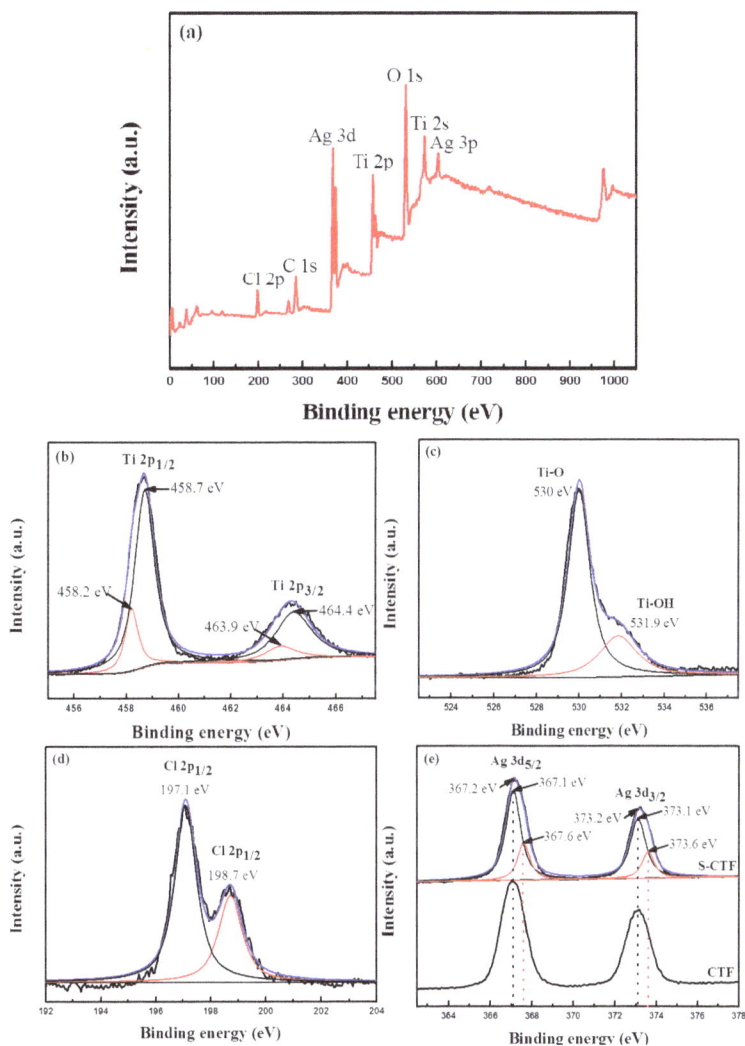

Figure 4. XPS spectra of all elements of S-CTF (**a**) and the band energy of Ti 2p (**b**), O 1s (**c**), Cl 2p (**d**) and Ag 3d (**e**).

The valence state of Ti, O, Cl and Ag in S-CTF are investigated by the high-resolution XPS spectrum (Figure 4b–e). As shown in Figure 4b, the binding energy at 458.6 and 464.6 eV of Ti $2p_{1/2}$ and Ti $2p_{3/2}$ are lower than those of pure TiO_2 (459.4 and 464.7 eV) [39]. The curve resolution of the Ti 2p signal of S-CTF indicates four different peaks. The peaks at binding energies of 458.2 and 463.9 eV belong to Ti $2p_{1/2}$ and Ti $2p_{3/2}$ of Ti^{3+}, while the other two peaks at 458.7 and 464.4 eV belong to Ti $2p_{1/2}$ and Ti $2p_{3/2}$ of Ti^{4+} [40]. Ti^{3+} (and oxygen vacancy) in S-CTF was produced during the heat treatment by the reduction of TF (the removal of partial lattice oxygen) with organic matters in steam as a reducing atmosphere. The O 1s region is decomposed into two peaks at 530 and 531.9 eV (Figure 4b). The two kinds of oxygen contributions are ascribed to Ti–O in TiO_2 and the OH in Ti–OH (adsorbed water on the surface of S-CTF) [35]. In Figure 4c, two signal peaks of Cl $2p_{1/2}$ and Cl $2p_{3/2}$ at binding energy of 197.1 and 198.7 eV with a separation of 1.6 eV confirmed the Cl^- from AgCl [41]. The Ag 3d region is decomposed into four peaks (Figure 4e). Binding energy peaks at 367.6 and 373.6 eV are ascribed to Ag^0, while 367.1 and 373.1 eV correspond to Ag^+ in AgCl particles [26,42]. The difference of binding energy (0.1 eV) between CTF and S-CTF also indicates the formation of Ag^0. According to the results of XPS, the detail atomic percentages of Ag, AgCl, and TiO_2 on the surface of S-CTF were about 8.5%, 40.7% and 51.8%, respectively. Moreover, the interaction among Ag, AgCl and TiO_2 [43] is further confirmed by the binding energy shifts peaks for all elements compared with standard values.

The crystal structure and chemical component of S-CTF are investigated by XRD and XPS analysis. The formation of Ag^0 and Ti^{3+} would enhance the optical property of S-CTF.

2.4. UV-Vis DRS

Compared with TF, S-CTF exhibits a slight enhancement in the UV light region (<400 nm) and considerable enhancement in the visible light region (>400 nm) (Figure 5). CTF shows the highest absorption in the UV light region among the samples, even higher than S-CTF. With the photoreduction of AgCl and the formation of Ag^0, the absorption of UV light decreases but is still higher than TF. Therefore, the enhancement of absorption in the UV light region can be ascribed to the combination of TF and AgCl with excellent UV light absorption. Because of the formation of Ti^{3+} and oxygen vacancy, TF possesses localized states below the conduct bind(CB) minimum and shows expected light absorption in the visible light region [39,44,45]. TF shows an absorption band edge at about 400 nm, and the absorption band edge of CTF and S-CTF (at about 420 nm) shows an expected red-shift, in comparison to TF. Compared with Ag/AgCl, S-CTF almost retains the same absorption intensity in the visible light region, illustrating that the composition of TF and Ag/AgCl has no influence on the light absorption of Ag/AgCl. Obviously, an additional absorption for S-CTF which was observed in the region of 400–600 nm, which is attributed to the SPR of Ag^0 nanoparticles [41]. The broad and enhanced absorption in the visible light region is also confirmed the formation of Ag^0 [46,47]. These evidences of Ag^0 formation are consistent with the results of XPS.

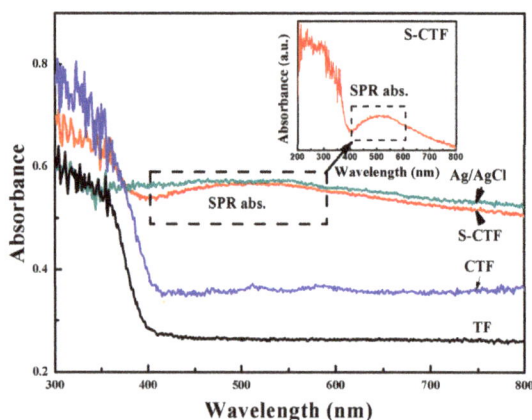

Figure 5. UV-vis diffuse reflectance spectrum of Ag/AgCl, TF, CTF and S-CTF.

2.5. Photocatalytic Activity

To evaluate the effect of the interaction between Ag/AgCl and the TF surface, Ag/AgCl/precursor fibers (S-CPF) were prepared. To investigate the influence of photosensitization and to confirm the photoactivity of TF, pure TiO$_2$ fibers (pure-TF) were also prepared by PF through heat treatment (the heating program was the same as that used for TF preparation) in O$_2$ atmosphere with the flow rate at 0.5 L/min. The photocatalytic activities of pure-TF, TF, CTF, S-CPF and S-CTF toward X-3B degradation under visible and UV light are shown in Figure 6.

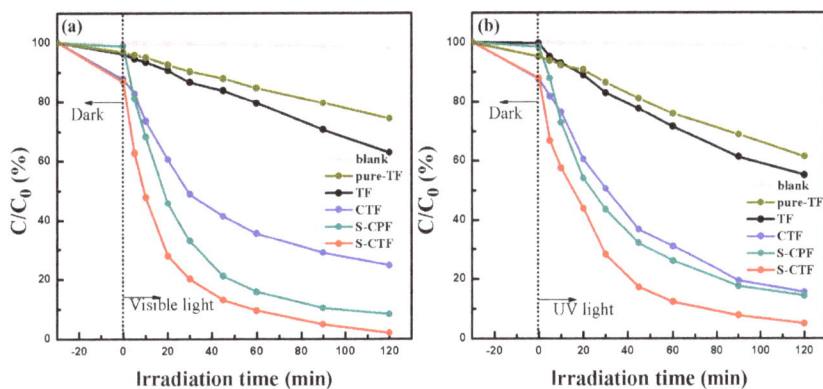

Figure 6. Photocatalytic degradation of 100 mL X-3B (20 mg L^{-1}) over pure-TF TF, CTF, S-CPF and S-CTF (0.05 g) under visible light (**a**) and ultrasonic light (**b**) with blank control groups.

After dark treatment for 30 min (Figure 6), S-CTF and CTF (about 12%) have higher X-3B adsorption than pure-TF and TF (about 4%). Nearly no adsorption is observed for S-CPF. The enhancement for X-3B adsorption is a result of the enlarged surface area and narrowed bimodal mesoporous structure after AgCl deposition.

The degradation efficiencies within 120 min under visible light (Figure 6a) follow an order of pure $-$ TF (25.6%) $<$ TF (37.1%) $<$ CTF (73.4%) $<$ S $-$ CPF (91.8%) $<$ S $-$ CTF (98.1%). After 120 min, the degradation efficiency by pure-TF was 25.6%, which was lower than TF (37.1%). The degradation of X-3B by pure-TF can be attributed to the photosensitization of dye, and the degradation elevation of TF can be attributed to the existence of Ti^{3+} and the optical absorption of visible light. CTF shows high photoactivity due to the reduction of AgCl and the formation Ag^0 as the photoreaction progressed [32]. Because of the strong SPR of Ag^0, S-CPF and S-CTF exhibit much higher degradation efficiencies than TF and CTF. The highest efficiency by S-CTF is ascribed to the better separation capacity of photo-generated carriers than S-CPF.

Under UV light, the degradation efficiencies of X-3B (Figure 6b) for 120 min are displayed as following sequence: pure $-$ TF (38.6%) $<$ TF (44.8%) $<$ CTF (84.5%) $<$ S $-$ CPF (85.7%) $<$ S $-$ CTF (95.1%). The higher efficiency of TF than pure-TF can be attributed to the more efficient carrier separation with oxygen vacancy in TF. CTF and S-CPF exhibit much higher activity than TF. Compared with X-3B degradation efficiency under visible light, the efficiency elevation by CTF results from the excitement of AgCl and the higher activity of TF. Since Ag^0 possesses the capacity to trap electrons [48], S-CTF shows the highest degradation efficiency.

The stability of S-CTF is further measured by a recycling experiment. After five runs, the degradation efficiencies of S-CTF under visible and UV light are still as high as 90.8% and 87.5%. The catalyst is stable enough, even considering the inevitable loss of weight during each cycle.

2.6. Photocatalytic Mechanism

2.6.1. Photoluminescence (PL)

The recombination of the photoinduced electrons and holes is a crucial factor in a photocatalytic reaction for the decrease of quantum yield. The efficiencies of photo-generated carriers [12,49,50] are explained by PL analysis. A strong emission peak of TF is observed at about 400 nm with the excitation wavelength at 325 nm (Figure 7), which indicates the direct recombination of electrons and holes. No obvious emission peak is observed in PL spectra of Ag/AgCl and S-CTF, while S-CTF shows lower PL intensity than Ag/AgCl, indicating its better separation capacity caused by the interaction among Ag, AgCl and TiO_2. The results of PL analysis are in accordance with the results of XRD and XPS. Otherwise, an additional peak of TF (at about 470 nm), belonging to oxygen vacancy [51], further confirms the formation of oxygen vacancy and Ti^{3+}.

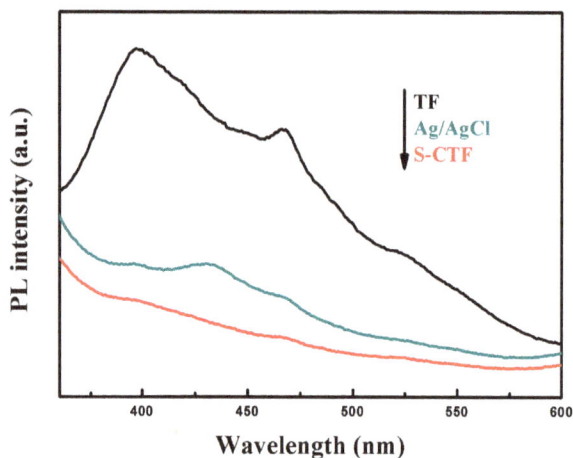

Figure 7. Photoluminescence spectra of TF, Ag/AgCl and S-CTF.

2.6.2. Photocatalytic Mechanism

To better understand the photocatalytic mechanisms of S-CTF and to investigate the main oxidative species during the photodegradation of X-3B, tert-butyl alcohol (TBA), $AgNO_3$, disodium ethylenediaminetetraacetate (EDTA-Na_2) and benzoquinone (BQ) are used as $\bullet OH$, electrons, holes and $\bullet O_2^-$ quenchers. Under visible light irradiation, the degradation efficiency of X-3B decreases from 98.0% to 87.2% (TBA), 74.6% ($AgNO_3$), 43.9% (EDTA-Na_2) and 36.7% (BQ). Under UV light irradiation, the degradation efficiency of X-3B reduces from 95.1% to 85.5% (TBA), 82.2% ($AgNO_3$), 65.5% (EDTA-Na_2) and 40.3% (BQ). These results illustrate that the critical species under visible light are holes and $\bullet O_2^-$, while the most critical specie under UV light is $\bullet O_2^-$.

The photocatalytic mechanism of S-CTF under visible light is shown in Figure 8a. Ag^0 exhibits strong SPR and generates electron-hole pairs under visible irradiation [52]. TF with the existence of Ti^{3+} is also excited under visible light. With the efficient interaction of Ag^0, AgCl and TF, the plasmon-excited electrons are injected into the conduction band (CB) of TF [20,45] easily, and subsequently produce $\bullet O_2^-$ with O_2. Meanwhile, the photo-excited holes are transferred to the surface of AgCl, X-3B molecules [20,32] and the value band (VB) of TF. The efficient separation of carriers among Ag, AgCl and TF is confirmed by PL spectra. Holes participate in the degradation of organic dyestuff by formation of Cl^0 and $\bullet OH$. Firstly, the holes transfer to AgCl and convert Cl^- ions into Cl^0 atoms [53]. The Cl^0 species oxidize the X-3B molecules and form AgCl. These reactions ensure that Ag/AgCl and S-CTF remain stable during the photodegradation process [16,20,37]. Secondly, the holes generate by TF yield •OH radicals or degrade organic pollutants directly. The photosensitization of X-3B [54] (generated electrons can inject into the CB of TF from excited dyes) also can be regarded as an advantage in its degradation.

(a)   **(b)**

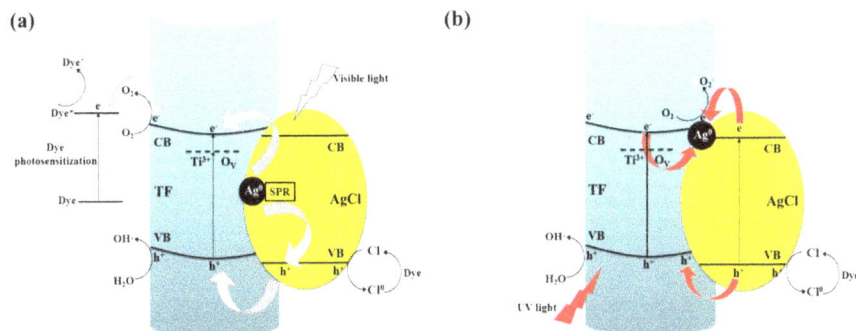

Figure 8. Photocatalytic mechanisms of S-CTF under visible (**a**) and UV (**b**) light irradiation.

Figure 8b shows the photocatalytic mechanism of S-CTF under UV light. The transformation of electron-hole pairs is different from visible irradiation. According to the CB and VB edge potentials of TiO_2 (-0.29 and 2.91 eV [40]) and AgCl (0.09 and 3.16 [55]), both TiO_2 and AgCl are excited under UV irradiation. The generated holes of AgCl can easily inject into the VB of TF [41], and then yield •OH with adsorbed water molecules, or directly degrade organic pollutants such as the holes generated by TF. The holes also yield Cl^0 and reduce this to Cl^- via the same route under visible irradiation. The electrons generated from the CB of TF and AgCl transfer to Ag^0, which serves as an electron trap under UV irradiation [48,56]. These electrons participate in the formation of $•O_2^-$ radicals. Moreover, the oxygen vacancy in TF is also beneficial to trap electrons and promote their separation under visible or UV light irradiation [40].

3. Materials and Methods

3.1. Chemicals

$TiCl_4$, acetylacetone (AcAc), triethylamine (TEA), methyl alcohol (MeOH), tetrahydrofuran (THF), NaOH and $AgNO_3$ were purchased from Sinopharm International Co. Ltd. (Shanghai, China). Brilliant red X-3B was purchased from Shanghai Dyestuff Chemical Plant. All of the experiments in this research used deionized water. All of the chemical reagents were analytical grade, and were used without further purification.

3.2. Preparation of Catalysts

TiO_2 fibers (TF) were synthesized by a sol-gel method combined with centrifugal spinning technique, followed by stage-temperature-programmed heat treatment in a steam atmosphere. The hydrolysis of $TiCl_4$ (10 mL in MeOH) was carried out by mixing 4 mL NaOH aqueous solution (0.5 g mL^{-1}) and 25 mL MeOH under ice-cold conditions. The reaction can be expressed as: $TiCl_4 + H_2O \rightarrow TiO_2 + HCl$. During the reaction, NaOH neutralized the produced HCl to form NaCl, and the formed NaCl was supersaturated and precipitated due to its relatively small saturation in MeOH. Then, NaCl was filtered and collected for Ag/AgCl deposition. The TiO_2 was chelated by AcAc to form a spinning solution. After the centrifugal spinning, precursor fibers (PF) were obtained and converted to TF by the heat treatment in steam as a reducing atmosphere. The detailed heat treatment program is shown in Table 2.

Table 2. Program of Segmented Heating.

Temperature (°C)	Heating Rate (°C min^{-1})	Time (min)
25–95	1.67	42
95–250	1.29	120
250–350	0.83	120
350–550	5.00	40
550–600	0.83	60
600–600	-	120

Ag/AgCl@TiO$_2$ fibers (S-CTF) were prepared by the deposition of Ag/AgCl on TF. The recycled NaCl and AgNO$_3$ were prepared into impregnation liquids. TF was soaked in 1 M NaCl and 0.1 M AgNO$_3$ solution for 10 min prior for AgCl deposition, and washed by deionized water to rinse the excess elements after each soak. This process was repeated for five trials. After being dried at 353 K, the prepared AgCl@TiO$_2$ fibers (CTF) were irradiated under a 1000 W xenon lamp (50 cm away from the light source) for 10 min to reduce partial Ag$^+$ to Ag^0 on the surface of the AgCl particles.

3.3. Characterization of Catalysts

To observe the structure and morphology of the prepared samples, scanning electron microscopy (SEM, JSM 6700F, JEOL, Tokyo, Japan) was employed. A Quadrasorb SI-MP system (Quantachrome, Boynton Beach, FL, USA) was used to acquire the N$_2$ adsorption-desorption isotherms. The Brunauer-Emmett-Teller (BET) specific surface area was detected by a multipoint BET method, and the adsorption data was acquired in the relative pressure (P/P_0) range of 0.05–0.3. The pore size distribution was calculated by employing the Density Functional Theory (DFT). The average pore size and pore volume data were obtained by the N$_2$ adsorption volume with the relative pressure at 0.991. An X-ray diffractometer (XRD, D8 Advance, Bruker AXS GmbH, Karlsruhe, Germany) was employed to determine the crystal structure of the prepared photocatalysts. An ESCALAB 250 spectrometer (Thermo Electron Corp., Cramlington, UK) with an Al KR source and a charge neutralizer was used to perform the X-ray photoelectron spectroscopy (XPS). All the spectra were calibrated to the C 1s peak at 284.6 eV. A UV-vis spectrophotometer (UV-3100, Shimadzu, Kyoto, Japan) was employed to investigate the light absorption property of the photocatalysts samples. To measure the solid-state UV-vis diffuse reflectance spectra, an integrating sphere attachment was equipped with BaSO$_4$ as a background in room temperature and air conditions. A fluorescence spectrometer (F-4600, Hitachi, Tokyo, Japan) was used to measure the photoluminescence (PL) spectra of all the samples.

3.4. Photoactivity Studies

The photocatalytic performance of S-CTF was evaluated by the decolorization of X-3B under visible and UV light. A 1000 W Xe lamp inserted by a glass optical filter to cut off short wavelength components ($\lambda < 420$ nm) and a 250 W high pressure mercury lamp with a maximum wavelength at 365 nm served as visible and UV light sources, respectively. Before the photocatalytic reaction, 50 mg of each sample was added to 200 mL X-3B aqueous solution (20 mg/L) in an open fixed-bed photo-reactor [57] which was cooled through the circulation of water at 293 K. To establish the adsorption/desorption equilibrium, the photo-reactor was stirred in dark conditions for 30 min. Then, the reactor was placed under visible and UV irradiation with the irradiation intensity at the surface of the dye solution being 250 W m^{-2} and 15 W m^{-2}, respectively, and cycled by a peristaltic pump with a flow rate of 20 mL min^{-1}. During the photocatalytic reaction, 3 mL reaction solution was taken at each interval and filtrated through a 0.45 μm syringe filter. A UV-vis 1601 spectrophotometer (Shimadzu, Japan) was used to measure the concentration of X-3B by the variation of the maximum absorption wavelength (536 nm) of filtrate which was determined. The degradation efficiency was calculated by the formula $R = (1 - C/C_0) \times 100\%$, in which C and C_0 stood for the initial concentration and the concentration of dye at each moment in the reaction, respectively.

Species capture experiments were performed to investigate the mechanism of TF. 10.0 mM of TBA), BQ, AgNO$_3$ and EDTA-Na$_2$ were added into X-3B solution acting as •OH, •O$_2$$^-$, electrons and holes scavengers. The reaction processes are same as that of the degradation reaction by S-CTF.

4. Conclusions

In summary, the plasmonic photocatalyst S-CTF was prepared by the gel-sol method to synthesize TF, and the impregnation-precipitation-photoreduction method to deposit Ag/AgCl particles on the produced TF. AgCl particles were deposited on TF with recycled NaCl as Cl sources, which were produced in the hydrolysis of TiCl$_4$ by NaOH aqueous solution. The improved dye adsorption compared to TF was due to the enlarged specific surface area and narrowed pore-size distribution after AgCl deposition. The increased photocatalytic efficiency resulted from the SPR of Ag^0, enhanced light absorption, narrowed band gap and efficient charge separation. With its outstanding photocatalytic performance and high stability, S-CTF is a promising application in organic pollutant treatment and wastewater purification. Furthermore, this mild, less-toxic and lower-cost synthesis process is accordance with environmental protection concepts.

Acknowledgments: This work was supported by the Natural Science Foundation of Shandong Province, China (No. ZR2011BM005) and the Fundamental Research Funds of Shandong University (No. 2015JC022).

Author Contributions: Xinhan Miao and Xiuyan Jie prepared the catalysts; Xinhan Miao and Xinde Hu performed the experiment; Xinhan Miao and Xiyue Zheng prepared the manuscript; Qingzhe Zhang directed the manuscript drafting; Nan Bao directed the project.

Conflicts of Interest: The authors declare no conflict of interest.

References

1. Legrini, O.; Oliveros, E.; Braun, A.M. Photochemical processes for water treatment. *Chem. Rev.* **1993**, *93*, 671–698. [CrossRef]
2. Chong, M.N.; Jin, B.; Chow, C.W.K.; Saint, C. Recent developments in photocatalytic water treatment technology: A review. *Water Res.* **2010**, *44*, 2997–3027. [CrossRef] [PubMed]
3. Liu, G.; Yin, L.-C.; Wang, J.; Niu, P.; Zhen, C.; Xie, Y.; Cheng, H.-M. A red anatase TiO$_2$ photocatalyst for solar energy conversion. *Energy Environ. Sci.* **2012**, *5*, 9603–9610. [CrossRef]
4. Kudo, A. Development of photocatalyst materials for water splitting with the aim at photon energy conversion. *J. Ceram. Soc. Jpn.* **2001**, *109*, S81–S88. [CrossRef]
5. Pelaez, M.; Nolan, N.T.; Pillai, S.C.; Seery, M.K.; Falaras, P.; Kontos, A.G.; Dunlop, P.S.M.; Hamilton, J.W.J.; Byrne, J.A.; O'Shea, K.; et al. A review on the visible light active titanium dioxide photocatalysts for environmental applications. *Appl. Catal. B Environ.* **2012**, *125*, 331–349. [CrossRef]
6. Choi, W.; Termin, A.; Hoffmann, M.R. The role of metal ion dopants in quantum-sized TiO$_2$: Correlation between photoreactivity and charge carrier recombination dynamics. *J. Phys. Chem.* **1994**, *98*, 13669–13679. [CrossRef]
7. Asahi, R.; Morikawa, T.; Ohwaki, T.; Aoki, K.; Taga, Y. Visible-light photocatalysis in nitrogen-doped titanium oxides. *Science* **2001**, *293*, 269–271. [CrossRef] [PubMed]
8. Liu, Y.; Liu, J.; Lin, Y.; Zhang, Y.; Wei, Y. Simple fabrication and photocatalytic activity of S-doped TiO$_2$ under low power led visible light irradiation. *Ceram. Int.* **2009**, *35*, 3061–3065. [CrossRef]
9. Chen, C.; Cai, W.; Long, M.; Zhou, B.; Wu, Y.; Wu, D.; Feng, Y. Synthesis of visible-light responsive graphene oxide/TiO$_2$ composites with p/n heterojunction. *ACS Nano* **2010**, *4*, 6425–6432. [CrossRef] [PubMed]
10. Bao, N.; Yin, Z.; Zhang, Q.; He, S.; Hu, X.; Miao, X. Synthesis of flower-like monoclinic BiVO$_4$/surface rough TiO$_2$ ceramic fiber with heterostructures and its photocatalytic property. *Ceram. Int.* **2016**, *42*, 1791–1800. [CrossRef]
11. Yang, Y.; Wen, J.; Wei, J.; Xiong, R.; Shi, J.; Pan, C. Polypyrrole-decorated Ag-TiO$_2$ nanofibers exhibiting enhanced photocatalytic activity under visible-light illumination. *ACS Appl. Mater. Interfaces* **2013**, *5*, 6201–6207. [CrossRef] [PubMed]
12. Li, F.B.; Li, X.Z. Photocatalytic properties of gold/gold ion-modified titanium dioxide for wastewater treatment. *Appl. Catal. A Gen.* **2002**, *228*, 15–27. [CrossRef]

13. Zhang, Q.; Thrithamarassery Gangadharan, D.; Liu, Y.; Xu, Z.; Chaker, M.; Ma, D. Recent advancements in plasmon-enhanced visible light-driven water splitting. *J. Materiomics* **2017**, *3*, 33–50. [CrossRef]

14. Kumar, R.; El-Shishtawy, M.R.; Barakat, A.M. Synthesis and characterization of Ag-Ag₂O/TiO₂@polypyrrole heterojunction for enhanced photocatalytic degradation of methylene blue. *Catalysts* **2016**, *6*, 76. [CrossRef]

15. Awazu, K.; Fujimaki, M.; Rockstuhl, C.; Tominaga, J.; Murakami, H.; Ohki, Y.; Yoshida, N.; Watanabe, T. A plasmonic photocatalyst consisting of silver nanoparticles embedded in titanium dioxide. *J. Am. Chem. Soc.* **2008**, *130*, 1676–1680. [CrossRef] [PubMed]

16. Wang, P.; Huang, B.; Qin, X.; Zhang, X.; Dai, Y.; Wei, J.; Whangbo, M.-H. Ag@AgCl: A highly efficient and stable photocatalyst active under visible light. *Angew. Chem. Int. Ed.* **2008**, *47*, 7931–7933. [CrossRef] [PubMed]

17. Jin, R.; Cao, Y.; Mirkin, C.A.; Kelly, K.L.; Schatz, G.C.; Zheng, J.G. Photoinduced conversion of silver nanospheres to nanoprisms. *Science* **2001**, *294*, 1901–1903. [CrossRef] [PubMed]

18. Jain, P.K.; Lee, K.S.; El-Sayed, I.H.; El-Sayed, M.A. Calculated absorption and scattering properties of gold nanoparticles of different size, shape, and composition: Applications in biological imaging and biomedicine. *J. Phys. Chem. B* **2006**, *110*, 7238–7248. [CrossRef] [PubMed]

19. Kakuta, N.; Goto, N.; Ohkita, H.; Mizushima, T. Silver bromide as a photocatalyst for hydrogen generation from CH₃OH/H₂O solution. *J. Phys. Chem. B* **1999**, *103*, 5917–5919. [CrossRef]

20. Yu, J.; Dai, G.; Huang, B. Fabrication and characterization of visible-light-driven plasmonic photocatalyst Ag/AgCl/TiO₂ nanotube arrays. *J. Phys. Chem. C* **2009**, *113*, 16394–16401. [CrossRef]

21. Guo, J.-F.; Ma, B.; Yin, A.; Fan, K.; Dai, W.-L. Highly stable and efficient Ag/AgCl@TiO₂ photocatalyst: Preparation, characterization, and application in the treatment of aqueous hazardous pollutants. *J. Hazard. Mater.* **2012**, *211–212*, 77–82. [CrossRef] [PubMed]

22. Hu, C.; Lan, Y.; Qu, J.; Hu, X.; Wang, A. Ag/AgBr/TiO₂ visible light photocatalyst for destruction of azodyes and bacteria. *J. Phys. Chem. B* **2006**, *110*, 4066–4072. [CrossRef] [PubMed]

23. Nunes, D.; Pimentel, A.; Santos, L.; Barquinha, P.; Fortunato, E.; Martins, R. Photocatalytic TiO₂ nanorod spheres and arrays compatible with flexible applications. *Catalysts* **2017**, *7*, 60. [CrossRef]

24. Liao, W.; Zhang, Y.; Zhang, M.; Murugananthan, M.; Yoshihara, S. Photoelectrocatalytic degradation of microcystin-LR using Ag/AgCl/TiO₂ nanotube arrays electrode under visible light irradiation. *Chem. Eng. J.* **2013**, *231*, 455–463. [CrossRef]

25. Wang, D.; Li, Y.; Li Puma, G.; Wang, C.; Wang, P.; Zhang, W.; Wang, Q. Ag/AgCl@helical chiral TiO₂ nanofibers as a visible-light driven plasmon photocatalyst. *Chem. Commun.* **2013**, *49*, 10367–10369. [CrossRef] [PubMed]

26. Wang, P.; Huang, B.; Lou, Z.; Zhang, X.; Qin, X.; Dai, Y.; Zheng, Z.; Wang, X. Synthesis of highly efficient Ag@AgCl plasmonic photocatalysts with various structures. *Chem. Eur. J.* **2010**, *16*, 538–544. [CrossRef] [PubMed]

27. Ma, B.; Guo, J.; Dai, W.-L.; Fan, K. Ag-AgCl/WO₃ hollow sphere with flower-like structure and superior visible photocatalytic activity. *Appl. Catal. B Environ.* **2012**, *123–124*, 193–199. [CrossRef]

28. Di Paola, A.; Bellardita, M.; Ceccato, R.; Palmisano, L.; Parrino, F. Highly active photocatalytic TiO₂ powders obtained by thermohydrolysis of TiCl₄ in water. *J. Phys. Chem. C* **2009**, *113*, 15166–15174. [CrossRef]

29. Zhu, Y.; Zhang, L.; Gao, C.; Cao, L. The synthesis of nanosized TiO₂ powder using a sol-gel method with TiCl₄ as a precursor. *J. Mater. Sci.* **2000**, *35*, 4049–4054. [CrossRef]

30. Xu, H.; Gao, L.; Guo, J. Hydrothermal synthesis of tetragonal barium titanate from barium chloride and titanium tetrachloride under moderate conditions. *J. Am. Ceram. Soc.* **2002**, *85*, 727–729. [CrossRef]

31. Bao, N.; Yin, G.-B.; Wei, Z.-T.; Li, Y.; Ma, Z.-H. Preparation of TiO₂ continuous fibers with oxygen vacancies and photocatalytic activity. *Integr. Ferroelectr.* **2011**, *127*, 97–105. [CrossRef]

32. Zhou, J.; Cheng, Y.; Yu, J. Preparation and characterization of visible-light-driven plasmonic photocatalyst Ag/AgCl/TiO₂ nanocomposite thin films. *J. Photochem. Photobiol. A Chem.* **2011**, *223*, 82–87. [CrossRef]

33. Dai, Y.; Lu, X.; McKiernan, M.; Lee, E.P.; Sun, Y.; Xia, Y. Hierarchical nanostructures of K-birnessite nanoplates on anatase nanofibers and their application for decoloration of dye solution. *J. Mater. Chem.* **2010**, *20*, 3157–3162. [CrossRef]

34. Zhuang, X.; Wan, Y.; Feng, C.; Shen, Y.; Zhao, D. Highly efficient adsorption of bulky dye molecules in wastewater on ordered mesoporous carbons. *Chem. Mater.* **2009**, *21*, 706–716. [CrossRef]

35. Bao, N.; Li, Y.; Wei, Z.; Yin, G.; Niu, J. Adsorption of dyes on hierarchical mesoporous TiO_2 fibers and its enhanced photocatalytic properties. *J. Phys. Chem. C* **2011**, *115*, 5708–5719. [CrossRef]

36. Xie, C.; Yang, S.; Shi, J.; Niu, C. Highly crystallized C-doped mesoporous anatase TiO_2 with visible light photocatalytic activity. *Catalysts* **2016**, *6*, 117. [CrossRef]

37. Xu, H.; Li, H.; Xia, J.; Yin, S.; Luo, Z.; Liu, L.; Xu, L. One-pot synthesis of visible-light-driven plasmonic photocatalyst Ag/AgCl in ionic liquid. *ACS Appl. Mater. Interfaces* **2011**, *3*, 22–29. [CrossRef] [PubMed]

38. Ghicov, A.; Yamamoto, M.; Schmuki, P. Lattice Widening in Niobium-Doped TiO_2 Nanotubes: Efficient Ion Intercalation and Swift Electrochromic Contrast. *Angew. Chem. Int. Ed.* **2008**, *47*, 7934–7937. [CrossRef] [PubMed]

39. Qiu, B.; Zhou, Y.; Ma, Y.; Yang, X.; Sheng, W.; Xing, M.; Zhang, J. Facile synthesis of the Ti^{3+} self-doped TiO_2-graphene nanosheet composites with enhanced photocatalysis. *Sci. Rep.* **2015**, *5*, 8591. [CrossRef] [PubMed]

40. Li, K.; Gao, S.; Wang, Q.; Xu, H.; Wang, Z.; Huang, B.; Dai, Y.; Lu, J. In-situ-reduced synthesis of Ti^{3+} self-doped TiO_2/g-C_3N_4 heterojunctions with high photocatalytic performance under led light irradiation. *ACS Appl. Mater. Interfaces* **2015**, *7*, 9023–9030. [CrossRef] [PubMed]

41. Bu, Y.; Chen, Z.; Feng, C.; Li, W. Study of the promotion mechanism of the photocatalytic performance and stability of the Ag@AgCl/g-C_3N_4 composite under visible light. *RSC Adv.* **2014**, *4*, 38124–38132. [CrossRef]

42. Wang, P.; Huang, B.; Qin, X.; Zhang, X.; Dai, Y.; Whangbo, M.-H. Ag/AgBr/WO_3·H_2O: Visible-light photocatalyst for bacteria destruction. *Inorg. Chem.* **2009**, *48*, 10697–10702. [CrossRef] [PubMed]

43. Xu, Y.; Xu, H.; Li, H.; Xia, J.; Liu, C.; Liu, L. Enhanced photocatalytic activity of new photocatalyst Ag/AgCl/ZnO. *J. Alloys Compd.* **2011**, *509*, 3286–3292. [CrossRef]

44. Wang, Z.; Wen, B.; Hao, Q.; Liu, L.-M.; Zhou, C.; Mao, X.; Lang, X.; Yin, W.-J.; Dai, D.; Selloni, A.; et al. Localized excitation of Ti^{3+} ions in the photoabsorption and photocatalytic activity of reduced rutile TiO_2. *J. Am. Chem. Soc.* **2015**, *137*, 9146–9152. [CrossRef] [PubMed]

45. Yin, H.; Wang, X.; Wang, L.; Nie, Q.; Zhang, Y.; Yuan, Q.; Wu, W. Ag/AgCl modified self-doped TiO_2 hollow sphere with enhanced visible light photocatalytic activity. *J. Alloys Compd.* **2016**, *657*, 44–52. [CrossRef]

46. Han, L.; Wang, P.; Zhu, C.; Zhai, Y.; Dong, S. Facile solvothermal synthesis of cube-like Ag@AgCl: A highly efficient visible light photocatalyst. *Nanoscale* **2011**, *3*, 2931–2935. [CrossRef] [PubMed]

47. Ma, B.; Guo, J.; Dai, W.-L.; Fan, K. Highly stable and efficient Ag/AgCl core-shell sphere: Controllable synthesis, characterization, and photocatalytic application. *Appl. Catal. B Environ.* **2013**, *130*, 257–263. [CrossRef]

48. Sung-Suh, H.M.; Choi, J.R.; Hah, H.J.; Koo, S.M.; Bae, Y.C. Comparison of ag deposition effects on the photocatalytic activity of nanoparticulate TiO_2 under visible and UV light irradiation. *J. Photochem. Photobiol. A Chem.* **2004**, *163*, 37–44. [CrossRef]

49. Yu, J.-G.; Yu, H.-G.; Cheng, B.; Zhao, X.-J.; Yu, J.C.; Ho, W.-K. The effect of calcination temperature on the surface microstructure and photocatalytic activity of TiO_2 thin films prepared by liquid phase deposition. *J. Phys. Chem. B* **2003**, *107*, 13871–13879. [CrossRef]

50. Cheng, H.; Huang, B.; Dai, Y.; Qin, X.; Zhang, X. One-step synthesis of the nanostructured AgI/BiOI composites with highly enhanced visible-light photocatalytic performances. *Langmuir* **2010**, *26*, 6618–6624. [CrossRef] [PubMed]

51. Klaysri, R.; Tubchareon, T.; Praserthdam, P. One-step synthesis of amine-functionalized TiO_2 surface for photocatalytic decolorization under visible light irradiation. *J. Ind. Eng. Chem.* **2017**, *45*, 229–236. [CrossRef]

52. Zhu, M.; Chen, P.; Liu, M. Ag/Agbr/graphene oxide nanocomposite synthesized via oil/water and water/oil microemulsions: A comparison of sunlight energized plasmonic photocatalytic activity. *Langmuir* **2012**, *28*, 3385–3390. [CrossRef] [PubMed]

53. Sun, L.; Zhang, R.; Wang, Y.; Chen, W. Plasmonic Ag@AgCl nanotubes fabricated from copper nanowires as high-performance visible light photocatalyst. *ACS Appl. Mater. Interfaces* **2014**, *6*, 14819–14826. [CrossRef] [PubMed]

54. Bao, N.; Li, Y.; Yu, X.-H.; Niu, J.-J.; Wu, G.-L.; Xu, X.-H. Removal of anionic azo dye from aqueous solution via an adsorption–photosensitized regeneration process on a TiO_2 surface. *Environ. Sci. Pollut. Res.* **2013**, *20*, 897–906. [CrossRef] [PubMed]

55. Ye, L.; Liu, J.; Gong, C.; Tian, L.; Peng, T.; Zan, L. Two different roles of metallic Ag on Ag/AgX/BiOX (X = Cl, Br) visible light photocatalysts: Surface plasmon resonance and Z-scheme bridge. *ACS Catal.* **2012**, *2*, 1677–1683. [CrossRef]

56. Hirakawa, T.; Kamat, P.V. Charge separation and catalytic activity of Ag@TiO$_2$ core-shell composite clusters under UV-irradiation. *J. Am. Chem. Soc.* **2005**, *127*, 3928–3934. [CrossRef] [PubMed]

57. Zhang, Q.; Bao, N.; Wang, X.; Hu, X.; Miao, X.; Chaker, M.; Ma, D. Advanced fabrication of chemically bonded graphene/TiO$_2$ continuous fibers with enhanced broadband photocatalytic properties and involved mechanisms exploration. *Sci. Rep.* **2016**, *6*, 38066. [CrossRef] [PubMed]

© 2017 by the authors. Licensee MDPI, Basel, Switzerland. This article is an open access article distributed under the terms and conditions of the Creative Commons Attribution (CC BY) license (http://creativecommons.org/licenses/by/4.0/).

catalysts

MDPI

Article

Photocatalytic Graphene-TiO₂ Thin Films Fabricated by Low-Temperature Ultrasonic Vibration-Assisted Spin and Spray Coating in a Sol-Gel Process

Fatemeh Zabihi [1,2], Mohammad-Reza Ahmadian-Yazdi [1] and Morteza Eslamian [1,3,*]

[1] University of Michigan-Shanghai Jiao Tong University Joint Institute, Shanghai 200240, China; fzabihi@dhu.edu.cn (F.Z.); 1143709026@sjtu.edu.cn (M.-R.A.-Y.)
[2] State Key Laboratory for Modification of Chemical Fibers and Polymer Materials, College of Materials Science and Engineering, Donghua University, Shanghai 201620, China
[3] State Key Laboratory for Composite Materials, School of Materials Science and Engineering, Shanghai Jiao Tong University, Shanghai 200240, China
* Correspondence: morteza.eslamian@sjtu.edu.cn or morteza.eslamian@gmail.com;
  Tel.: +86-21-3420-7249; Fax: +86-21-3420-6525

Academic Editors: Vladimiro Dal Santo and Alberto Naldoni
Received: 29 March 2017; Accepted: 27 April 2017; Published: 2 May 2017

Abstract: In this work, we communicate a facile and low temperature synthesis process for the fabrication of graphene-TiO₂ photocatalytic composite thin films. A sol-gel chemical route is used to synthesize TiO₂ from the precursor solutions and spin and spray coating are used to deposit the films. Excitation of the wet films during the casting process by ultrasonic vibration favorably influences both the sol-gel route and the deposition process, through the following mechanisms. The ultrasound energy imparted to the wet film breaks down the physical bonds of the gel phase. As a result, only a low-temperature post annealing process is required to eliminate the residues to complete the conversion of precursors to TiO₂. In addition, ultrasonic vibration creates a nanoscale agitating motion or microstreaming in the liquid film that facilitates mixing of TiO₂ and graphene nanosheets. The films made based on the above-mentioned ultrasonic vibration-assisted method and annealed at 150 °C contain both rutile and anatase phases of TiO₂, which is the most favorable configuration for photocatalytic applications. The photoinduced and photocatalytic experiments demonstrate effective photocurrent generation and elimination of pollutants by graphene-TiO₂ composite thin films fabricated via scalable spray coating and mild temperature processing, the results of which are comparable with those made using lab-scale and energy-intensive processes.

Keywords: photocatalysis; spray coating; ultrasonic vibration; graphene-TiO₂; sol-gel; microstreaming

1. Introduction

A photocatalyst performs catalytic activity using incident photons as the driving force for a chemical reaction, without being consumed or chemically altered as a result of the reaction. Photocatalysts are low-cost, efficient and environmentally-favored alternatives to commonly used industrial catalysts [1–3]. Photocatalyst works based on oxidative surface decomposition of the reactants are typically used for the removal of residual oils and solvents and for inhibiting the growth of microorganisms on the surface [2–4]. Some metal oxides, such as TiO₂, with inherent resistance to oxidation and hydration exhibit photocatalytic properties at room temperature [4–6]. TiO₂ is a large band gap semiconductor that absorbs high energy UV photons to generate electron and hole pairs. As Figure 1a depicts, the holes may react with the hydroxyl ions from the adsorbed surface water molecules to form highly reactive but neutral hydroxyl radicals. Airborne or aqueous pollutants may be readily adsorbed on the TiO$_2$ surface and react with these hydroxyl radicals, and reduced to minerals and small molecules [7].

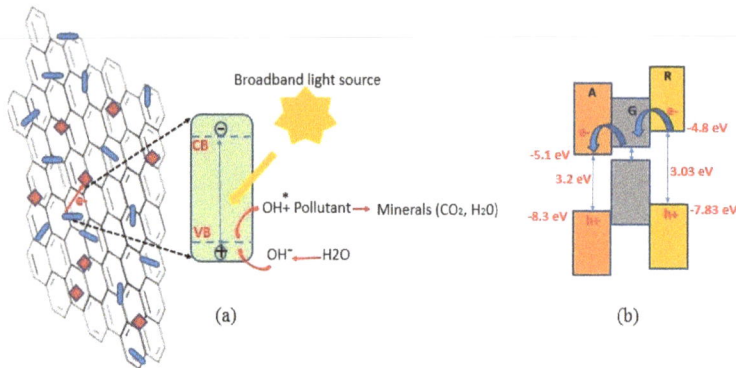

Figure 1. (a) Structure and photocatalytic mechanism in graphene-TiO$_2$ thin films. The rectangular and rod-like features on graphene illustrate anatase and rutile TiO$_2$, respectively; (b) energy band alignment of graphene-TiO$_2$. A, R and G denote anatase, rutile and graphene, respectively.

The photocatalytic performance of TiO$_2$ depends on its crystalline form. The differences in spatial coordination and chemical bonding result in far different ionization potentials, and therefore different electrical affinities [8–10]. Anatase is famous for its size-dependent physical properties and fast photoresponse [6,8]. On the other hand, rutile is more stable, and the difference between its direct and indirect band gap energies is favorably small (quasi-direct band gap) [8]. Therefore, application of mixed phases of rutile and anatase is a more desirable state for photoreaction purposes [3,8]. According to the literature reports, configuration of amorphous TiO$_2$ to a regulated crystalline form requires Ti-O$_2$ cleavage at elevated temperatures [9,11], and this requirement raises the production cost and limits its applications. Therefore, fabrication of multiphase crystalline TiO$_2$ via a low temperature process is desirable but challenging. This has been achieved in this work.

Carbon-based materials may be combined with TiO$_2$, to alleviate the fast recombination of the excited electron-hole pairs and to serve as supporting matrix for TiO$_2$ [8,11–13]. Compared to 3D carbon materials, graphene nanosheets with 2D structure are a better alternative, in that the incorporation and entrapment of TiO$_2$ nanoparticles into 2D graphene nanosheets is readily achieved. In addition, graphene-TiO$_2$ hybrid compound, in the form of powders or thin films, enables an extended light harvesting capability, owing to Ti-O-C bonding. Also, graphene-TiO$_2$ interfaces provide effective charge transfer junctions, which help the injection of electrons from TiO$_2$ to graphene sheets leading to prolonged recombination [7,13–16]. Beside coordination with inorganic materials, graphene provides a strong chemical affinity with organic materials, in particular with the organic dyes [17]. Figure 1b illustrates the band gap alignment of graphene-TiO$_2$ hybrid thin films. Recent electron paramagnetic resonance analyses ascertain that the electrical band alignment of rutile/anatase bi-morph allows electrons to flow from rutile into anatase [8]. This is ascribed to the work function offset, placing the conduction band of anatase about 0.3 eV more negative relative to that of rutile. The work function of few-layered graphene (~−5.0 eV) lies between the conduction bands of rutile and anatase. Therefore, a graphene lattice accommodated between rutile and anatase phases favorably serves as an electron shuttle, prolonging charge recombination. On the other hand, the valence band of graphene stands much higher than those of anatase and rutile, inhibiting unwanted hole transfer, thus favoring photocatalytic function of the composite graphene-TiO$_2$ structure.

The oxidative nature of the composite photocatalyst will be discounted, if the TiO$_2$ nanoparticles are agglomerated or improperly dispersed in the graphene matrix. To alleviate this complication, several strategies have been suggested, such as using TiO_2 nanowires instead of nanoparticles [7], and using layer-by-layer assembly of TiO_2 and graphene nanosheets [13]. In a study conducted by Cheng et al. [14], graphene-TiO_2 composite was synthesized by solvothermal reaction, using various graphene to TiO_2 ratios. Rahimi et al. [15] studied the role of graphene content on light absorption and photoactivity of graphene-TiO_2 blend made by solvothermal method. Xia et al. [3] used chemisorption assembly in which titanium (IV) isopropoxide (TIP) was added to functionalized graphene oxide suspension, followed by an intensive thermal treatment. In another study, incorporation of TiO_2 nanoparticles into graphene sheets was conducted by electrospinning [16]. Posa et al. [18] used graphene oxide and titanium isopropoxide to grow anatase on reduced graphene oxide nanosheets. Chemisorption was carried out in an acid-catalyzed sol-gel process which resulted in graphene oxide-TiO_2, demonstrating superior photocatalytic response. In a recent work by Hu et al. [19], graphene-TiO_2 thin films were synthesized by electrostatical self-assembly of graphene oxide on a cellulose-TiO_2 film under an annealing temperature of 500 °C. Gopalakrishnan et al. [20] reported in-situ solvothermal preparation of graphene-TiO_2 nanocomposite powder and its photocatalytic activity.

The abovementioned representative works show the great potential of graphene-TiO_2 for photocatalytic applications. Issues such as the presence of toxic hydrazine in the solvothermal method, the high-temperature processing required for crystallization of TiO_2, and the development and application of low-cost and scalable manufacturing methods have yet to be addressed. In this work, to obtain functional graphene-TiO_2 photocatalysts at low temperatures, we employ the sol-gel route, combined with ultrasonic substrate vibration-assisted spray [21] and spin [22] coating methods. Ultrasonic substrate-vibration-assisted spray coating (SVASC) [21] is a novel and more controllable version of spray coating, which can be used to manufacture films with large areas in a low-cost industrial process. The employed method has resulted in intact, uniform, and high quality graphene thin films, e.g., [23,24]. Moreover, uniform and high performance spun-on functional thin films, such as polymers, perovskite and graphene-polymer hybrid, subjected to ultrasonic substrate vibration post treatment (SVPT) have been previously developed [22,25,26]. Based on the hydrodynamic and instability analysis of thin liquid solution films subjected to ultrasonic vibration, Rahimzadeh and Eslamian [27] concluded that the imposed vibration has a destabilizing effect on the liquid film. However, if the vibration power and amplitude are kept low, the destabilizing effect is moderate or insignificant; therefore, if the liquid film can resist the destabilizing effect of vibration and remains intact, the circulating motion or microstreaming created within the film as a result of the imposed vibration will actually stir and mix the precursors, a process that results in preparation of uniform and homogenized composite thin solid films, after solvent evaporation. This simple mechanical technique is therefore able to replace some tedious and energy intensive chemical and thermal treatments traditionally used for the fabrication of thin films. In this work, we prepare graphene-TiO_2 composite thin films, where both anatase and rutile coexist, using a sol-gel chemical route assisted with ultrasonic vibration, in which we show that vibration significantly reduces the required heat treatment temperature. We will elaborate later on the fact that the imposed ultrasonic vibration on the wet films assists the chemical conversion in the sol-gel process as well. In the following sections, the physical and optoelectronic and photocatalytic performance of the developed graphene-TiO_2 thin films are presented and discussed.

2. Results and Discussion

Chemical composition of graphene disperse (GD) and a mixture of GD and the TiO_2 precursor solution (titanium isopropoxide bis(acetylacetonate) solution), abbreviated as TS, was studied using liquid-phase Fourier transform infrared spectroscopy (FTIR), shown in Figure 2. Typical graphene features appearing in both spectra are as follows: superimposed sharp peaks at 950–1100 cm^{-1} reflect C-O stretching on graphene surface, due to the presence of a small percentage of oxygen in the graphene used in this study. The bold signals at 1250, 1327 and 1385 cm^{-1} represent shifted C-O-C, C-O···H or C-O bindings, and imply interlinking of unsaturated –C and –OH groups in alcohols [28].

Signals at 1430, 1507 and 1580 cm^{-1} are related to bending vibration of H-C-H and C=O, perhaps formed during the long term dispersion of graphene in dimethylformamide (DMF). The weak reflection at 3450 cm^{-1} shows –OH stretching due to the hydroxyl groups attached to graphene planes [17,28]. The FTIR spectrum of the TS:GD solution presents some additional peaks (Figure 2b). Ti-O vibration is identified at 670 cm^{-1}. The left shoulder absorption band at 807 cm^{-1} and the minor peaks around 2800–3100 cm^{-1} are consistent with Ti-O-C binding, showing the chemisorption between TS and GD solutions [15,28]. The same peaks (2800–3100 cm^{-1}) may be attributed to metal (in this case, Ti) and methyl groups (Ti-CH$_x$), as well [29].

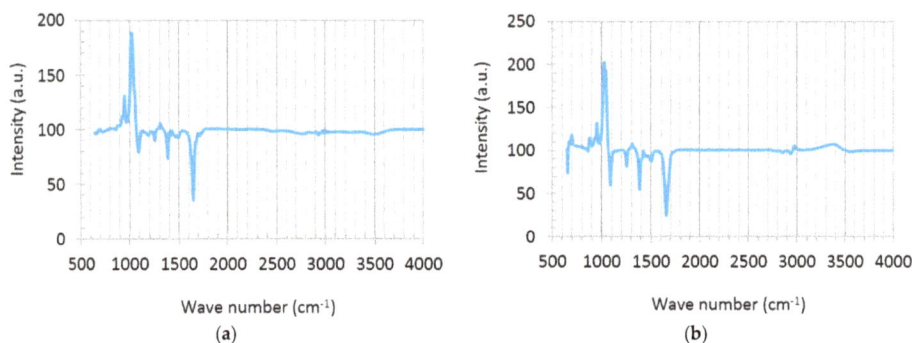

Figure 2. Liquid phase FTIR spectra of precursor solutions. (**a**) Graphene disperse (GD); and (**b**) hybrid TS:GD solution with volume ratio of 1:4.

Figure 3 shows scanning electron microscope (SEM) images of the surface morphology of graphene and graphene-TiO$_2$ thin films made using spin-SVPT, and SVASC. The effect of the volume ratio of TS:GD solutions and annealing temperature is also investigated. Figure 3 evidences the improving role of TiO$_2$ content in surface topography and therefore quality of the composite thin films, in that a higher fraction of TiO$_2$ in graphene-TiO$_2$ films results in better uniformity, owing to the reinforcing effect of TiO$_2$ in graphene matrix. Moreover, at identical precursor solutions and annealing temperatures, application of SVASC results in the formation of slightly more uniform films compared to spin-SVPT (images (c) vs. (d), and (g) vs. (h)), perhaps due to the detrimental effect of centrifugal forces of spin coating applied to graphene nanosheets and the titanium gel. Comparison of the upper and lower panels of Figure 3 reveals that annealing at moderate temperature of 150 °C compared to high temperature of 450 °C results in a more uniform and intact structure, owing to gradual drying and reduced thermal stresses. The surface wrinkles are attributed to the flexible nature of graphene-TiO$_2$ thin films [3,11]. To further demonstrate the remarkable effect of the imposed ultrasonic vibration, in Figure S1 we have shown the SEM images of selected thin films prepared without substrate vibration, i.e., by conventional spin and spray coating, where the non-uniform surface of the films are evidenced. Figure S2 shows the effect of the temperature and TiO$_2$ content on graphene-TiO$_2$ film thickness (~10–50 nm). An increase in the TS to GD volume ratio results in a decrease in the film thickness. This may be attributed to the higher density of TiO$_2$ compared to graphene. It is observed that the films annealed at 150 °C are thinner than those annealed at 450 °C. This is because at lower temperatures, the solvent vapor diffuses away from the wet film more effectively, leaving behind a denser film with less voids. A high temperature may result in fast drying of the film surface and entrapment of the moisture within the film, leading to a thicker and porous film.

Figure 4 displays the X-ray diffraction (XRD) patterns of graphene and graphene-TiO$_2$ thin films. Four selected samples are compared to elucidate the effect of the annealing temperature and precursor composition on the crystalline structure of the ensuing thin films. The typical XRD of graphene is comprised of a wide background with a sharp peak at 26.6° [23,24,30,31]. This sharp signal is present in all graphene-TiO$_2$ spectra, except for one case, i.e., the composite thin film prepared using the precursor solution of TS:GD = 1:4 (lowest graphene content) and annealed at 150 °C, which implies homogenous dispersion of graphene nanosheets [6]. It is found that the abovementioned conditions lead to the same XRD patterns independent of the casting method (spin-SVPT or SVASC). The signals at 31.8° and 34.6° represent the graphene oxide and graphene hydroxide perhaps formed during dispersion in organic and oxidative media [2,3,18]. These unwanted bindings deteriorate the optoelectronic performance of the graphene-TiO$_2$ thin films. Nevertheless, these two peaks only appear in XRD patterns of the samples annealed at 450 °C. The peak at 36.4 °, assigned to 004 anatase and 101 rutile planes, appear in TiO$_2$-rich samples and is intensified when the film is annealed at 150 °C. The peaks at 44.7°, associated with the 105 plane of anatase is present in all composite films, but is weak and slightly shifted in the films with low TiO$_2$ content and annealed at 450 °C [18]. The other peak at 45.4° is due to 211 anatase plane and appears when the film is deposited from the solution with TS:GD of 1:4 and annealed at 150 °C [18]. The reflection peak at 56.6° is assigned to 211 anatase and 105 rutile planes [3,13,15]. These signals are weak in the composite films with low TiO$_2$ content, but are quite strong in the TiO$_2$-rich film annealed at 150 °C. Another signature of TiO$_2$, 200 anatase plane at 48° only appears in the rich-TiO$_2$ film annealed at 150°C. Therefore, a TiO$_2$-rich composite film annealed at 150 °C shows ideal transformation of titanium precursors to crystalline TiO$_2$. It is deduced that the imposed ultrasonic vibration has significantly reduced the required annealing temperature to achieve desired crystalline TiO$_2$ film. The explicit peaks of the rutile and anatase TiO$_2$ phases in XRD patterns indicate that TiO$_2$ was physically combined with the graphene lattice, and no chemical binding has occurred between graphene and TiO$_2$.

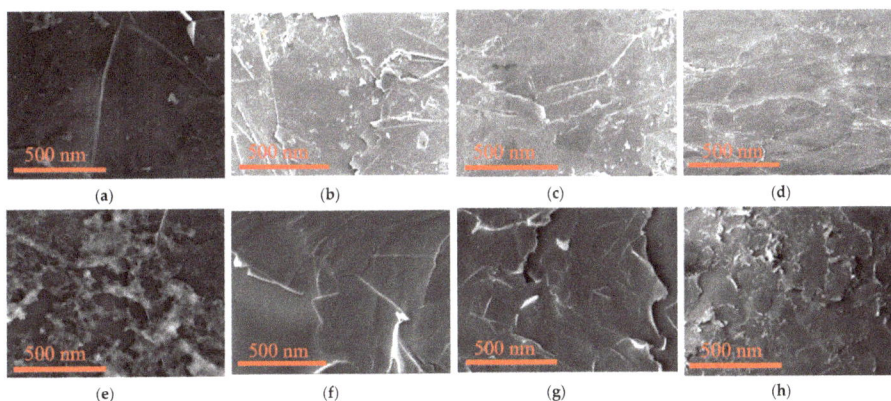

Figure 3. SEM surface topography images of graphene and graphene-TiO$_2$ nanocomposite thin films, made by spin-SVPT and SVASC at various TS:GD volume ratios (for composite films) and annealing temperatures. (**a**) Pristine graphene, SVPT, 150 °C; (**b**) graphene-TiO$_2$, SVPT, TS:GD = 1:9, 150 °C; (**c**) graphene-TiO$_2$, SVPT, TS:GD = 1:4, 150 °C; (**d**) graphene-TiO$_2$, SVASC, TS:GD = 1:4, 150 °C; (**e**) pristine graphene, SVPT, 450 °C; (**f**) graphene-TiO$_2$, SVPT, TS:GD = 1:9, 450 °C; (**g**) graphene-TiO$_2$, SVPT, TS:GD = 1:9, 450 °C; (**h**) graphene-TiO$_2$, SVPT, TS:GD = 1:9, 450 °C. The films on the upper panel were annealed at 150 °C, while those on the lower panel were annealed at 450 °C. Images (**a**,**e**) show pristine graphene films, whereas others are images of graphene-TiO$_2$ composite films prepared at various TS:GD volume ratios. The films associated with images (**d**,**h**) were made by SVASC, whereas the rest of the films were made using spin-SVPT.

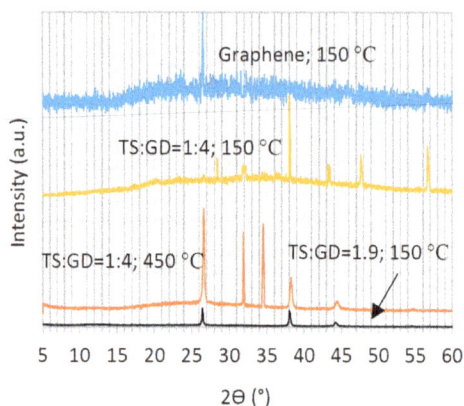

Figure 4. XRD patterns of graphene and graphene-TiO$_2$ thin films prepared at various compositions (TS:GD volume ratios) and annealing temperatures.

The transmittance spectra of graphene and graphene-TiO$_2$ thin films are presented in Figure 5. In general, it is evidenced that the films with higher TiO$_2$ content, deposited by SVASC, and annealed at 150 °C are more transparent. TiO$_2$ is unable to absorb the photons in the visible range, due to its large band gap. Thus, it is expected that a higher TiO$_2$ content in the thin film results in better transparency in the visible range [3,11]. The films annealed at 450 °C show low transparency, presumably due to their larger thickness, as shown in Figure S2, and the defective porous structure. The SVASC films show a relatively better transmittance, compared to spin-SVPT films, perhaps due to the destructive effect of the centrifugal forces that may cause detachment of titanium in the form of hydrogels from the graphene network in the wet films. Therefore, even when deposited from the same precursor solution, the spray-on thin films contain larger amount of TiO$_2$, thus showing higher transparency in the visible range.

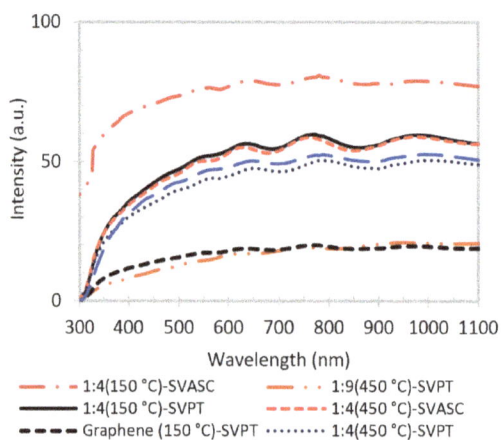

Figure 5. UV-visible transmission of graphene and graphene-TiO$_2$ thin films fabricated at various TiO$_2$ contents and annealing temperatures. The first number in the labels is the TS:GD volume ratio.

Raman spectra of graphene-TiO$_2$ thin films present various patterns depending on the annealing temperature and the casting method. Here, we only display the Raman spectra of the film fabricated using precursor solution with TS:GD volume ratio of 1:4, fabricated by spin-SVPT and SVASC, and annealed at 150 °C (Figure 6). The Raman spectra of the films fabricated using the same precursor solution, but annealed at 450 °C are presented in Figure S3 of the Supporting Information. Raman spectra of the films deposited from the solution with TS:GD volume ratio of 1:9 showed no clear TiO_2 peaks, due to the low content of TiO_2 in the composite films and high intensity of graphene bands, which obscure the TiO_2 peaks. The prominent peaks at 1570 cm^{-1} (G), 2700 cm^{-1} (G': the unique feature of few-layered graphene) and the weak peak at 1350 cm^{-1} (D) are graphene reflections [23,24,32–34]. The sharp and symmetric graphene peaks indicate the small size, few-layered, nanoscale and homogenous form of graphene sheets [33–35]. We attribute the formation of this favorable structure to ultrasonic vibration, which homogenizes the film nanostructure. Another evidence of the smaller, few-layered configuration of graphene is the high intensity ratio of D to G peaks ($I_D/I_G > 0.23$). Small values of I_D/I_G (<0.2) suggest the presence of large graphite segments in the domain [34–37]. According to the Raman spectra of Figure 6, in both SVPT and SVASC films this value is above 0.24. The signals appearing between the graphene peaks are related to different TiO_2 phases. Both spectra of Figure 6 demonstrate the known 440 cm^{-1} vibrational band of rutile and the 391 cm^{-1} band of anatase phases. The known peaks at 144 and 236 cm^{-1} associated with rutile have been combined and circumvented by a wide peak at the left shoulder of the Raman spectra. The other known vibrational bands of rutile at 580, 613 and 769 cm^{-1} and conjugated signals of anatase at 520–640 cm^{-1} [1,10–12] are somewhat detected, although the TiO_2 peaks are weak, due to the strong peaks of graphene. It is noted that in Figure 6 there is no footprint of amorphous TiO_2 in the samples annealed at 150 °C. Moreover, in both cases rutile is the dominant phase, which is chemically and thermodynamically stable and is a stronger charge carrier with lower band gap compared to anatase [3,9,10]. It is also deduced that the high temperature processing at 450 °C disrupts the Raman peaks of graphene (Figure S3), as discussed before. Amorphous TiO_2 has reflections at 1061, 1100 and 1342 cm^{-1} [9,10,12], and Figure S3a clearly shows the peaks at 1100 and 1342 cm^{-1}, substantiating that the films annealed at 450 °C contain amorphous TiO_2 phase.

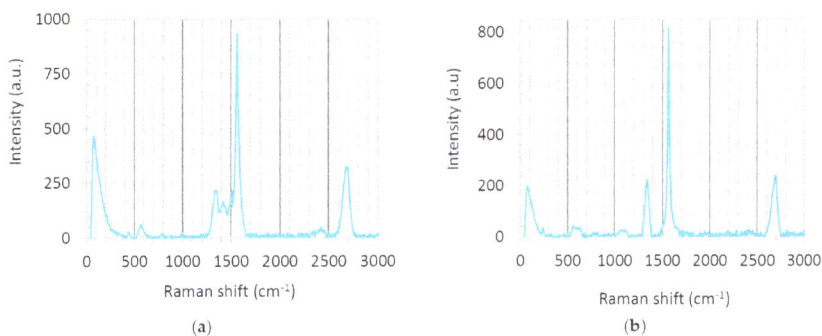

(a) (b)

Figure 6. Raman spectra of graphene-TiO_2 thin films, deposited from precursor solution with TS:GD volume ratio of 1:4: (**a**) spin-SVPT; and (**b**) SVASC films annealed at 150 °C.

According to the discussed XRD and Raman spectroscopy results, the ultrasonic vibration-assisted sol-gel followed by spin or spray coating and a mild heat treatment results in the formation of anatase/rutile polymorph. As substantiated in our previous works [21,25,26] and observed in our ongoing studies, a very striking effect of the excitation by ultrasonic vibration is its controlling effect for preferential orientation of crystallization (nucleation and growth), and a decrease in the activation energy of crystallization, leading to the formation of crystalline phases of titanium at a lower temperature.

The dark current-voltage (I-V) and sheet resistance curves are affected by the presence of TiO_2 in graphene-TiO_2 thin films [2,38], because TiO_2 behaves as an insulator in the dark. The I-V curves shown in Figure 7a confirm that TiO_2 thin film creates no current in the dark, whereas graphene thin film is highly conductive. The Hall measurements of Figure 7b show that the sheet resistance of graphene-TiO$_2$ increases by increasing the TiO$_2$ content, as expected. The sheet resistance also increases in the films annealed at 450 °C, due to the formation of defects and voids in the film structure as a result of rapid drying, as discussed before. However, due to the synergic light-reactive function of TiO$_2$ and graphene, a different optoelectronic behavior is observed when the films are exposed to broadband illumination.

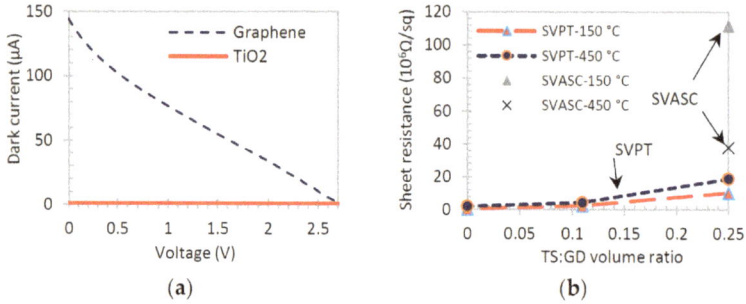

Figure 7. (a) Typical dark I-V curves of graphene and TiO$_2$ films; (b) influence of TiO$_2$ content and annealing temperature on the sheet resistance of graphene-TiO$_2$ thin films.

Figure 8 shows the photoinduced current of graphene-TiO$_2$ thin films compared with those of pristine TiO$_2$ and graphene thin films. The illumination source was blocked after about 80 s. TiO$_2$ and graphene films show negligible photoinduced current response, individually (overlapped in the x axis of the graph), whereas graphene-TiO$_2$ films demonstrate a good synergic photocurrent generation. The SP^2 barrier effect of –O and –OH groups in graphene networks may be responsible for its negligible photocurrent activity [39]. Also, the large band gap and extremely short lifetime of excitons in TiO$_2$ result in poor photoinduced current. On the other hand, when combined in the form of graphene-TiO$_2$ thin film with a well-structured architecture, the composite film benefits from the high photoinduced charge dissociation of TiO$_2$ at the presence of graphene, increased absorption range, particularly at Ti-O-C bonds, and prolonged recombination provided by graphene. The larger photoinduced responses of the thin films fabricated by SVASC are consistent with the aforementioned characterization results, i.e., the improved structural arrangement and the good uniformity and higher content of TiO$_2$ in spray-on thin films, compared to those of the spun-on thin films.

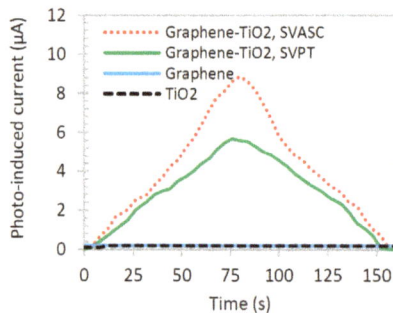

Figure 8. Time-resolved photocurrent generation of TiO$_2$, graphene, and graphene-TiO$_2$ thin films under on and off broadband illumination and 3 V bias. The graphene-TiO$_2$ thin films were deposited from precursor solution with TS:GD volume ratio of 1:4 and annealed at 150 °C. The illumination was blocked after about 80 s.

We also investigated the distribution of surface potential and phase images by atomic force microscopy (AFM) (Figure 9). The potential roughness (Rq) indicates the deviation in distribution of the surface potential. Figure 9 shows that an increase in the TiO_2 content results in a decrease in Rq, because as shown in SEM images of Figure 3 and AFM phase images of Figure 9, the addition and increase of the TiO_2 content has a positive effect on the surface uniformity and structural homogeneity. The potential roughness of the spin-SVPT graphene film decreases from 14.3 to 12 mV by adding a small amount of TiO_2 (TS:GD volume ratio of 1:9). A further increase in the TiO_2 content (TS:GD volume ratio of 1:4) results in further reduction of potential roughness to 8.16 mV. With the same precursor solution, the SVASC film shows the lowest potential roughness of 6 mV. The potential profiles and the peak-to-valley roughness values along the lines shown on the potential maps are displayed in Figure S4 of the Supplementary Materials. The variation of peak-to-valley roughness is consistent with the Rq roughness of Figure 9. The size of the graphene nanosheets and TiO_2 particles (bright spots) can be inferred from some of the AFM phase images.

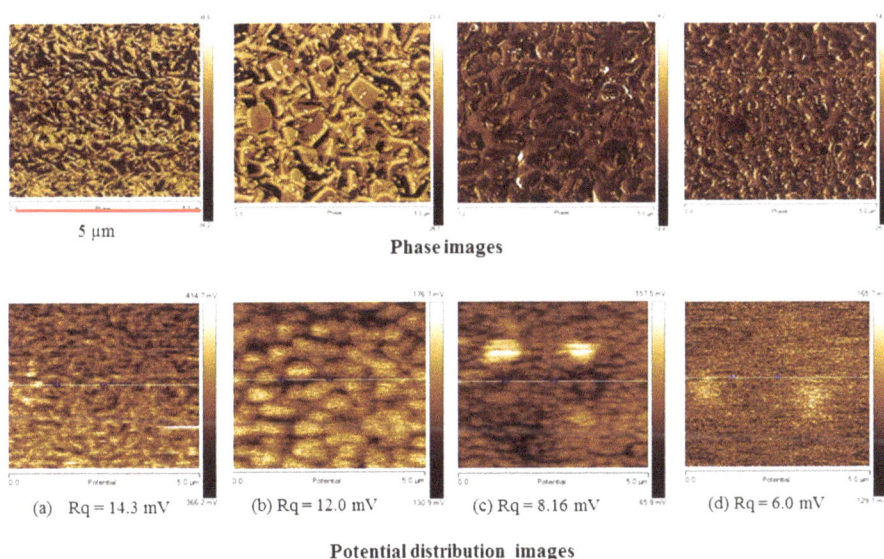

Phase images

5 µm

(a) Rq = 14.3 mV   (b) Rq = 12.0 mV   (c) Rq = 8.16 mV   (d) Rq = 6.0 mV

Potential distribution images

Figure 9. AFM phase (upper panel) and potential mapping (lower panel) images of graphene-TiO_2 films with various TiO_2 contents (TS:GD volume ratio), annealed at 150 °C. (a) spin-SVPT graphene; (b) spin-SVPT graphene-TiO_2 at TS:GD = 1:9; (c) spin-SVPT graphene-TiO_2 at TS:GD = 1:4; and (d) SVASC graphene-TiO_2 at TS:GD = 1:4. Note that letters (a) to (d) under each column in the lower panel refers to both phase (upper panel) and potential distribution (lower panel) images.

The contact angle of water droplets on graphene and graphene-TiO_2 films are shown in Figure S5, where it is found that graphene-TiO_2 film is highly wetting (small contact angle) and therefore superhydrophilic. This is due to the hydrogen bonding and strong interlinking between Ti–O and unsaturated –O and –OH groups on graphene [12,40–42]. It is known that TiO_2 surface is self-cleaning, meaning that it removes the surface dirt, such as organic compounds. This may be related to superhydrophilicity and/or photocatalysis capability of TiO_2 surface [43]. Here we show that the graphene-TiO_2 film has a similar capability.

To test the photocatalytic activity of graphene-TiO_2 thin films, the best graphene-TiO_2 films (spin-SVPT and SVASC, deposited from the solution with TS:GD volume ratio of 1:4 and annealed at 150 °C) in terms of functionality, composition and uniformity, as well as pristine TiO_2 and graphene thin films, were subjected to photodegradation analysis. Figure 10a,b show the time-varying and maximum degradation performance (after 60 min) of the abovementioned films, respectively. The TiO$_2$ thin film shows a poor photodegradation of MB in water, less than 8% after 60 min. This low photocatalytic performance may be attributed to the fast charge recombination [44,45]. Graphene thin film also shows a poor involvement in chemical conversion, as observed by others also, e.g., [46], here about 5% after 60 min. In the case of graphene-TiO$_2$ thin films, it is observed that under illumination, the spin-SVPT and SVASC samples linearly degrade the contaminant, and after 60 min about 77% and 84% of MB in water is removed, respectively. In graphene-TiO$_2$ thin films, the photoinduced carriers from TiO$_2$ are conducted within the graphene network, leading to longer carrier lifetime. Moreover, the increased photoconversion might be due to better trapping of the contaminants, owing to the large surface area of the 2D structure of graphene, and the unsaturated functional groups on graphene sheets [7,19]. The photoconversion performance and degradation rates observed in Figure 10 are consistent with photoinduced current results of Figure 8. It is worth noting that the obtained photoinduced reaction rates of the composite films are comparable with few similar works available in the literature [3,7,18], but in the present study, the films contain both TiO$_2$ phases and are fabricated using scalable and facile spray coating at much reduced annealing temperatures, achieved by ultrasonic vibration.

Figure 11 shows the coordination of graphene with a small amount of oxygen with DMF, reaction of titanium precursor with ethanol and water under acid catalyzing by HCl, and a suggested route for the sol-gel conversion process and transformation of the amorphous TiO$_2$ to crystalline TiO$_2$. In this process, it is speculated that the –OH groups for the formation of the gel phase in the hydrolysis step of the sol-gel process is supplied by ethanol, as well as water. The ultrasonic vibration and a mild heat treatment convert the titanium gel into crystalline TiO$_2$. It is deduced and supported by the characterization results that the imposed ultrasonic vibration helps disruption of the gel phase, and therefore a lower annealing temperature is required to remove the residuals to convert the titanium precursors to TiO$_2$. Figure 12 illustrates the abovementioned explanation. It is noted that in this work, ultrasonic vibration is used to assist the sol-gel conversion process, as well as to assist the formation of a uniform composite film, in one fabrication step. Ultrasonic vibration has been used by others to boost the chemical conversion, e.g., powder synthesis by sol-gel [47] and hydrothermal [48] processes, but not thin films.

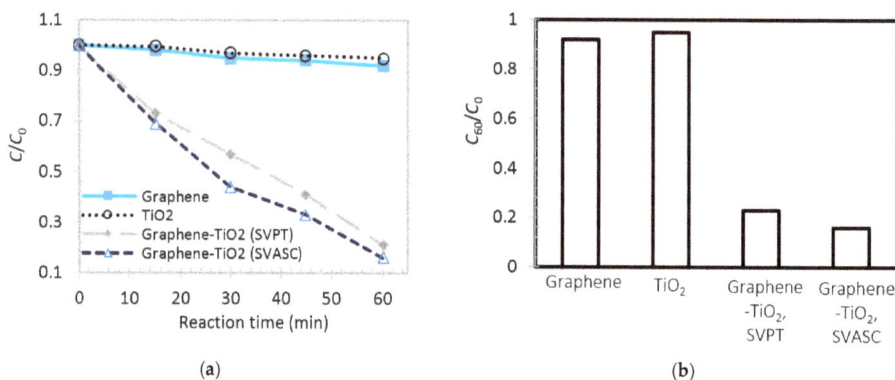

Figure 10. (a) Time varying photocatalytic performance of graphene, TiO$_2$ and graphene-TiO$_2$ thin films fabricated by spin-SVPT and SVASC; (b) Maximum photodegradation performance of the same films as those in (a) after 60 min under broadband illumination. The model reaction medium is methylene blue (MB) in deionized water. C_0 is the initial concentration of MB in water, and C_{60} is the concentration after 60 min. All samples were annealed at 150 °C and the graphene-TiO$_2$ samples were deposited from the precursor solution with a TS:GD volume ratio of 1:4.

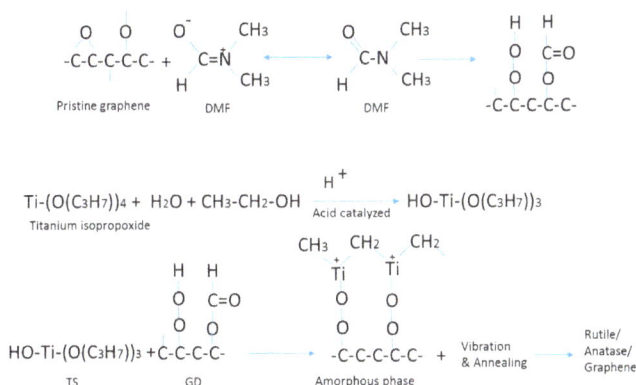

Figure 11. The top reaction shows the dispersion of graphene nanosheets in DMF. The middle reaction shows the coordination between TiO_2 precursors. The bottom reaction shows mixing of graphene and TiO_2 precursors and the sol-gel process assisted by ultrasonic vibration.

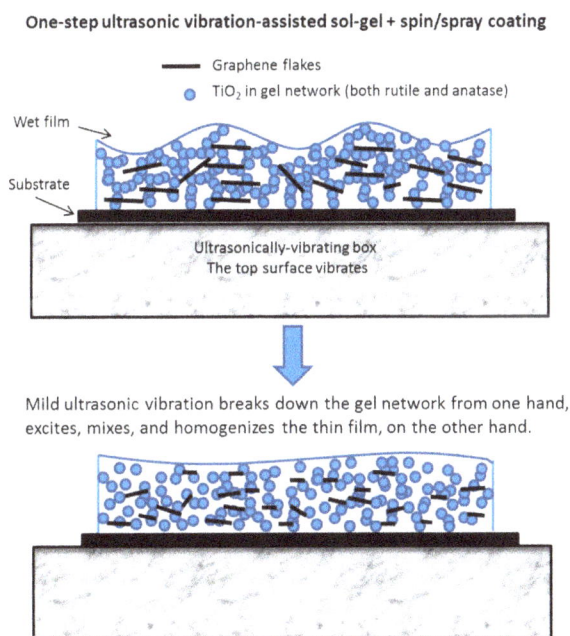

Figure 12. The process of graphene-TiO_2 thin film formation by low-temperature ultrasonic vibration-assisted spin and spray coating in a sol-gel route.

3. Materials and Methods

Few-layered graphene nano-sheets (FLGNS, 3 stacked layers, 1 nm average thickness of each layer with 1.5% oxygen content, and average surface area of 1960 m^2/g) were provided by Hengqiu Graphene Technology Co., Ltd., Suzhou, China. HCl aqueous solution (36.5%), dimethylformamide (DMF, 99.5%), ethanol (99.5%), 2-propanol (99.8%), acetone (99.5%) and titanium isopropoxide bis(acetylacetonate) (75% in 2-propanol) were purchased from Sigma-Aldrich, St. Louis, MO, USA.

Bare glass and indium-doped tin oxide (ITO)-coated glass substrates (1.5 cm × 1.5 cm) were purchased from Nanbo Display Technology, Shenzhen, China.

The glass substrates were washed in an ultrasonic bath using detergent, 2-propanol, and deionized water, sequentially, for 30 min and dried in a vacuum oven and placed in an ultraviolet cleaner for 12 min. Precursor solution of graphene-TiO_2 thin films was composed of titanium isopropoxide bis(acetylacetonate) solution (TS) and graphene disperse (GD) mixed with TS:GD volume ratios of 1:4 and 1:9. For preparation of GD, 50 mg of graphene nanosheets was added to 10 mL of DMF, supersonicated for 6 h, and agitated overnight on a magnetic stirrer. For preparation of TS, 35 μL of HCl solution mixed with ethanol (1:50 volume ratio, respectively) was gradually added to diluted titanium isopropoxide bis(acetylacetonate) (0.07 volume ratio in ethanol, stirred for 10 min) and the mixture was stirred for 30 min. The GD and TS solutions were then mixed, sonicated for 6 h and stirred overnight before casting. The TS precursor solution was converted into TiO_2 in a sol-gel process. The graphene-TiO_2 composite thin films were deposited by spin coating followed by ultrasonic substrate vibration post treatment, called spin-SVPT or simply SVPT and ultrasonic substrate vibration-assisted spray coating (SVASC). Table S1 in the Supporting Information lists the experimental conditions. In the case of SVASC, the substrate is ultrasonically vibrated by placing it on a vibrating metal box. A piezoelectric ceramic (5 W and 40 kHz) is mounted inside the top surface of the metal box, which vibrates the substrate in the vertical direction. SVASC was performed using an air-assist spray nozzle mounted on a 3D traverser arm. Back pressure of the atomizing air was set to 0.3 MPa and the distance between the nozzle tip and the substrate was kept constant at 80 mm. Nozzle speed, spray flow rate and number of spray passes were set to 10 mm/s, 200 μL/min, and single spray pass, respectively. In spin-SVPT experiments, the precursor solution was spun at 2000 rpm for 60 s. The as-spun wet films were immediately placed on the surface of the vibrating box for 10 s. As-sprayed and as-spun wet thin films were annealed at 150 or 450 °C, for 45 min. TiO_2 forms in a sol-gel process assisted by energy impartment to the solution by ultrasonic vibration, and is completed by removal of the solvents after heat treatment (c.f. Figures 11 and 12).

The intermolecular bindings in composite films were characterized using Raman spectroscopy (Horiba Jobin Yvon LabRam model HR800, Horiba Scientific, Kyoto, Japan). The Raman spectra were recorded at room temperature with a micro-Raman system equipped with a CCD camera, using 514 nm laser line, under attenuated power of 5 mW. Raman shifts were calibrated at 521 cm^{-1}. Liquid phase Fourier transmission infrared spectroscopy (FTIR) was performed on the precursor solutions, using a Smart iTR accessory connected to a Nicolet 6700 FTIR spectrometer (Thermo Fisher Scientific, Waltham, MA, USA). DMF:ethanol:HCl mixture (10:2:1 volume ratio) was used as the background medium. The transmission UV-vis spectra of the thin films were recorded using a Shimadzu UV-3101PC UV-Vis-NIR spectrophotometer, Shimadzu, Kyoto, Japan. Samples were prepared on bare glass and a second bare glass was used as the background. Surface morphology of the thin films was studied by scanning electron microscopy (SEM, Hitachi, Model S-3400 N, Tokyo, Japan). The local surface potential and the phase images were obtained by atomic force microscopy (AFM, Dimension Icon & FastScan Bio, Brucker, Bremen, Germany), while the thin films were deposited on ITO-coated glass. Surface potential was determined based on the local differences of the electrical potential between the sample and a Cr/Pt coated conductive tip (Multi75E-G, BudgetSensors, Sofia, Bulgaria) at constant force of 3 N/m, positioned at 2 μm distance from the surface. Electrical resistivity of thin films was obtained using a Hall measurement instrument (MMR Technologies, San Jose, CA, USA), at room temperature, based on the van der Pauw four-point method. The samples were cast on bare glass. X-ray diffraction (XRD) patterns of the films fabricated on bare glass were obtained using XRD spectroscopy (XPert3 MRD (XL), PANalytical, Westborough, MA, USA). The surface profilometry (KLA-Tencor P7, Milpitas, CA, USA) was used for thickness measurements. The dark current-to-voltage trend and the photoresponse under illumination were measured in a standard probe station, using a Keithley source meter Model 2602A, Gorinchem, The Netherlands. The broadband irradiation was generated using a solar simulator Xenon lamp with an intensity of 100 mW·cm^{-2}. Equilibrium contact angles of a 20 μL deionized water droplet were measured using a Theta Lite Optical Tensiometer, Biolin Scientific AB, Gothenburg, Sweden. Photocatalytic performance of the thin films was evaluated based on photodegradation of methylene blue (MB) in water under broadband illumination. For each measurement, 1 mL of MB solution (2 ppm in deionized water) was dispensed on the thin films and the films were placed under a Xenon lamp (100 mW·cm^{-2}). The distance between the light source and sample was set to 5 cm in all experiments and 50 μL of reaction fluid was used in 15 min time intervals, to study the time-resolved catalytic performance, in 60 min. Concentration of MB in water was measured by UV-vis absorption at 663 nm.

4. Conclusions

In this study, graphene-TiO$_2$ photocatalytic thin films were fabricated via the sol-gel method, as the chemical route for the formation of TiO$_2$, and spin and spray coating as the casting methods. The wet films were excited by imposing ultrasonic vibration on the substrate. As a result, rutile and anatase TiO$_2$ crystalline phases formed in a low-temperature (150 °C) annealing process. Therefore, it is deduced that the ultrasonic vibration assists the conversion of titanium precursors to TiO$_2$ and facilitates the breakdown of the physical bonds of the gel phase. This ultrasonic vibration-assisted sol-gel process for casting thin films requires only a mild annealing step, resulting in significant energy savings compared to the conventional sol-gel process. It is also noted that the spray coating process is a scalable, fast, and low-cost casting method, suitable for large-scale manufacturing of the developed photocatalysts.

Morphology, optical, electrical and optoelectronic properties of the composite thin films were studied by varying the content of TiO$_2$ and the annealing temperature. The best film was obtained with the highest TiO$_2$ content used in this study (volume ratio of TS to GD precursor solutions of 1:4), and a mild annealing temperature of 150 °C. A higher annealing temperature of 450 °C deteriorated the film characteristics, perhaps due to rapid drying and the formation of voids in the film, thermal sintering, etc. The method of deposition was also found to be a determining factor. The composite films made by ultrasonic substrate vibration-assisted spray coating (SVASC) outperformed spin coating followed by ultrasonic substrate vibration post treatment (spin-SVPT). In this case, the centrifugal forces acting on titanium-based sol and/or gel phases during the spinning process are presumably responsible for the removal of titanium from the matrix, and disruption of the film structure.

The characterization techniques showed that the optimized graphene-TiO$_2$ thin film is comprised of rutile and anatase particles uniformly embedded in a matrix of few-layered graphene thin film. The composite thin films demonstrated significant photoinduced current generation and photocatalytic activity. This enhancement was attributed to the advantages of graphene and TiO$_2$, collectively, as follows. TiO$_2$ can generate photoinduced current in the UV range, but suffers from fast recombination, due to negligible electrical conductivity. Graphene is highly conductive and a strong charge carrier, and facilitates charge dissociation in TiO$_2$. A well-structured graphene-TiO$_2$ thin film could ideally exploit the traits of both graphene and TiO$_2$, offering favorable photoresponse and photocatalytic functions.

Supplementary Materials: The following are available online at www.mdpi.com/2073-4344/7/5/136/s1. Figure S1: Surface morphology of graphene-TiO$_2$ thin films prepared by conventional spin (a) and spray (b) coating. Both films were deposited from precursor solution with TS:GD volume ratio of 1:4, and the films were annealed at 150 °C, Figure S2: Effect of precursor composition and annealing temperature on the thickness of graphene-TiO$_2$ composite thin films, Figure S3: Raman spectra of graphene-TiO$_2$ thin films, deposited from the precursor solution with TS:GD vol. ratio of 1:4. (a) Spin-SVPT, and (b) SVASC, annealed at 450 °C, Figure S4: Line potential profiles of graphene-TiO$_2$ films with various TiO$_2$ contents (TS:GD vol. ratio), annealed at 150 °C. (a) spin-SVPT graphene, (b) spin-SVPT graphene-TiO$_2$ at TS:GD = 1:9, (c) spin-SVPT graphene-TiO$_2$ at TS:GD = 1:4, and (d) SVASC graphene-TiO$_2$ thin film at TS:GD = 1:4. The line profiles were obtained along the lines shown on the AFM potential images of Figure 9, Figure S5: Contact angle measurement tests of water droplets on (a) graphene-TiO$_2$ thin films fabricated by spin-SVPT using a solution with TS:GD volume ratio of 1:4 and annealed at 150 °C, and (b) graphene deposited by spin-SVPT and annealed at 150 °C, Table S1: Experimental conditions used for the fabrication of graphene, TiO$_2$, and graphene-TiO$_2$ thin films via spin coating followed by substrate vibration post treatment (spin-SVPT) and substrate vibration-assisted spray coating (SVASC). GD stands for graphene disperse and TS stands for TiO$_2$ precursor solution.

Acknowledgments: Financial support from the Shanghai Municipal Education Commission in the framework of the oriental scholar and distinguished professor designation and funding from the National Natural Science Foundation of China (NSFC) is acknowledged.

Author Contributions: Fatemeh Zabihi and Morteza Eslamian conceived and designed the experiments; Fatemeh Zabihi performed the experiments and characterizations and analyzed the data; Mohammad-Reza Ahmadian-Yazdi assisted with some of the experiments and characterizations; Fatemeh Zabihi and Morteza Eslamian wrote the paper. All authors read and approved the paper.

Conflicts of Interest: The authors declare no conflict of interest.

References

1. Hardcastle, F.D. Raman spectroscopy of titania (TiO_2) nanotubular water-splitting catalysts. *J. Ark. Acad. Sci.* **2011**, *65*, 43–48.

2. Zhou, K.; Zhu, Y.; Yang, X.; Jiang, X.; Li, C. Preparation of graphene-TiO_2 composites with enhanced photocatalytic activity. *New J. Chem.* **2011**, *35*, 353–359. [CrossRef]

3. Xia, H.Y.; He, G.Q.; Min, Y.L.; Liu, T. Role of the crystallite phase of TiO_2 in graphene/TiO_2 photocatalysis. *J. Mater. Sci. Mater. Electron.* **2015**, *26*, 3357–3363. [CrossRef]

4. Hashizume, M.; Kunitake, T. Preparations of self-supporting nanofilms of metal oxides by casting processes. *Soft Matter* **2006**, *2*, 135–140. [CrossRef]

5. Wang, J.T.W.; Ball, J.M.; Barea, E.M.; Abate, A.; Alexander-Webber, J.A.; Huang, J.; Saliba, M.; Mora-Sero, I.; Bisquert, J.; Snaith, J.H.; et al. Low-temperature processed electron collection layers of graphene/TiO_2 nanocomposites in thin film perovskite solar cells. *Nano Lett.* **2014**, *14*, 724–730. [CrossRef] [PubMed]

6. Keun, L.Y.; Choi, H.; Lee, H.; Lee, C.; Choi, J.S.; Choi, C.G; Hwang, E.; Young Park, J. Hot carrier multiplication on graphene/TiO_2 Schottky nanodiodes. *Sci. Rep.* **2016**, *6*, 27549.

7. Pan, X.; Zhao, Y.; Wang, S.; Fan, Z. TiO_2/graphene nanocomposite for photocatalytic application. In *Materials and Processes for Energy: Communicating Current Research and Technological Developments*; Méndez-Vilas, A., Ed.; Formatex Research Center: Extremadura, Spain, 2013; pp. 913–920.

8. Scanlon, D.O.; Dunnill, C.W.; Buckeridge, J.; Shevlin, S.A.; Logsdail, A.J.; Woodley, S.M.; Catlow, C.R.A.; Powell, M.J.; Palgrave, R.G.; Parkin, I.P.; et al. Band alignment of rutile and anatase TiO_2. *Nat. Mater.* **2013**, *12*, 798–801. [CrossRef] [PubMed]

9. Scepanović, M.J.; Grujic-Brojcin, M.; Dohcevic-Mitrovic, Z.D.; Popovic, Z.V. Characterization of anatase TiO_2 nanopowder by variable-temperature raman spectroscopy. *Sci. Sinter.* **2009**, *41*, 67–73. [CrossRef]

10. Castrejon-Sanchez, V.H.; Enrique, C.; Camacho-Lopez, M. Quantification of phase content in TiO_2 thin films by raman spectroscopy. *Superf. Y Vacio* **2014**, *27*, 88–92.

11. Sayilkan, F.; Ilturk, M.A.; Sayilkan, H.; Onal, Y.; Akarsu, M.; Arpac, E. Characterization of TiO_2 synthesized in alcohol by a sol-gel process: The effects of annealing temperature and acid catalyst. *Turk. J. Chem.* **2005**, *29*, 697–706.

12. Cravanzola, S.; Jain, S.M.; Cesano, F.; Damin, A.; Scarano, D. Development of a multifunctional TiO_2/MWCNT hybrid composite grafted on a stainless steel grating. *RSC Adv.* **2015**, *5*, 103255–103264. [CrossRef]

13. Zhu, J.; Cao, Y.; He, J. Facile fabrication of transparent, broadband photoresponse, self-cleaning multifunctional graphene—TiO_2 hybrid films. *J. Colloid Interface Sci.* **2014**, *420*, 119–126. [CrossRef] [PubMed]

14. Cheng, P.; Yang, Z.; Wang, H.; Cheng, W.; Chen, M.; Shangguan, W.; Ding, G. TiO_2-graphene nanocomposites for photocatalytic hydrogen production from splitting water. *Int. J. Hydrogen Energy* **2012**, *37*, 2224–2230. [CrossRef]

15. Rahimi, R; Zargari, S.; Sadat Shojaei, Z. Photoelectrochemical investigation of TiO_2-graphene nanocomposites. In Proceedings of the 18th International Electronic Conference on Synthetic Organic Chemistry, Basel, Switzerland, 1–30 November 2014. [CrossRef]

16. Peining, Z.; Sreekumaran, N.A.; Shengjie, P.; Shengyuan, Y.; Ramakrishna, S. Facile fabrication of TiO_2 graphene composite with enhanced photovoltaic and photocatalytic properties by electro-spinning. *ACS Appl. Mater, Interfaces* **2012**, *4*, 581–585.

17. Cravanzola, S.; Cesano, F.; Magnacca, G.; Zecchina, A.; Scarano, D. Designing rGO/MoS_2 hybrid nanostructures for photocatalytic applications. *RSC Adv.* **2016**, *6*, 59001–59008. [CrossRef]

18. Posa, V.R.; Annavaram, V.; Reddy Koduru, J.; Bobbala, P.; Madhavi, V.; Reddy Somala, A. Preparation of graphene-TiO$_2$ nanocomposite and photocatalytic degradation of Rhodamine-B under solar light irradiation. *J. Exp. Nano Sci.* **2016**, *11*, 722736.

19. Hu, L.; Zhang, L.; Zhang, S.; Li, B. A transparent TiO$_2$-C@TiO$_2$-graphene free-standing film with enhanced visible light photocatalysis. *RSC Adv.* **2016**, *6*, 43098–43103. [CrossRef]

20. Gopalakrishnan, A.; Binitha, N.N.; Yaakob, Z.; Mohammed Akbar, P.; Padikkaparambi, S. Excellent photocatalytic activity of titania—graphene nanocomposites prepared by a facile route. *J. Sol. Gel. Sci. Technol.* **2016**, *80*, 189–200. [CrossRef]

21. Zabihi, F.; Eslamian, M. Substrate vibration-assisted spray coating (SVASC): Significant improvement in nano-structure, uniformity, and conductivity of PEDOT:PSS thin films for organic solar cells. *J. Coat. Technol. Res.* **2015**, *12*, 711–719. [CrossRef]

22. Wang, Q.; Eslamian, M. Improving uniformity and nanostructure of solution-processed thin films using ultrasonic substrate vibration post treatment (SVPT). *Ultrasonics* **2016**, *67*, 55–64. [CrossRef] [PubMed]

23. Chen, Q.; Zabihi, F.; Eslamian, M. Improved functionality of PEDOT:PSS thin films via graphene doping, fabricated by ultrasonic substrate vibration-assisted spray coating. *Synth. Met.* **2016**, *222*, 309–317. [CrossRef]

24. Soltani-Kordshuli, F.; Zabihi, F.; Eslamian, M. Graphene-doped PEDOT:PSS nanocomposite thin films fabricated by conventional and substrate vibration-assisted spray coating (SVASC). *Eng. Sci. Technol.* **2016**, *19*, 1216–1223. [CrossRef]

25. Zabihi, F.; Chen, Q.; Xie, Y.; Eslamian, M. Fabrication of efficient graphene-doped polymer/fullerene bilayer organic solar cells in air using spin coating followed by ultrasonic vibration post treatment. *Superlattices Microstruct.* **2016**, *100*, 1177–1192. [CrossRef]

26. Eslamian, M.; Zabihi, F. Ultrasonic substrate vibration-assisted drop casting (SVADC) for the fabrication of solar cell arrays and thin film devices. *Nanoscale Res. Lett.* **2015**, *10*, 462. [CrossRef] [PubMed]

27. Rahimzadeh, A.; Eslamian, M. Stability of thin liquid films subjected to ultrasonic vibration and characteristics of the resulting thin solid films. *Chem. Eng. Sci.* **2017**, *158*, 587–598. [CrossRef]

28. Amiri, A.; Shanbedi, M.; Ahmadi, G.; Eshghi, H.; Kazi, S.N.; Chew, B.T.; Savari, M.; Mohd Zubir, M.N. Mass production of highly-porous graphene for high-performance supercapacitors. *Sci. Rep.* **2016**, *6*, 32686. [CrossRef] [PubMed]

29. Morrow, B.A.; Beauchamp, Y. Infrared spectra of some alkyl platinum compounds. Part II. Assignment of the CH stretching modes of a methyl group. *Can. J. Chem.* **1970**, *4*, 2921–2926. [CrossRef]

30. Batmunkh, M.; Shearer, C.J.; Biggs, M.J.; Shapter, J.G. Solution processed graphene structures for perovskite solar cells. *J. Mater. Chem. A* **2016**, *4*, 2605–2616. [CrossRef]

31. Neill, A.O.; Khan, U.; Nirmalraj, P.N.; Boland, J.; Coleman, J.N.J. Graphene dispersion and exfoliation in low boiling point solvents. *Phys. Chem. C* **2011**, *115*, 5422–5428.

32. Naik, G.; Krishnaswamy, S. Room-temperature humidity sensing using graphene oxide thin films. *Graphene* **2016**, *5*, 1–13. [CrossRef]

33. El Gemayel, M.; Narita, A.; Dossel, L.F.; Sundaram, R.S.; Kiersnowski, A.; Pisula, W.; Hansen, M.R.; Ferrari, A.C.; Orgiu, E.; Feng, X.; et al. Graphene nanoribbon blends with P3HT for organic electronics. *Nanoscale* **2014**, *6*, 6301–6314. [CrossRef] [PubMed]

34. Dang, T.T.; Pham, V.H.; Hur, S.H.; Kim, E.J.; Kong, B.S.; Chung, J.S.J. Superior dispersion of highly reduced graphene oxide in N,N-dimethylformamide. *J. Colloid Interface Sci.* **2012**, *376*, 91–96. [CrossRef] [PubMed]

35. Kim, K.; Bae, S.H.; Toh, C.T.; Kim, H.; Cho, J.H.; Whang, D.; Lee, T.W.; Ozyilmaz, B.J.; Ahn, J.H. Ultrathin organic solar cells with graphene doped by ferroelectric polarization. *ACS Appl. Mater. Interfaces* **2014**, *6*, 3299–3304. [CrossRef] [PubMed]

36. Sharma, S.; Klita, G.; Hirano, R.; Hayashi, Y.; Tanemura, M. Influence of gas gomposition on the formation of graphene domain synthesized from camphor. *Mater. Lett.* **2013**, *93*, 18258–18262. [CrossRef]

37. Acik, M.; Darling, S.B. Graphene in perovskite solar cells: Device design, characterization and implementation. *J. Mater. Chem. A* **2016**, *4*, 6185–6235. [CrossRef]

38. Zhao, Y.L.; Lv, M.W.; Liu, Z.Q.; Zeng, S.W.; Motapothula, M.; Dhar, S. Variable range hopping in TiO$_2$ insulating layers for oxide electronic devices. *AIP Adv.* **2012**, *2*, 012129. [CrossRef]

39. Seo, H.; Ahn, S.; Kim, J.; Lee, Y.A.; Chung, K.H.; Jeon, K.J. Multi-Resistive reduced graphene oxide diode with reversible surface electrochemical reaction induced carrier control. *Sci. Rep.* **2012**, *4*, 5642. [CrossRef] [PubMed]

40. Pal, S.; Contaldi, V.; Licciulli, A.; Marzo, F. Self-cleaning mineral paint for application in architectural heritage. *Coatings* **2016**, *6*, 48. [CrossRef]

41. Chen, M.; Straatsma, T.P.; Dixon, D.A. Molecular and dissociative adsorption of water on (TiO$_2$) clusters, n = 1–4. *J. Phys. Chem. A* **2015**, *119*, 11406–11421. [CrossRef] [PubMed]

42. Ansari, M.; Mansoob Khan, M.; Ansari, S.A.; Cho, M.H. Electrically conductive polyaniline sensitized defective-TiO$_2$ for improved visible light photocatalytic and photoelectrochemical performance: A synergistic effect. *New J. Chem.* **2015**, *39*, 8381–8388. [CrossRef]

43. Kommireddy, D.S.; Patel, A.; Shutava, T.G.; Mills, D.K.; Lvov, Y.M. Layer-by-layer assembly of TiO$_2$ nanoparticles for stable hydrophilic biocompatible coatings. *J. Nanosci. Nanotechnol.* **2005**, *5*, 1081–1087. [CrossRef] [PubMed]

44. Srivastava, A.K.; Deepa, M.; Bhandari, S.; Fuess, H. Tunable nanostructures and crystal structures in titanium oxide films. *Nanoscale Res. Lett.* **2009**, *4*, 54–62. [CrossRef] [PubMed]

45. Li, J.; Wang, Z.; Wang, J.; Sham, T.K. Unfolding the anatase-to-rutile phase transition in TiO$_2$ nanotubes using X-ray spectroscopy and spectromicroscopy. *J. Phys. Chem. C* **2016**, *120*, 22079–22087. [CrossRef]

46. Boyd, D.A.; Lin, W.H.; Hsu, C.C.; Teague, M.L.; Chen, C.C.; Lo, Y.Y.; Chan, W.Y.; Su, W.B.; Cheng, T.C.; Chang, C.S.; et al. Single-step deposition of high-mobility graphene at reduced temperatures. *Nat. Commun.* **2015**, *6*, 6620. [CrossRef] [PubMed]

47. Meskin, P.E.; Ivanov, V.K.; Barantchikov, A.E.; Churagulov, B.R.; Tretyakov, Y.D. Ultrasonically assisted hydrothermal synthesis of nanocrystalline ZrO$_2$, TiO$_2$, NiFe$_2$O$_4$ and Ni$_{0.5}$Zn$_{0.5}$Fe$_2$O$_4$ powders. *Ultrason. Sonochem.* **2006**, *13*, 47–53. [CrossRef] [PubMed]

48. Yi, T.; Hu, X.; Gao, K. Synthesis and physicochemical properties of LiAl$_{0.05}$Mn$_{1.95}$O$_4$ cathode material by the ultrasonic-assisted sol-gel method. *J. Power Sources* **2016**, *162*, 36–643. [CrossRef]

© 2017 by the authors. Licensee MDPI, Basel, Switzerland. This article is an open access article distributed under the terms and conditions of the Creative Commons Attribution (CC BY) license (http://creativecommons.org/licenses/by/4.0/).

catalysts

MDPI

Article

Flame-Made Cu/TiO₂ and Cu-Pt/TiO₂ Photocatalysts for Hydrogen Production

Massimo Bernareggi [1], Maria Vittoria Dozzi [1], Luca Giacomo Bettini [2,3], Anna Maria Ferretti [4], Gian Luca Chiarello [1] and Elena Selli [1,3,*]

[1]   Dipartimento di Chimica, Università degli Studi di Milano, via Golgi 19, 20133 Milano, Italy; massimo.bernareggi@unimi.it (M.B.); mariavittoria.dozzi@unimi.it (M.V.D.); gianluca.chiarello@unimi.it (G.L.C.)

[2]   Dipartimento di Fisica, Università degli Studi di Milano, via Celoria 16, 20133 Milano, Italy; lucagiacomo.bettini@unimi.it

[3]   CIMAINA, Università degli Studi di Milano, via Celoria 16, 20133 Milano, Italy

[4]   ISTM-CNR Lab Nanotechnology, Via Fantoli 16/15, 20138 Milano, Italy; anna.ferretti@istm.cnr.it

*   Correspondence: elena.selli@unimi.it; Tel.: +39-02-503-14237

Academic Editor: Bunsho Ohtani
Received: 8 September 2017; Accepted: 11 October 2017; Published: 13 October 2017

Abstract: The effect of Cu or Cu-Pt nanoparticles in TiO₂ photocatalysts prepared by flame spray pyrolysis in one step was investigated in hydrogen production from methanol photo-steam reforming. Two series of titanium dioxide photocatalysts were prepared, containing either (i) Cu nanoparticles (0.05–0.5 wt%) or (ii) both Cu (0 to 0.5 wt%) and Pt (0.5 wt%) nanoparticles. In addition, three photocatalysts obtained either by grafting copper and/or by depositing platinum by wet methods on flame-made TiO₂ were also investigated. High hydrogen production rates were attained with copper-containing photocatalysts, though their photoactivity decreased with increasing Cu loading, whereas the photocatalysts containing both Cu and Pt nanoparticles exhibit a bell-shaped photoactivity trend with increasing copper content, the highest hydrogen production rate being attained with the photocatalyst containing 0.05 wt% Cu.

Keywords: photocatalysis; flame-spray pyrolysis; TiO₂ modification; Cu and Pt nanoparticles; photocatalytic hydrogen production; methanol photo-steam reforming

1. Introduction

The continued use of fossil fuels led to an increased greenhouse effect; thus, cleaner and renewable energy sources are urgently required. Hydrogen is considered the main alternative to fossil fuels and technologies based on hydrogen exploitation as an energy vector are already mature, such as fuel cells or internal hydrogen combustion engines [1,2]. Photocatalysis can provide a straightforward route to hydrogen production from water solutions, possibly converting solar light into chemical energy in the form of H-H bond.

Many photocatalysts have been proposed and tested in recent years, for both thermodynamically up-hill (e.g., hydrogen production from aqueous solutions) and down-hill reactions (photodegradation of organic pollutants), but titanium dioxide still remains the most widely investigated one, due to its advantageous physical and chemical properties [3,4]. One of its main drawbacks consists in the fast recombination of photoproduced electron-hole couples. This drawback can be overcome by modifying TiO₂ with noble metal (Au, Ag, Pd, Pt) nanoparticles (NPs), which improve the separation of photoproduced charge carriers and, thus, the quantum efficiency of photocatalytic processes [5]. Furthermore, the rate of photocatalytic hydrogen production from water is largely increased by performing the reaction in the presence of organic compounds which scavenge the holes photoproduced in the semiconductor valence band (VB) more efficiently than water, making the reaction irreversible [6–11].

Aiming at increasing the photoactivity of titanium dioxide by modification with non-noble metals (e.g., Ni, Cu, Co), thus reducing the photocatalyst costs, Irie et al. [12] deposited copper on titanium dioxide powders by grafting. This led to successful visible light activation of the so obtained photocatalysts both in oxidation reactions (photodegradation of organic pollutants, e.g., 2-propanol) in the presence of O_2 and in hydrogen production from water solutions under anaerobic conditions [13]. In fact, Cu^{2+} ions are able (i) to accept electrons from the conduction band (CB) of TiO_2, since the redox potential of the Cu^{2+}/Cu^+ couple is more positive than the CB edge of TiO_2; and (ii) to accept photoexcited electrons directly from the VB of TiO_2, also under visible light irradiation. Both these electron transfer paths contribute to increase the separation of photoproduced charge carriers, with a consequent photoactivity improvement [10].

To further increase the photoactivity of TiO_2, modification of the oxide surface by non-noble metals, such as copper or nickel, has been coupled with noble metal (i.e., Au, Pt, Ag) NPs deposition, attaining remarkable improvement in terms of photocatalytic hydrogen production rates and efficiency in solar light exploitation [14–18]. In this context, synergistic effects in photoactivity have been demonstrated in the case of copper-platinum co-modified TiO_2 when small amounts of Cu were deposited together with Pt NPs, under both aerobic [19] and anaerobic conditions [20,21]. A strong synergistic effect between copper and platinum NPs deposited over TiO_2 has been recently demonstrated in hydrogen production by methanol photo-steam reforming by our research group [22]. The NPs of the two metals were deposited on commercial titanium dioxide in subsequent steps, i.e., Cu(II) was pre-grafted on the oxide surface, followed by Pt NP deposition and chemical reduction.

In the present work, we investigate another, potentially less time consuming, synthetic route to produce Cu and Cu-Pt co-modified TiO_2 photocatalysts in one step, i.e., flame spray pyrolysis (FSP), and report on the structural characterization of the so obtained materials in relation to their photoactivity in methanol photo-steam reforming. The FSP technique proved to be an effective method to synthesize TiO_2 photocatalyst powders containing noble metal NPs with high anatase content and crystallinity, high surface area and excellent metal dispersion [23–26], which are particularly suitable for photocatalytic hydrogen production from water solutions. In addition, three photocatalysts obtained either by grafting copper and/or by depositing platinum by wet methods on FSP-made TiO_2 were also investigated for comparison.

2. Results and Discussion

2.1. Photocatalyst Characterization

2.1.1. XRPD and BET Analyses

As shown in Figure 1, the X-ray powder diffraction (XRPD) pattern of pure titanium dioxide FP-T showed a biphasic crystalline composition (89% anatase, 11% rutile) with no evidence of brookite reflections; the mean crystallite size of the anatase phase was 14 nm (Table 1). The XRPD analysis of both FP-(X)Cu-T and FP-(X)Cu/Pt-T series indicate that the anatase crystallite mean size appear to increase with increasing copper content, as well as the rutile/anatase ratio, a phenomenon which has already been observed in previous studies [27]. Nevertheless, no reflections due to the metals or to copper oxides were detected, suggesting that they are finely dispersed in small NPs.

As reported in Table 1, the specific surface area (SSA) of the FP-(X)Cu-T series was only slightly higher than that of the FSP-made pure TiO_2 ($110 \, m^2 \, g^{-1}$), while the whole FP-(X)Cu/Pt-T series showed a somewhat higher SSA with respect to the other samples, with the FP-(0.05)Cu/Pt-T photocatalyst possessing the highest SSA, i.e., $153 \, m^2 \, g^{-1}$.

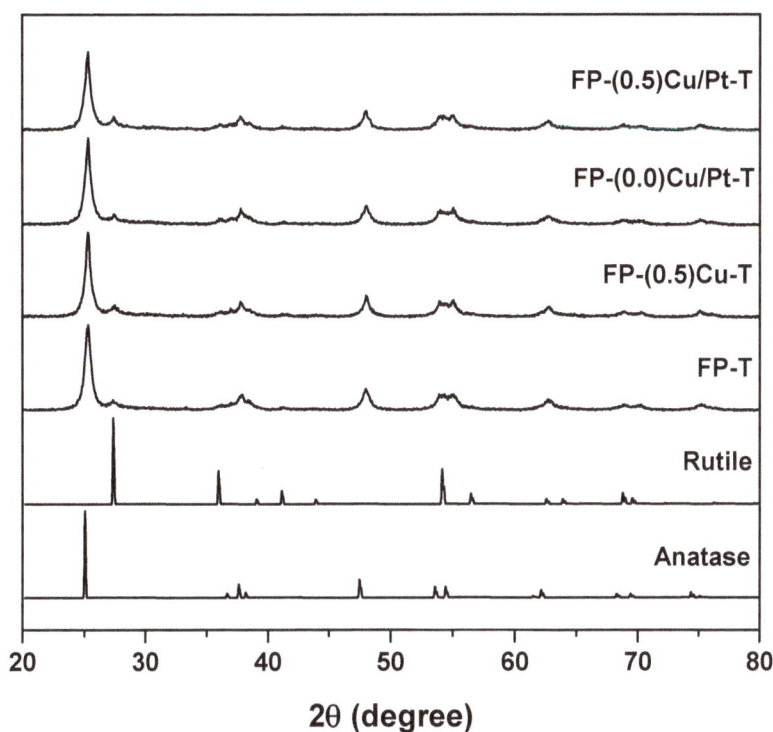

Figure 1. XRPD pattern of selected photocatalyst samples, with standard reference patterns of the anatase and rutile phases.

Table 1. Crystal phase composition, average anatase particles diameter d_A, and specific surface area SSA of the FSP-made photocatalysts.

Photocatalyst	Anatase (%)	Rutile (%)	d_A (nm)	SSA ($m^2 g^{-1}$)
FP-T	88.6	11.4	13.9	110
FP-(0.05)Cu-T	83.2	16.8	13.8	116
FP-(0.1)Cu-T	91.8	8.2	14.3	119
FP-(0.5)Cu-T	78.9	21.1	17.1	115
FP-(0.0)Cu/Pt-T	87.3	12.7	14.0	131
FP-(0.05)Cu/Pt-T	91.9	8.1	12.8	153
FP-(0.1)Cu/Pt-T	87.7	12.3	13.4	130
FP-(0.2)Cu/Pt-T	82.4	17.6	14.9	127
FP-(0.3)Cu/Pt-T	83.5	16.5	14.9	129
FP-(0.5)Cu/Pt-T	82.0	18.0	15.2	121

2.1.2. UV-VIS Absorption Properties

The ultraviolet-visible (UV-VIS) absorption spectra of the FP-(X)Cu-T series together with that of bare FP-T are collected in Figure 2a. First of all, reference FP-T sample shows an important absorption tail in the whole visible light range originated from the carbonaceous impurities typical of FSP-made samples [26,28,29]. In the presence of copper the absorption of the materials increases, all FP-(X)Cu-T samples showing an extra absorption contribution in the 400–500 nm region, to be ascribed to the direct interfacial charge transfer (IFCT) of electrons from the VB of TiO_2 to Cu(II) surface species [12]. In addition, specific Cu(II) d-d transitions, evidenced by the absorption in the 700–800 nm region in the spectrum of FP-(0.5)Cu-T (see Figure 2b), confirm that the here employed single-step FSP synthesis of Cu-containing TiO$_2$ stabilizes surface Cu$_x$O species, with copper in an oxidized state.

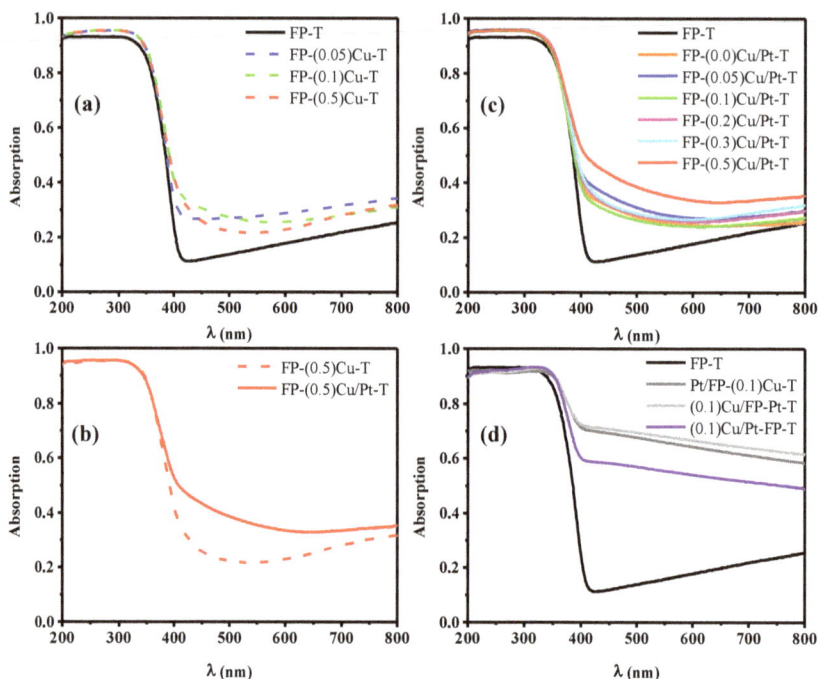

Figure 2. UV-VIS absorption spectra of (**a**) the FP-(X)Cu-T series; (**b**) FP-(0.5)Cu-T in comparison with FP-(0.5)Cu/Pt-T; (**c**) the FP-(X)Cu/Pt-T series and (**d**) the hybrid Pt and Cu co-modified TiO$_2$ samples prepared by combining FSP with Cu grafting and/or Pt NP deposition through the DP route.

All Pt-containing TiO$_2$ samples showed enhanced absorption with respect to the corresponding FP-(X)Cu-T photocatalysts (Figure 2c vs. Figure 2a), as clearly evidenced in Figure 2b, where the absorption spectrum of FP-(0.5)Cu/Pt-T is compared with that of FP-(0.5)Cu-T. However, the materials of the FP-(X)Cu/Pt-T series (characterized by a light grey color) showed a much lower absorption with respect to the Pt-Cu/TiO$_2$ "hybrid" samples prepared by combining the FSP technique with alternative TiO$_2$ surface modification routes implying post-deposition metal reduction (Figure 2c vs. Figure 2d). This indirectly confirms that, in FSP-made Pt/TiO$_2$ samples, platinum is also mostly present in oxidized, rather than in metallic, form [30] and that post-deposition chemical reduction may promote the reduction of both metal co-catalysts into metallic NPs (Figure 2d). Importantly, hybrid materials with identical co-catalyst content showed the same optical absorption profiles, independently of their preparation sequence (see Pt/FP-(0.1)Cu-T and (0.1)Cu/FP-Pt-T in Figure 2d). Compared to such materials, (0.1)Cu/Pt-FP-T, prepared by directly contacting the two metal precursors with the FP-T powder in subsequent steps, absorbs less light in the visible range and shows a UV-VIS absorption spectrum comparable to that of the photocatalyst obtained by applying exactly the same two-step deposition procedure to commercial P25 [22].

2.1.3. XPS Analysis

X-Ray photoelectron spectroscopy (XPS) analysis (see for example Figure 3 and Table 2) confirms the presence of ca. 20 at% of carbon in FSP-made materials. The C 1s signal exhibits a band peaking at ca. 284.8 eV, which can be attributed to organic carbon, and a second peak at ca. 288 eV, ascribable to carbonaceous traces. The O 1s signal consists of a main peak at ca. 530.3 eV, originating from oxygen linked to titanium (Ti–O bonds) and a minor peak at ca. 532.5 eV compatible with oxygen in carbonate species, CO, CO_2, and in physisorbed water. The Ti 2p doublet signal is almost identical for all samples (Figure 3), with the main peak at ca. 458.8 eV and the second one at 464.5 eV, both typical of Ti^{4+} in TiO_2 [31]. The absence of shoulders at lower energy points to a negligible contribution of sub-stoichiometric titanium dioxide (TiO_{2-x}) or of Ti–OH surface groups [32–34]. This is also confirmed by the O/Ti ratio greater than 2 (Table 2). No signals originated from platinum and copper photoemission could be detected, their intensity possibly being below the detection limit.

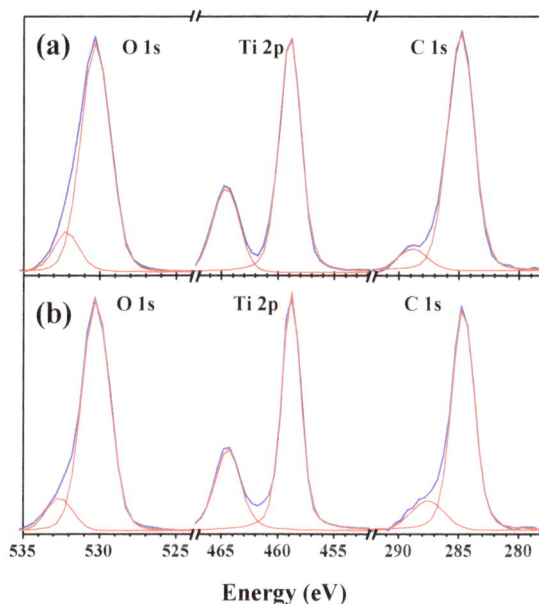

Figure 3. XPS spectra of the O 1s, Ti 2p, and C 1s regions for (**a**) FP-(0.0)Cu/Pt-T and (**b**) FP-(0.5)Cu/Pt-T.

Table 2. Results of the XPS analysis for 3 selected FP-(X)Cu/Pt-T photocatalysts.

Photocatalyst	Concentration (at%)			O/Ti Ratio
	O 1s	Ti 2p	C 1s	
FP-(0.0)Cu/Pt-T	55.4	23.5	19.7	2.36
FP-(0.05)Cu/Pt-T	54.6	23.1	21.0	2.36
FP-(0.5)Cu/Pt-T	53.5	22.2	23.4	2.41

2.1.4. HRTEM Analysis

Transmission electron microscopy (TEM) investigation of FP-(0.0)Cu/Pt-T (Figure 4a) and of FP-(0.5)Cu/Pt-T confirms the typical morphology of the flame made powder consisting of micro-aggregates of spherical nanocrystals. The TiO_2 particle size distribution obtained by counting 170 NPs (Figure 4b) is in the 4–28 nm range, with an average value of 11 nm and a standard deviation of 5 nm.

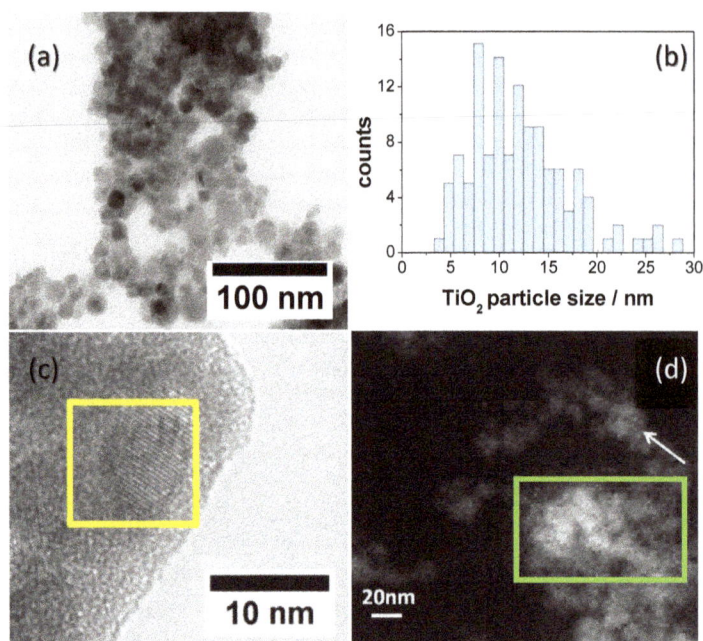

Figure 4. (**a**) TEM; (**b**) TiO_2 particle size distribution; (**c**) HRTEM; and (**d**) STEM-HAADF investigation of FP-(0.0)Cu/Pt-T. The green rectangle in (**d**) shows the acquisition area of the EDX spectrum. The white arrow points a surface Pt NP appearing as a bright dot due to the Z-contrast.

The high resolution TEM (HRTEM) image shown in Figure 4c confirms that the TiO_2 NPs are monocrystalline and that their crystal structure corresponds to the anatase phase. Indeed, the plane distance calculated by FFT analysis of the nanocrystals appearing within the yellow frame in Figure 4c is 3.5 Å, corresponding to the d-spacing of the [101] plane of anatase.

Pt NPs can be revealed by scanning transmission electron spectroscopy—high angular annular dark field (STEM-HAADF) analysis as bright dots on the TiO_2 surface because of its higher Z-contrast compared to the lighter Ti and O elements. However, Pt NPs can hardly be distinguished in Figure 4d (see for example the point indicated by the white arrow), in line with the fact that Pt is finely and homogeneously dispersed on the TiO_2 surface in the form of approximately 1 nm sized NPs. The presence of Pt was confirmed by energy dispersive X-ray spectroscopy (EDX) analysis, giving a 0.4 wt% of Pt for FP-(0.0)Cu/Pt-T, in good agreement with its nominal 0.5 wt% content. Similarly, EDX analysis of the FP-(0.5)Cu/Pt-T (dispersed on a molybdenum grid to avoid artefact on the Cu signal) confirmed the 0.5 wt% Cu content and its homogeneous dispersion on TiO_2.

2.2. Photocatalytic Activity

In the photocatalytic steam reforming of methanol the alcohol acts as an efficient hole scavenger, thus decreasing the electron-hole recombination rate and making conduction band electrons more readily available for H^+ reduction. Hydrogen production is, thus, coupled with methanol oxidation up to CO_2. Several intermediates, such as carbon monoxide, formic acid, or formaldehyde, are produced together with other side products, such as methane or ethane. H_2, CO_2, and CO accumulate at a constant rate in the closed recirculation system during the photocatalytic tests, according to pseudo-zero order kinetics.

The results obtained in methanol photosteam reforming photocatalytic tests are reported in Figure 5, in terms of hydrogen production rate, r_{H2}, and selectivity to CO_2 and CO, S_{CO2} and S_{CO}, as in previous studies [35]. We note, first of all, that the higher rate of hydrogen production r_{H2} obtained with FP-T with respect to P25 TiO_2 (3.6 vs. 2.7 mmol h^{-1} g$_{cat}$$^{-1}$) can be ascribed to the higher surface area and to the larger anatase content. On the other hand, these two reference samples behave in the same way regarding side-product formation, with similar selectivities towards carbon dioxide and monoxide (Figure 5). The r_{H2} values obtained with the photocatalysts of the FP-(X)Cu-T series were all around 7 mmol h^{-1} g$_{cat}$$^{-1}$, more than twice of those obtained with the bare materials, with a slightly decreasing rate with increasing copper loading. In addition to the beneficial effect on H_2 production rate, the presence of Cu species on the TiO_2 surface also influences the selectivity to CO in methanol photosteam reforming, the higher the amount of this metal, the lower being the selectivity to carbon monoxide. Thus, in the presence of copper as co-catalyst, preferential complete oxidation of methanol to carbon dioxide occurs, rather than to carbon monoxide, with a consequent higher rate of hydrogen production. In fact, full oxidation of one methanol molecule produces three H_2 molecules, while incomplete methanol oxidation to carbon monoxide produces two molecules of hydrogen per methanol molecule [35].

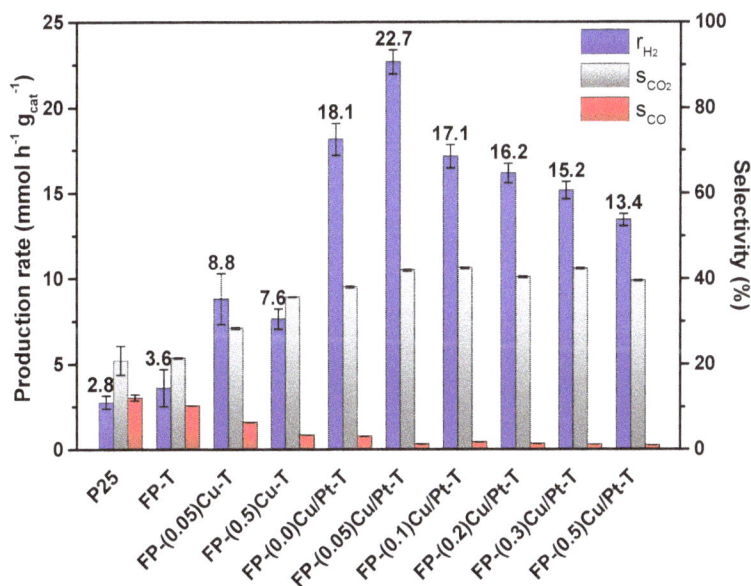

Figure 5. H_2 production rate (left ordinate) and selectivity to CO_2 and CO (right ordinate) obtained with Cu-containing TiO_2 and Cu/Pt-containing TiO_2 photocatalysts prepared by FSP in one step.

With respect to the Cu/TiO_2 photocatalysts obtained by grafting Cu on P25 TiO_2 [22] the photocatalysts of the FP-(X)Cu-T series showed a lower selectivity towards CO and a higher rate of hydrogen production. Considering two photocatalysts with the same 0.1 wt% Cu nominal content, the hydrogen production rate obtained with the FSP-made one is almost double (6.9 vs. 3.8 mmol h^{-1} g$_{cat}$$^{-1}$), with a halved selectivity to CO (6.1% vs. 10.8%). This might be a consequence of the formation of small NPs of crystalline copper oxides during the FSP synthesis, acting as semiconductors and thus forming a heterojunction with TiO_2, which improves the separation of the photoproduced charge couples [36]. In Cu/TiO_2 photocatalysts produced by the grafting technique the oxidized copper species on the TiO_2 surface are expected to be in amorphous form, as grafting is carried out at room temperature and, thus, their action mechanism, consisting of switching between the Cu^{2+} and Cu^{+} oxidation states, may be different [12].

The presence of Pt on the FP-(X)Cu/Pt-T photocatalysts led to much higher photoactivity in terms of hydrogen production rate with respect to pure TiO_2 and to a ca. doubled photoactivity with respect to the FP-(X)Cu-T series. Platinum, due to its high work function (5.93 eV for the 111 crystal plane) [37], is a well-known, very efficient co-catalyst of TiO_2, particularly contributing to increased H^+ photocatalytic reduction leading to molecular hydrogen evolution [21,31,35,38,39]. As shown in Figure 5, the rate of photocatalytic hydrogen production obtained with the FSP-made Cu-Pt co-modified photocatalysts showed a bell-shaped trend, with the best performance in terms of r_{H2} (22.7 mmol h^{-1} g_{cat}^{-1}) being achieved with FP-(0.05)Cu/Pt-T, followed by the Pt-only containing sample FP-(0.0)Cu/Pt-T. Further increase of the copper content of the FSP-made photocatalysts had instead detrimental effects on the hydrogen production rate, with r_{H2} values lower than that obtained with copper-free FP-(0.0)Cu/Pt-T. The formation of an alloy between Cu and Pt during the FSP synthesis with the resulting decrease in the total work function of metal NPs and consequent decreased efficiency in H^+ reduction in comparison to pure platinum NPs, may be at the origin of such a behavior [35]. At the same time, the increase of nominal Cu content employed during the FSP synthesis may promote the formation of larger copper oxide domains, implying reduced Cu-TiO_2 interactions, which possibly determine a decrease of photocatalytic H_2 evolution [40].

Anyway, with increasing copper content in the FSP-made Cu-Pt/TiO_2 photocatalysts the selectivity towards carbon dioxide reached values up to 42% in the case of Cu- and Pt-containing photocatalysts, while the selectivity towards CO was significantly reduced. In fact, S_{CO} dropped from 3.1% and 3.4% for FP-(0.0)Cu/Pt-T and FP-(0.5)Cu-T, respectively, to 1.0% for FP-(0.5)Cu/Pt-T.

Concerning the photoactivity achieved by the "hybrid" samples, as shown in Figure 6, a limited increase in r_{H2} up to 10.6 mmol h^{-1} g_{cat}^{-1} was attained upon platinum deposition by the DP method (see Pt/FP-(0.1)Cu-T vs. FP-(0.1)Cu-T), as a consequence of the above mentioned positive role played by platinum NPs in favoring electron-hole separation. On the other hand, Pt/FP-(0.1)Cu-T exhibits a photoactivity very similar to that obtained with (0.1)Cu/FP-Pt-T, in terms of both hydrogen production rate and selectivity to by-products. The obtained r_{H2} value is lower than that attained with the corresponding unmodified photocatalyst, i.e., FP-(0.0)Cu/Pt-T, and points to a negative effect of copper grafting on the photoactivity of Pt-containing FSP-made TiO_2. Surprisingly, (0.1)Cu/Pt-FP-T, obtained by Cu grafting followed by Pt deposition using the DP method on flame-made bare TiO_2, exhibits a hydrogen production rate much higher than that of the other two "hybrid" samples, containing the same nominal amount of metals.

Differently from the results obtained in our previous work on Cu-Pt modified TiO_2 photocatalysts prepared starting from commercial TiO_2 [22], with the presently investigated FSP-made Cu and Pt-containing TiO_2 photocatalysts no synergistic effect between the two metal co-catalysts was observed. In fact, with none of them was a r_{H2} value was attained greater than the sum of those observed with the corresponding single metal (Pt or Cu)-modified TiO_2 photocatalyst. Nevertheless, a limited improvement in photoactivity was observed for very low copper content, i.e., in the case of FP-(0.05)Cu/Pt-T.

Thus, FSP proves to be an effective way to synthesize single metal-containing TiO_2-based photocatalysts, since both Cu-only and Pt-only containing FSP-made TiO_2 samples showed very high photoactivity in hydrogen production, performing better than TiO_2-based photocatalysts with analogous composition produced through wet-phase techniques, such as grafting of Cu or Pt deposition through the DP method.

Importantly, the beneficial effects induced in H_2 production by combining Cu(II) grafting with Pt NPs deposition on the TiO_2 surface by means of the DP procedure has been confirmed also in the case of bare FSP-made TiO_2, i.e., not only for commercial P25 [22]. In fact, though implying a more time-consuming procedure, the mild modification conditions ensured by these wet-phase techniques may avoid the undesired formation of a Cu-Pt alloy, with the consequent stabilization of

Cu nanoclusters, able to promote the transfer of photoexcited electrons from TiO_2 towards Pt NPs, where H_2 evolution occurs [22].

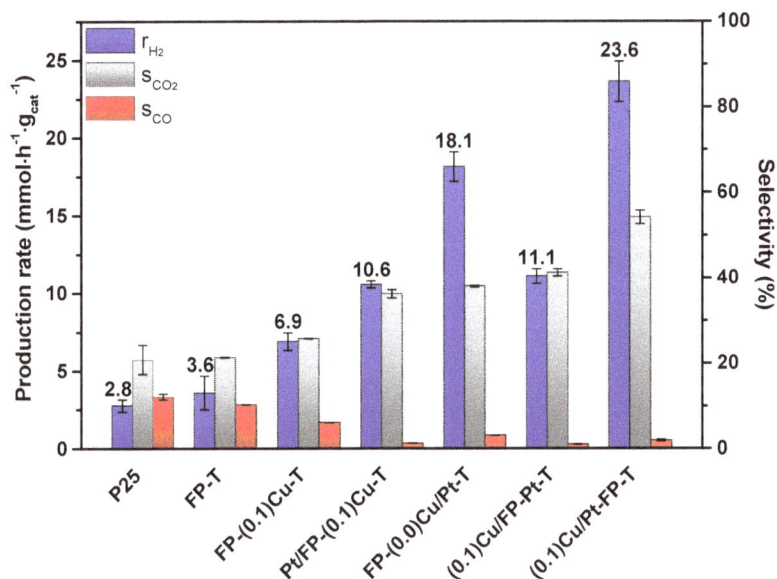

Figure 6. H_2 production rate (left ordinate) and selectivity to CO_2 and CO (right ordinate) obtained with photocatalysts prepared by combining different preparation techniques (see text).

3. Materials and Methods

3.1. Synthesis of the Photocatalysts

Except for commercial P25 TiO_2 from Degussa (Evonik, Essen, Germany), all investigated photocatalysts were home-prepared by FSP in a single step [41], employing a commercial FSP system (NPS10 Tethis S.p.A., Milano, Italy). The precursor solutions to be burned were prepared by mixing 25 mL of a 1.2 M titanium(IV)-tetraisopropoxide (TTIP) solution in xylene with a fixed volume (i.e., 5 mL) of a Pt-containing mother solution and/or variable volumes of a Cu-containing solution. This latter was prepared by dissolving 0.381 g of copper acetate monohydrate into 100 mL of propanoic acid. The Pt-containing mother solution was prepared by dissolving 0.135 g of hexachloroplatinic acid (30 wt% actual Pt content) into 50 mL of propanoic acid.

In order to maintain the combustion enthalpy constant in all synthesis, the xylene to propanoic acid volume ratio was, thus, kept constant (7:4) by diluting all solutions with 10 mL of xylene and with the required volume of propanoic acid, up to a 50 mL constant final volume.

The so-obtained solutions were injected into the burner at 4 mL min^{-1} by means of a syringe pump through a capillary tube and dispersed by pure oxygen (5.0 L min^{-1}). The spray was ignited by a methane/oxygen flamelet ring surrounding the nozzle. The methane and oxygen flow rates were 1.0 L min^{-1} and 2.0 L min^{-1}, respectively, and the pressure drop across the nozzle was kept constant at 2 bar. The powders were collected on glass fiber filters (Whatman, Maidstone, UK, model GF6, 257 mm in diameter) positioned 64 cm over the burner, on top of a steel vessel connected to a vacuum pump (Seco SV 1040C by Busch, Magden, Switzerland).

Two photocatalyst series were synthesized: the first series, labeled FP-(X)Cu-T, consisted of copper containing titanium dioxide powders; the second series, labeled FP-(X)Cu/Pt-T, consisted of platinum and copper containing powders. X corresponds to the Cu/Ti nominal weight percent ratio, ranging from 0 to 0.5, while the Pt/Ti nominal weight percent ratio in the second series was fixed at 0.5. Bare titanium dioxide containing no Cu and Pt NPs was also produced by FSP and named FP-T.

Three additional samples were prepared by combining different metal NPs deposition techniques. Sample (0.1)Cu/Pt-FP-T was obtained from FP-T, by grafting 0.1 wt% of copper [42] and 0.5 wt% of platinum by the deposition-precipitation (DP) method [43], in two consecutive steps. Sample (0.1)Cu/FP-Pt-T was obtained by grafting Cu on the surface of FP-(0.0)Cu/Pt-T, followed by the addition of an aqueous NaBH$_4$ solution in slight excess. Sample Pt-FP-(0.1)Cu-T was obtained by DP of Pt NPs on FP-(0.1)Cu-T. Briefly, the grafting method [42] consists in drying under vigorous stirring and heating at 90 °C an aqueous suspension containing (Cu(NO$_3$)$_2$·3H$_2$O) and the dispersed TiO$_2$ powder. The DP technique [43] consists of stirring a heated aqueous suspension containing the starting photocatalyst and H$_2$PtCl$_6$ in the presence of urea to induce the precipitation of Pt NPs, followed by reduction of the so obtained powder with NaBH$_4$ in an aqueous dispersion.

All chemicals were purchased from Sigma-Aldrich (St. Louis, MO, USA) and used as received.

3.2. Characterization of the Photocatalysts

XRPD analyses were performed using a Philips PW3020 powder diffractometer (PANalytical, Almelo, The Netherlands), operating at 40 kV and 40 mA and exploiting copper Kα radiation (λ = 1.54056 Å) as X-ray source. The diffractograms were recorded by scanning between 5° and 80° 2θ angles, with a 0.05° step. Phase quantitative analysis was made by the Rietveld refinement method [44], using Quanto software (Ver. 1.0, free licence software) [45]; the mean anatase crystallite size was calculated by applying the Scherrer equation [46], from the width of the most intense reflection at 2θ = 25.4°.

The BET specific surface area (SSA) was measured by N$_2$ adsorption/desorption at liquid nitrogen temperature in a Micromeritics ASAP 2010 (Micromeritics, Norcross, GA, USA) apparatus after out-gassing in vacuo at 150 °C for 2 h. UV-VIS diffuse reflectance (R) analysis was performed with a Jasco V-670 spectrophotometer (Jasco, Easton, MD, USA) equipped with a PIN-757 integrating sphere, using barium sulfate as a reference. The results are presented as absorption (A) spectra (A = 1 − R).

XPS data were collected by a PHI-5500 Physical Electronics spectrometer (Physical Electronics, Chanhassen, MN, USA) equipped with an aluminum anode (Kα = 1486.6 eV) as the monochromatized source, operating at 200 W applied power, with a 58.7 eV pass energy, 0.5 eV energy step, and a 0.15 s step time. The vacuum level during the analyses was ca. 10^{-9} Torr and a neutralizer was used in order to avoid surface electrostatic charge accumulation on the nonconductive samples.

HRTEM analysis was carried out with a Zeiss LIBRA 200FE transmission electron microscope (Zeiss, Oberkochen, Germany), equipped with STEM—HAADF and EDX (Oxford X-Stream 2 and INCA software). The microscope has a 200 kV field emission gun-like source with an in-column second-generation omega filter for energy-selective spectroscopy. The sample was dispersed in isopropanol and then a drop of the suspension was deposited on a 300 mesh holey carbon copper or molybdenum grid.

3.3. Photocatalytic Tests

The photocatalytic activity in hydrogen production from methanol photo-steam reforming was tested in the already described stainless steel closed system [5], which was modified by substituting the Plexiglas photoreactor with a new stainless-steel photoreactor, with a front round hollow (4 mm thick and 63.5 mm in diameter), closed by a Pyrex glass window. The temperature of this new cell can be increased by means of four heating cartridges regulated by a thermocouple. The photocatalytic bed, placed in the front hollow of the photoreactor, was prepared by mixing 15 ± 2 mg of photocatalyst and 7.10 ± 0.05 g of 20–40 mesh (0.42–0.85 mm) quartz beads with few droplets of distilled water, followed by drying in an oven at 70 °C. Prior to any run, the whole system was purged in the dark with pure nitrogen at 110 mL min^{-1} for 40 min in order to remove any trace of oxygen. The temperature of the photoreactor was fixed at 40 °C and set at this value 30 min before the beginning of the experiment, with a heating rate of 10 °C min^{-1}. Then the gas phase, saturated with methanol and water vapors by bubbling nitrogen into a 20 vol% methanol aqueous solution kept at 30 °C, was continuously recirculated at a constant rate (60 mL min^{-1}) by means of a bellow pump for 15 min before starting the test. During the photocatalytic run, the gas flow was also set at 60 mL min^{-1}. The absolute pressure inside the system, initially 1.2 bar, gradually increased during the run due to the accumulation of gas products. The light source, always switched on 30 min before the beginning of the run and placed at ca. 20 cm from the photoreactor, was a 300 W xenon arc lamp (LSH302, LOT-Oriel, Darmstadt, Germany), emitting in the 350–400 nm range. The light intensity on the photocatalyst was 0.31 W cm^{-2}, as measured with an optical power meter (model PM200 by Thorlabs, Newton, NJ, USA) equipped with a thermal power sensor (Thorlabs S302C).

The gas-phase composition was analyzed on-line during irradiation by means of a gas chromatograph (GC, Agilent 6890N, by Agilent, Santa Clara, CA, USA) equipped with two capillary columns (HP-PlotU and Molesieve 5A), two detectors (thermo conductivity and flame ionization) and a Ni-catalyst system for CO_2 and CO methanation. The instrument was preliminary calibrated for H_2, CO_2, CO, CH_4, and H_2CO analyses. The amount of formic acid produced during the photoreaction and accumulated in the aqueous solution was determined by ion chromatography at the end of the run, employing a Metrohm 761 Compact IC instrument (by Metrohm AG, Herisau, Switzerland), equipped with an anionic Metrosep A column.

The results of photocatalytic tests are reported as H_2 production rate (r_{H2}), obtained as the slope of the straight lines of the produced hydrogen amount (normalized per unit catalyst weight) vs. the irradiation time plots. The selectivity in hydrogen production was calculated from the rates of CO_2 and CO formation (r_{CO2} and r_{CO}), as the ratio between these latter and the rate of H_2 production from methanol, by taking into account the stoichiometry of the CO_2 and CO formation reactions [35].

$$CH_3OH + H_2O \xrightarrow{hv,TiO_2} 3H_2 + CO_2$$

$$CH_3OH \xrightarrow{hv,TiO_2} 2H_2 + CO$$

$$S_{CO_2} = \frac{3r_{co_2}}{r_{H_2}}$$

$$S_{CO} = \frac{2r_{CO}}{r_{H_2}}$$

To ensure the reproducibility of the data, the photocatalytic tests were repeated at least twice with each sample, using the same photocatalytic bed; at the end of each run, the water/methanol solution in the flask was changed and the whole system was purged with N_2 in the dark for 30 min prior to start a new run.

4. Conclusions

In conclusion, FSP proves to be an effective way to synthesize highly-performing single metal-containing TiO$_2$-based photocatalysts for photocatalytic hydrogen production. However, the highest synergistic effect between the Cu and Pt co-catalysts was attained with photocatalysts prepared by wet-phase methods on bare FSP-made TiO$_2$, i.e., Cu(II) grafting followed by Pt NP deposition.

Copper overloading (up to 0.5 wt%) of FSP-made photocatalysts is detrimental for H_2 production rate, probably due to Cu-Pt alloying under the here employed harsh synthesis conditions. Nevertheless, the addition of very low amounts of copper (0.05 wt%) during the FSP synthesis of Pt/TiO$_2$ guarantees a ca. 20% improvement of the overall photocatalytic hydrogen production, together with a lower selectivity towards the less desired carbon monoxide side-product.

Acknowledgments: The collaboration of Laura Meda, Istituto ENI Donegani, Novara, in XPS analysis is gratefully acknowledged. This work received financial support from the Italian MIUR, through the PRIN 2015 SMARTNESS

(2015K7FZLH) project. The use of instrumentation purchased through the Regione Lombardia—Fondazione Cariplo joint SmartMatLab project (Fondazione Cariplo 2013-1766 project) is gratefully acknowledged.

Author Contributions: G.L.C., M.V.D., and E.S. conceived and designed the experiments and contributed in writing the manuscript; M.B. performed all experiments and data analyses, wrote a draft of the paper, and contributed to its revision; L.G.B. contributed to the FSP synthesis of the materials; A.M.F. performed and interpreted electron microscopy analyses; and M.V.D. and E.S. provided the final revision of the manuscript.

Conflicts of Interest: The authors declare no conflict of interest.

References

1. Cherry, R.S. A hydrogen utopia? *Int. J. Hydrogen Energy* **2004**, *29*, 125–129. [CrossRef]
2. Chaubey, R.; Sahu, S.; James, O.O.; Maity, S. A review on development of industrial processes and emerging techniques for production of hydrogen from renewable and sustainable sources. *Renew. Sustain. Energy Rev.* **2013**, *23*, 443–462. [CrossRef]
3. Rahimi, N.; Pax, R.A.; Gray, E.M.A. Review of functional titanium oxides. I: TiO$_2$ and its modifications. *Prog. Solid State Chem.* **2016**, *44*, 86–105. [CrossRef]
4. Ma, Y.; Wang, X.L.; Jia, Y.S.; Chen, X.B.; Han, H.X.; Li, C. Titanium Dioxide-Based Nanomaterials for Photocatalytic Fuel Generations. *Chem. Rev.* **2014**, *114*, 9987–10043. [CrossRef] [PubMed]
5. Chiarello, G.L.; Forni, L.; Selli, E. Photocatalytic hydrogen production by liquid- and gas-phase reforming of CH$_3$OH over flame-made TiO$_2$ and Au/TiO$_2$. *Catal. Today* **2009**, *144*, 69–74. [CrossRef]
6. Gomathisankar, P.; Noda, T.; Katsumata, H.; Suzuki, T.; Kaneco, S. Enhanced hydrogen production from aqueous methanol solution using TiO$_2$/Cu as photocatalysts. *Front. Chem. Sci. Eng.* **2014**, *8*, 197–202. [CrossRef]
7. Pérez-Larios, A.; Hernández-Gordillo, A.; Morales-Mendoza, G.; Lartundo-Rojas, L.; Mantilla, A.; Gomez, R. Enhancing the H$_2$ evolution from water-methanol solution using Mn^{2+}-Mn^{+3}-Mn^{4+} redox species of Mn-doped TiO$_2$ sol-gel photocatalysts. *Catal. Today* **2016**, *266*, 9–16. [CrossRef]
8. Wang, Q.; An, N.; Bai, Y.; Hang, H.; Li, J.; Lu, X.; Liu, Y.; Wang, F.; Li, Z.; Lei, Z. High photocatalytic hydrogen production from methanol aqueous solution using the photocatalysts CuS/TiO$_2$. *Int. J. Hydrogen Energy* **2013**, *38*, 10739–10745. [CrossRef]
9. Kawai, T.; Sakata, T. Conversion of carbohydrate into hydrogen fuel by a photocatalytic process. *Nature* **1980**, *286*, 474–476. [CrossRef]
10. Choi, H.J.; Kang, M. Hydrogen production from methanol/water decomposition in a liquid photosystem using the anatase structure of Cu loaded TiO$_2$. *Int. J. Hydrogen Energy* **2007**, *32*, 3841–3848. [CrossRef]
11. Chiarello, G.L.; Dozzi, M.V.; Selli, E. TiO$_2$-based materials for photocatalytic hydrogen production. *J. Energy Chem.* **2017**, *26*, 250–258. [CrossRef]
12. Irie, H.; Kamiya, K.; Shibanuma, T.; Miura, S.; Tryk, D.A.; Yokoyama, T.; Hashimoto, K. Visible light-sensitive Cu(II)-grafted TiO$_2$ photocatalysts: Activities and X-ray absorption fine structure analyses. *J. Phys. Chem. C* **2009**, *113*, 10761–10766. [CrossRef]
13. Liu, Y.; Wang, Z.; Huang, W. Influences of TiO$_2$ phase structures on the structures and photocatalytic hydrogen production of CuO$_x$/TiO$_2$ photocatalysts. *Appl. Surf. Sci.* **2016**, *389*, 760–767. [CrossRef]
14. Luna, A.L.; Novoseltceva, E.; Louarn, E.; Beaunier, P.; Kowalska, E.; Ohtani, B.; Valenzuela, M.A.; Remita, H.; Colbeau-Justin, C. Synergetic effect of Ni and Au nanoparticles synthesized on titania particles for efficient photocatalytic hydrogen production. *Appl. Catal. B Environ.* **2016**, *191*, 18–28. [CrossRef]
15. Barrios, C.E.; Albiter, E.; Gracia y Jimenez, J.M.; Tiznado, H.; Romo-Herrera, J.; Zanella, R. Photocatalytic hydrogen production over titania modified by gold—Metal (palladium, nickel and cobalt) catalysts. *Int. J. Hydrogen Energy* **2016**, *41*, 1–14. [CrossRef]
16. Kotesh Kumar, M.; Bhavani, K.; Naresh, G.; Srinivas, B.; Venugopal, A. Plasmonic resonance nature of Ag-Cu/TiO$_2$ photocatalyst under solar and artificial light: Synthesis, characterization and evaluation of H$_2$O splitting activity. *Appl. Catal. B Environ.* **2016**, *199*, 282–291. [CrossRef]

17. Nischk, M.; Mazierski, P.; Wei, Z.; Siuzdak, K.; Kouame, N.A.; Kowalska, E.; Remita, H.; Zaleska-Medynska, A. Enhanced photocatalytic, electrochemical and photoelectrochemical properties of TiO$_2$ nanotubes arrays modified with Cu, AgCu and Bi nanoparticles obtained via radiolytic reduction. *Appl. Surf. Sci.* **2016**, *387*, 89–102. [CrossRef] [PubMed]

18. Janczarek, M.; Wei, Z.; Endo, M.; Ohtani, B.; Kowalska, E. Silver-and copper-modified decahedral anatase titania particles as visible light-responsive plasmonic photocatalyst. *J. Photonics Energy* **2017**, *7*, 12008. [CrossRef]

19. Shiraishi, Y.; Sakamoto, H.; Sugano, Y.; Ichikawa, S.; Hirai, T. Pt-Cu bimetallic alloy nanoparticles supported on anatase TiO$_2$: Highly active catalysts for aerobic oxidation driven by visible light. *ACS Nano* **2013**, *7*, 9287–9297. [CrossRef] [PubMed]

20. Teng, F.; Chen, M.; Li, N.; Hua, X.; Wang, K.; Xu, T. Effect of TiO$_2$ Surface Structure on the Hydrogen Production Activity of the Pt@CuO/TiO$_2$ Photocatalysts for Water Splitting. *ChemCatChem* **2014**, *6*, 842–847. [CrossRef]

21. Jung, M.; Hart, J.N.; Boensch, D.; Scott, J.; Ng, Y.H.; Amal, R. Hydrogen evolution via glycerol photoreforming over Cu-Pt nanoalloys on TiO$_2$. *Appl. Catal. A Gen.* **2016**, *518*, 221–230. [CrossRef]

22. Dozzi, M.V.; Chiarello, G.L.; Pedroni, M.; Livraghi, S.; Giamello, E.; Selli, E. High Photocatalytic Hydrogen Production on Cu(II) Pre-grafted Pt/TiO$_2$. *Appl. Catal. B Environ.* **2017**, *209*, 417–428. [CrossRef]

23. Teoh, W.Y.; Mädler, L.; Beydoun, D.; Pratsinis, S.E.; Amal, R. Direct (one-step) synthesis of TiO$_2$ and Pt/TiO$_2$ nanoparticles for photocatalytic mineralisation of sucrose. *Chem. Eng. Sci.* **2005**, *60*, 5852–5861. [CrossRef]

24. Strobel, R.; Baiker, A.; Pratsinis, S.E. Aerosol flame synthesis of catalysts. *Adv. Powder Technol.* **2006**, *17*, 457–480. [CrossRef]

25. Pratsinis, S.E.; Vemury, S. Particle formation in gases: A review. *POWDER Technol.* **1996**, *88*, 267–273. [CrossRef]

26. Chiarello, G.L.; Selli, E.; Forni, L. Photocatalytic hydrogen production over flame spray pyrolysis-synthesised TiO$_2$ and Au/TiO$_2$. *Appl. Catal. B Environ.* **2008**, *84*, 332–339. [CrossRef]

27. Teleki, A.; Bjelobrk, N.; Pratsinis, S.E. Flame-made Nb- and Cu-doped TiO$_2$ sensors for CO and ethanol. *Sens. Actuators B Chem.* **2008**, *130*, 449–457. [CrossRef]

28. Chiarello, G.L.; Rossetti, I.; Lopinto, P.; Migliavacca, G.; Forni, L. Preparation by flame spray pyrolysis of ABO$_{3\pm\delta}$ catalysts for the flameless combustion of methane. *Catal. Today* **2006**, *117*, 549–553. [CrossRef]

29. Chiarello, G.L.; Rossetti, I.; Forni, L.; Lopinto, P.; Migliavacca, G. Solvent nature effect in preparation of perovskites by flame pyrolysis. 2. Alcohols and alcohols + propionic acid mixtures. *Appl. Catal. B Environ.* **2007**, *72*, 227–232. [CrossRef]

30. Chiarello, G.L.; Dozzi, M.V.; Scavini, M.; Grunwaldt, J.-D.; Selli, E. One step flame-made fluorinated Pt/TiO$_2$ photocatalysts for hydrogen production. *Appl. Catal. B Environ.* **2014**, *160–161*, 144–151. [CrossRef]

31. Dozzi, M.V.; Zuliani, A.; Grigioni, I.; Chiarello, G.L.; Meda, L.; Selli, E. Photocatalytic activity of one step flame-made fluorine doped TiO$_2$. *Appl. Catal. A Gen.* **2016**, *521*, 220–226. [CrossRef]

32. Reyes-Garcia, E.A.; Sun, Y.; Reyes-Gil, K.R.; Raftery, D. Solid-state NMR and EPR analysis of carbon-doped titanium dioxide photocatalysts (TiO$_{2-x}$C$_x$). *Solid State Nucl. Magn. Reson.* **2009**, *35*, 74–81. [CrossRef] [PubMed]

33. Yang, J.; Bai, H.; Tan, X.; Lian, J. IR and XPS investigation of visible-light photocatalysis-Nitrogen-carbon-doped TiO$_2$ film. *Appl. Surf. Sci.* **2006**, *253*, 1988–1994. [CrossRef]

34. Caruso, T.; Lenardi, C.; Agostino, R.G.; Amati, M.; Bongiorno, G.; Mazza, T.; Policicchio, A.; Formoso, V.; Maccallini, E.; Colavita, E.; et al. Electronic structure of cluster assembled nanostructured TiO$_2$ by resonant photoemission at the Ti L$_{2,3}$ edge. *J. Chem. Phys.* **2008**, *128*, 94704. [CrossRef] [PubMed]

35. Chiarello, G.L.; Aguirre, M.H.; Selli, E. Hydrogen production by photocatalytic steam reforming of methanol on noble metal-modified TiO$_2$. *J. Catal.* **2010**, *273*, 182–190. [CrossRef]

36. DeMeo, D.; MacNaughton, S.; Sonkusale, S.; Vandervelde, T. Electrodeposited Copper Oxide and Zinc Oxide Core-Shell Nanowire Photovoltaic Cells. In *Nanowires—Implementations and Applications*; Hashim, A., Ed.; InTech: Rijeka, Croatia, 2011; pp. 141–156, ISBN 978-953-307-318-7.

37. Lide, D.R.R.; Haynes, W.M.M.; Baysinger, G.; Berger, L.I.; Roth, D.L.; Zwillinger, D.; Frenkel, M.; Goldberg, R.N. *CRC Handbook of Chemistry and Physics*; Internet, V., Ed.; CRC Press: Boca Raton, FL, USA, 2005.

38. Vorontsov, A.V.; Dubovitskaya, V.P. Selectivity of photocatalytic oxidation of gaseous ethanol over pure and modified TiO$_2$. *J. Catal.* **2004**, *221*, 102–109. [CrossRef]

39. Chiarello, G.L.; Paola, D.; Selli, E. Effect of titanium dioxide crystalline structure on the photocatalytic production of hydrogen. *Photochem. Photobiol. Sci.* **2011**, *10*, 355–360. [CrossRef] [PubMed]
40. Jung, M.; Scott, J.; Ng, Y.H.; Jiang, Y.; Amal, R. CuOx dispersion and reducibility on TiO2 and its impact on photocatalytic hydrogen evolution. *Int. J. Hydrogen Energy* **2014**, *39*, 12499–12506. [CrossRef]
41. Chiarello, G.L.; Rossetti, I.; Forni, L. Flame-spray pyrolysis preparation of perovskites for methane catalytic combustion. *J. Catal.* **2005**, *236*, 251–261. [CrossRef]
42. Morikawa, T.; Irokawa, Y.; Ohwaki, T. Enhanced photocatalytic activity of TiO$_{2-x}$N$_x$ loaded with copper ions under visible light irradiation. *Appl. Catal. A Gen.* **2006**, *314*, 123–127. [CrossRef]
43. Dozzi, M.V.; Prati, L.; Canton, P.; Selli, E. Effects of gold nanoparticles deposition on the photocatalytic activity of titanium dioxide under visible light. *Phys. Chem. Chem. Phys.* **2009**, *11*, 7171–7180. [CrossRef] [PubMed]
44. Rietveld, H.M. A profile refinement method for nuclear and magnetic structures. *J. Appl. Crystallogr.* **1969**, *2*, 65–71. [CrossRef]
45. Altomare, A.; Burla, M.C.; Giacovazzo, C.; Guagliardi, A.; Moliterni, A.G.G.; Polidori, G.; Rizzi, R. Quanto: A Rietveld program for quantitative phase analysis of polycrystalline mixtures. *J. Appl. Crystallogr.* **2001**, *34*, 392–397. [CrossRef]
46. Scherrer, P. Estimation of the size and internal structure of colloidal particles by means of Röntgen rays. *Göttinger Nachrichten Math. Phys.* **1918**, *2*, 98–100.

© 2017 by the authors. Licensee MDPI, Basel, Switzerland. This article is an open access article distributed under the terms and conditions of the Creative Commons Attribution (CC BY) license (http://creativecommons.org/licenses/by/4.0/).

catalysts

MDPI

Article

Sulfur-Doped TiO$_2$: Structure and Surface Properties

Sara Cravanzola *, Federico Cesano *, Fulvio Gaziano and Domenica Scarano

Department of Chemistry, NIS (Nanostructured Interfaces and Surfaces) Inter-Departmental Centre and INSTM Centro di Riferimento, University of Torino, Via P. Giuria, 7, 10125 Torino, Italy; fulvio.gaziano@edu.unito.it (F.G.); domenica.scarano@unito.it (D.S.)
* Correspondence: sara.cravanzola@unito.it (S.C.); federico.cesano@unito.it (F.C.);
  Tel.: +39-011-670-7834 (S.C. & F.C.)

Academic Editors: Vladimiro Dal Santo and Alberto Naldoni
Received: 29 May 2017; Accepted: 11 July 2017; Published: 18 July 2017

Abstract: A comprehensive study on the sulfur doping of TiO$_2$, by means of H$_2$S treatment at 673 K, has been performed in order to highlight the role of sulfur in affecting the properties of the system, as compared to the native TiO$_2$. The focus of this study is to find a relationship among the surface, structure, and morphology properties, by means of a detailed chemical and physical characterization of the samples. In particular, transmission electron microscopy images provide a simple tool to have a direct and immediate evidence of the effects of H$_2$S action on the TiO$_2$ particles structure and surface defects. Furthermore, from spectroscopy analyses, the peculiar surface, optical properties, and methylene blue photodegradation test of S-doped TiO$_2$ samples, as compared to pure TiO$_2$, have been investigated and explained by the effects caused by the exchange of S species with O species and by the surface defects induced by the strong H$_2$S treatment.

Keywords: TiO$_2$; S-doping; sulfidation; HRTEM; XRD; UV-visible; FTIR

1. Introduction

Titanium dioxide (TiO$_2$) is widely used for photocatalysis. It has attracted considerable attention because of its characteristics, including optical properties, reactivity and chemical stability, as well as its non-toxicity [1,2]. In particular, TiO$_2$-based photocatalysts have been used for significant applications, such as antibacterial actions [3], medical research [4], drug delivery [5], and self-cleaning fields [6]. Most of all, this material is widely used in the degradation of pollutants in air and water by the decomposition of organic compounds [7,8].

Despite its outstanding photocatalytic properties, TiO$_2$ is only able to absorb a small range of the UV portion of the solar spectrum [9], because of its relatively high band gap. To solve this problem, the most-used strategy is the engineering and shift of the TiO$_2$ band gap to the visible light region, in such a way to enhance its photocatalytic activity. In this regard, the surface modification obtained by anchoring selected species, such as MoS$_2$ or graphene-like systems [10,11], or by the incorporation of metal or non-metal dopants into the TiO$_2$ structure [12], allows one to harvest the visible spectrum or to increase the reactivity in the UV spectrum. It has been found that metals are able to induce a desired band gap shift, but also induces recombination centers, thus reducing the photocatalysis capability in combination with thermal instability [13].

On the other hand, the incorporation of non-metals, including nitrogen, carbon, sulfur, fluorine, or iodine [14–21], possibly as quantum dots [22], was found to be a more efficient way to lower the band gap of TiO$_2$, thus obtaining a photocatalyst with higher activity.

Indeed, sulfur-doped TiO$_2$ has attracted much attention due to the fact that increasing quantities of S can reduce the band gap [23], as well as show a strong absorption in the visible light [24].

Many strategies have been adopted to synthesize S-doped TiO$_2$ nanocatalysts, from the oxidative annealing of TiS$_2$, to catalyzed hydrolysis, hydrothermal and solvothermal synthesis, as well as sol-gel and co-precipitation methods [25–28]. In this regard, H_2S can be used as a precursor to obtain an S-doped TiO_2 surface [29].

According to some authors, the mechanism of H_2S adsorption on the TiO_2 surface is explained with dissociative pathways, causing S to fill the O vacancies to obtain S-doped surfaces [29].

However, the integration mechanism of S in TiO_2, from a structural point of view, is still under debate, being to a great extent affected by the synthesis conditions. It has been reported that sulfur could be adsorbed predominantly in the form of $SO_4{}^{2-}$ species at the surface of TiO_2 nanoparticles [30], or could be embedded within the TiO_2 lattice, thus creating S-Ti-O bonds [31].

Therefore, to better understand and describe the properties and then the possible applications of the so-obtained S-doped TiO_2, it is fundamental to investigate the nature of the H_2S interaction with the TiO_2 surface.

Moreover, a further aspect to take into account is the role of H_2S as a reactant in catalytic hydro-treatment and Claus reactions [32,33]. It is noteworthy that, nowadays, the emission limits of SO_x are becoming very rigorous, because air pollution has become a serious global problem. Oil and gas extraction sites are one of the main sources of H_2S emissions, and they are usually removed by means of the well-known Claus process. The reaction occurs via dissociative adsorption of H_2S on a metal oxide [34], mainly on Al_2O_3 but also on TiO_2, used as catalyst.

In both processes, H_2S was found to modify the surface properties of the metal oxide support catalyst. According to the literature, many theoretical studies concerning H_2S reaction and adsorption on TiO_2 are known, particularly those on anatase and rutile phases [35–39].

To our knowledge, only a few experimental studies have focused on the effects originated by H_2S dosage on TiO_2 surface, concerning the relationship among surface, structure, and morphology properties. Our study aims to contribute to these themes, as it reports a quite extensive chemical and physical characterization of the surface properties of S-doped TiO_2, obtained after H_2S treatment. The samples are investigated by X-ray diffraction (XRD), and high resolution transmission electron microscopy (HRTEM), in addition to Raman, Fourier Transform-Infrared (FTIR) and UV-visible (UV-Vis) spectroscopies. The obtained results are compared to those of pure TiO_2.

2. Results and Discussion

2.1. Structure and Morphology by XRD, Raman, and HRTEM Analyses

2.1.1. XRD Analysis

The XRD patterns of TiO_2 before and after H_2S dosage at 673 K for 1 h are shown in Figure 1 (black and red lines, respectively), together with the typical crystalline features of anatase (PDF card # 21-1272) and rutile (PDF card # 21-1276) phases, as highlighted by blue and green lines (anatase and rutile, respectively).

From the pattern of TiO_2 after the sulfidation procedure, no considerable modifications of the peculiar peaks of anatase and rutile are observed, thus remarking that the amount of S species inside the S-doped TiO_2 does not affect the lattice under the adopted preparation conditions. We shall return to this point by analyzing the amount of sulfur via elemental analysis, as illustrated in Figure S1.

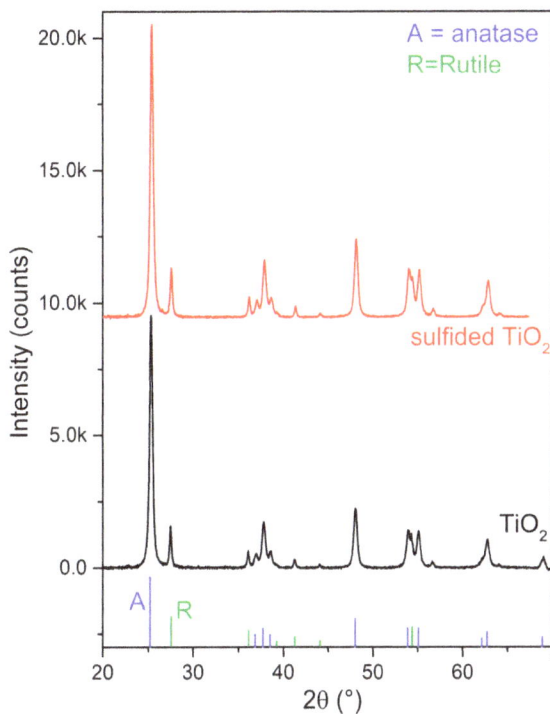

Figure 1. XRD patterns of TiO$_2$ (black line) and S-doped TiO$_2$ (red line). Anatase (PDF card # 21-1272) and rutile (PDF card # 21-1276) phases (blue and green lines, respectively) are shown for comparison.

2.1.2. Raman Spectroscopy

Figure 2 shows the Raman spectrum of S-doped TiO$_2$ compared with pure TiO$_2$, used as a reference material, both recorded with a 514 nm laser line.

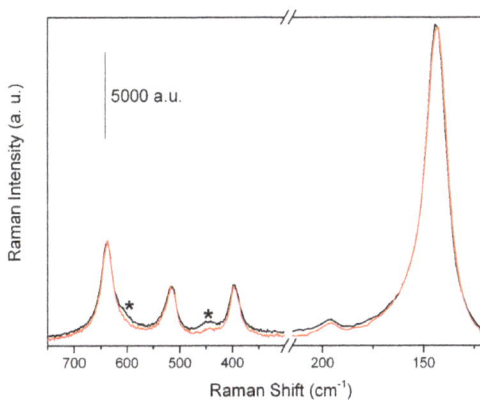

Figure 2. Raman spectra of TiO$_2$ and S-doped TiO$_2$ (black and red lines, respectively). Raman fingerprints of rutile are marked by asterisks.

In detail, concerning the spectrum of pure TiO_2 (black curve), the four bands at 144, 396, 514, and 636 cm^{-1} are ascribed to the E_g, B_{1g}, A_{1g}, E_g Raman active modes, respectively, of the anatase phase, as described in the literature [10,18,40]. Furthermore, the shoulder at 608 cm^{-1} and the small peak at 444 cm^{-1} (labelled by asterisks) are due to the A_{1g} and E_g modes of the rutile phase [40]. As for the S-doped TiO_2 sample, as obtained after the sulfidation step (red curve), the typical Raman fingerprints of TiO_2 are still present, even if a clear explanation of the erosion of the weak modes labelled by asterisks remains unclear [29,35].

2.1.3. Surface Area and HRTEM Analysis

Figure 3a,b depict HRTEM images of nanoparticles of ~15–50 nm in size, exposing lattice fringes spaced ~0.32 nm or 0.35 nm, corresponding to (110) and (101) planes of rutile and anatase, respectively, as confirmed by the fast Fourier transform (FFT) investigations (insets of Figure 3b,d).

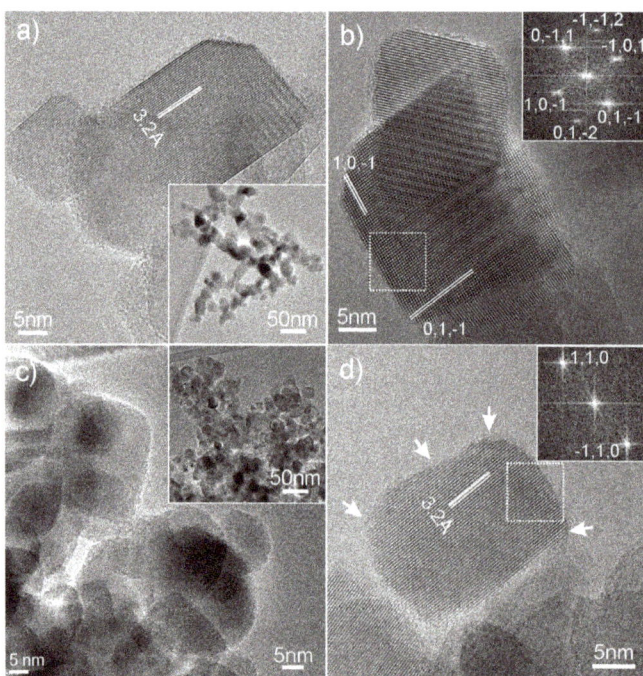

Figure 3. TEM images of TiO_2 (**a,b**) and S-doped TiO_2 (**c,d**). In the insets of (**a,c**) low resolution images of TiO_2 and S-doped TiO_2 are shown, while in the insets of (**b,d**) two selected regions are Fast Fourier Transform (FFT) imaged.

In particular, rutile (Figure 3a) and anatase (which reveals its diffraction pattern from [111] zone axis) nanoparticles (Figure 3b), which have a well-defined structure and shape, show extended faces with highly regular terminations, together with sharp corners and edges. Conversely, in Figure 3c,d, S-doped TiO_2 particles with rounded shapes are shown, as highlighted by the white arrows. From Figure 3d, it can be observed that the borders of the particle appear to be completely indented, with corners and edges sensitively smoothed, which is caused by the formation of local defective regions at the atomic level. The explanation of this phenomenon could be plausibly ascribed to the action of H_2S, whose strong acidic character could lead to remarkably defective surfaces of TiO_2 particles.

For a deeper understanding of the effects of H$_2$S treatment, the TiO$_2$ and S-doped TiO$_2$ samples were investigated by low resolution TEM (insets in Figure 3a,c, respectively). From the comparison of the related particle size distributions shown in Figure S2 (Supplementary Materials), a slight increment of particle dimensions for S-doped TiO$_2$ is detectable, as confirmed by the mean crystal sizes provided by the anatase (101) and rutile (110) XRD peak broadenings (Scherrer's equation) shown in Table S1 (Supplementary Materials).

Moreover, the slight decreasing of the specific surface area (38 m^2/g) of S-doped TiO$_2$ observed, as compared with TiO$_2$ (50 m^2/g), in addition to the slight increasing of the particle size in the case of S-doped TiO$_2$ samples, can be explained by the moderate sintering effect due to annealing treatment conditions occurring at 673 K under an H$_2$S atmosphere.

Furthermore, the impressive action of H$_2$S on the nature of TiO$_2$ particles has been also confirmed by FTIR investigation (*vide infra*).

2.2. Surface Properties by FTIR and UV-Vis Spectroscopies

2.2.1. FTIR Spectroscopy

FTIR spectra collected at 100 K at decreasing CO coverage on TiO$_2$ were compared to a similar sequence on the S-doped TiO$_2$ sample (Figure 4a,b). The spectra were acquired by increasing an initial CO dose (70 Torr) to reach equilibrium conditions, i.e., the maximum CO coverage, after which the system was progressively outgassed up to the complete removal of the adsorbed CO molecules.

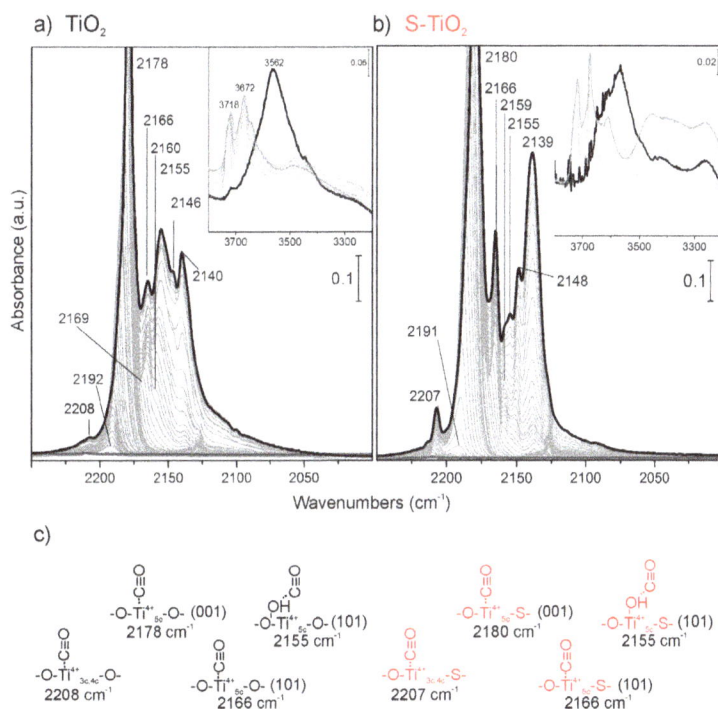

Figure 4. FTIR spectra of CO adsorbed at 100 K on (a) TiO$_2$ and (b) S-doped TiO$_2$ at decreasing coverages (black bold line, maximum coverage: p$_{CO}$ = 70 Torr, grey bold line, minimum coverage: p$_{CO}$ → 0); (c) scheme modelling of Ti sites on two different faces interacting with CO molecules and the frequency values of the corresponding bands at maximum coverage (black bold line).

TiO$_2$ spectra show the typical features due to the adsorption of CO on the different Ti sites on activated TiO$_2$ surfaces, as discussed in the literature (Figure 4a) [41–43].

As can be seen from Figure 4a, the intense main peak observed at 2179 cm^{-1} can be explained by the building up of lateral interactions among an array of parallel CO oscillators, adsorbed on Ti^{4+} sites on flat (101) surfaces [41], while that at 2155 cm^{-1} is due to CO molecules interacting by hydrogen bonds with residual OH groups. The main band, in the 2178 cm^{-1}–2192 cm^{-1} range (Figure 4a), undergoes a frequency shift upon decreasing CO pressure, due to the changing of the lateral interactions between CO oscillators on the TiO$_2$ surface [43,44]. Notice that a higher frequency shift is indicative of a highly regular and extended face.

Coming back to OH groups, which remain on the surface despite the activation treatment at 673 K, their presence is confirmed by the stretching modes observed at 3718 cm^{-1} and 3672 cm^{-1} (inset of Figure 4a). These features are shifted to lower frequency as a consequence of CO adsorption, thus giving rise to a broad and more intense feature centered at 3562 cm^{-1}. Notice that the spectral features of the OH groups are then completely restored at a lower CO pressure together with the progressively disappearing and shifting of the 2155 cm^{-1} band to 2160 cm^{-1}. These two events indicate that the complete and reversible CO desorption from OH groups is occurring [42,45].

The sharp band at 2140 cm^{-1} is easily associated to physically adsorbed CO, which forms a multi-layer surface when the liquid nitrogen temperature brings CO to a "liquid-like" state [46]. On the other hand, the band at 2146 cm^{-1} that, by decreasing CO coverage, gradually shifts upward and merges with the 2155 cm^{-1} band, is due to the CO species on rutile facelets present in TiO$_2$ P25 [41].

The weak feature at 2166 cm^{-1} was assigned to CO adsorbed on Ti Lewis centers on (001) surfaces, where Ti sites along Ti-O rows are strongly bound to two oxygens, which cause a screened electrostatic potential at these Ti sites, as concluded by Mino et al. [47] in a combined FTIR/Density Functional Theory (DFT) study on the CO adsorption on anatase (001) and (101) facets.

The weak band at 2208 cm^{-1} was assigned to CO adsorbed on highly acid Ti Lewis sites located on defective situations such as edges, steps, and corners, thus exhibiting very low coordination [41].

After H$_2$S dosage (Figure 3b), some modifications can be highlighted. In particular, a slightly wider Full Width at Half Maximum (FWHM) for the main feature at 2180 cm^{-1} is observed due to a more disordered system, caused by the presence of S species interrupting the regularity of the extended faces.

Notice that the 2180 cm^{-1}–2191 cm^{-1} main band for S-doped TiO$_2$ (Figure 4b) undergoes a quite similar frequency shift upon decreasing CO pressure, as compared to pure TiO$_2$ (Figure 4a).

The 2166 cm^{-1} band, previously assigned to Ti Lewis centers on flat (001) faces and strongly bound to O anions, presents an increased intensity and can be ascribed to a reduced screening electrostatic potential due to the O $\rightarrow$ S exchange. In fact, the presence of sulfur could cause an increased acidity followed by the observed increased intensity (Figure 4b). Conversely, the reduced intensity of the 2160–2155 cm^{-1} envelope can be explained by the weaker interaction of CO molecules with the residual OH groups (Figure 4b).

Moreover, notice that the band at 2207 cm^{-1}, ascribed to CO on highly cus (coordinatively unsaturated) Ti sites on S-doped TiO$_2$ [41], shows higher intensity with respect to pure TiO$_2$ and is the last one to disappear by outgassing. It can be hypothesized that the larger S atoms replace the smaller O ones, thus favoring the formation of defects such as edges, steps, and corners.

These results can be confirmed by the wide band observed in the 3400–3200 cm^{-1} region, before CO dosage, due to the formation of surface H$_2$O as a consequence of oxygen-sulfur exchange (inset of Figure 4b), as explained in the next paragraph.

On the basis of studies concerning the dissociative adsorption of H$_2$S on TiO$_2$ [48], it was found that above a temperature of 593 K, the Ti–SH bond is still strong, being –SH irreversibly adsorbed on the surface of TiO$_2$. However, by increasing the temperature, S-H bonds become weaker and finally break. H moves to a neighboring O, forming –OH groups. Finally H$_2$O is formed at the surface and S moves to oxygen vacancy positions [29].

Moreover, besides the dissociative adsorption, H_2S molecules can also interact with surface –OH groups (bonded to Ti^{4+}) to give rise to hydrogen bonds.

2.2.2. UV-Vis Spectroscopy

The UV-Vis spectrum of S-doped TiO_2 (red curve) as compared to pure TiO_2 (black curve), is shown in Figure 5. From this, a wide shift of the absorption edge for the S-doped TiO_2 sample towards higher wavelengths (lower energies) is clearly detectable, together with an additional wide absorption centered at around 390 nm.

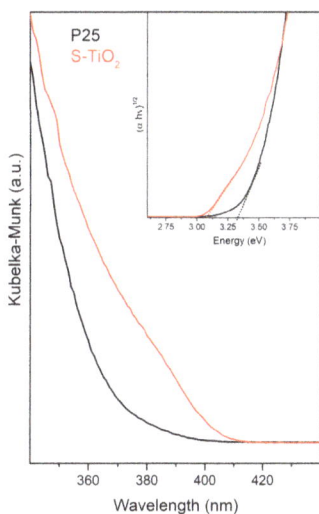

Figure 5. UV-Vis spectra of S-doped TiO_2 (red curve) and pure TiO_2 (black curve) used as a reference material. The Tauc plot of both samples is shown in the inset. An example of a Tauc Plot is reported in ref. [49].

In particular, the band gap shift is emphasized by the Tauc plot (see inset in Figure 5), where the intercepts with the abscissa axis of the extrapolation of the linear part of the curves clearly highlight a red shift of the absorption edge for S-doped TiO_2 (see black and red dotted lines for TiO_2 and S-doped TiO_2, respectively).

In general, this energy red-shift is associated with a change of TiO_2 electronic structure, after the treatment with H_2S, which causes the S → O exchange at the surface of TiO_2 [50,51]. Taking into consideration this aspect, it is well known that the formation of doping states involves additional electronic levels that can be formed between the valence (VB) and the conduction (CB) bands, thus reducing the electron transition energy [21,52]. Along this line, the features in the UV-visible range can be ascribed to the presence of additional electronic states above the valence band edge of pure TiO_2, due to S species, as well supported by XPS measurements [52,53]. These states can be attributed to sulfur 3p atomic orbitals mixing with the VB of TiO_2. In addition, another plausible mechanism to explain the observed absorption shift could also involve O vacancies caused by the thermal treatment at 673 K under vacuum conditions [54].

The S-doped TiO_2 sample was tested for the photodegradation of methylene blue (MB) in water solution under solar light irradiation as compared to the TiO_2 P25 benchmark (Figure S4, Supplementary Materials). In this figure, the MB C/C_0 vs. time plots of S-TiO_2 and of the TiO_2 P25 are shown. Although the MB photodegradation of S-doped TiO_2 is definitely high, under solar light irradiation its photocatalytic activity is lower than that of pure TiO_2 P25. We think that the explanation for this lower photocatalytic performance in the degradation of MB is the balance of two opposite effects. The first one, associated with the sulfur doping which allows the solar light harvesting, is beneficial. However, the defective surface, as a result of the heavy H_2S treatment, has a strong, definitely detrimental effect in the charge separation efficiency. It is safely concluded that, upon the adopted H_2S conditions (673 K, vacuum), the non-reversible surface modification does not help to increase the photocatalytic efficiency with respect to the TiO_2 P25 powder used as a starting material.

3. Materials and Methods

TiO_2 (P25, Evonik), in pellet form, was activated at 673 K for 30 min under dynamic vacuum, then oxidized in an oxygen (40 Torr) atmosphere at the same temperature for 30 min, twice. By keeping the temperature at 673 K, the obtained sample was sulfided in an H_2S atmosphere (30 Torr) for 1 h, twice, and then outgassed. The sample was further sulfided following the same method.

FTIR spectra of CO adsorbed at 100 K on TiO_2 and S-doped TiO_2 at decreasing coverages were obtained in an IR cell designed for liquid nitrogen flowing, and were recorded by means of a Bruker IFS-28 spectrometer, equipped with a Mercury Cadmium Telluride (MCT) cryogenic detector, with a resolution of 4 cm^{-1} (64 interferograms were averaged for each spectrum). The spectra were acquired in the 4000–400 cm^{-1} interval, where the fundamental vibration modes are observed.

Raman spectra were recorded using a Renishaw Raman InVia Reflex spectrophotometer equipped with an Ar+ laser emitting at 514 nm, using both static and rotating configurations.

UV-Vis measurements were collected by using a UV-Vis-NIR spectrophotometer (Varian Cary 5000, equipped with a reflectance sphere. Due to their strong optical absorption, the samples were diluted in $BaSO_4$ powder.

X-ray diffraction patterns were collected by means of a diffractometer (PANalytical PW3050/60 X'Pert PRO MPD) with a Ni-filtered Cu anode, working with a reflectance Bragg-Brentano geometry, by using the spinner mode. The mean crystal sizes were calculated from XRD measurements by Scherrer's equation: $L = K\lambda/\beta \cos\theta$ (λ is the X-ray wavelength, β is the full width at half maximum (FWHM) of the diffraction line corrected by the instrumental broadening, θ is the diffraction angle, and K is a constant assumed to be 0.9). Peak fitting of XRD patterns was adopted, using the Pseudo-Voigt function of anatase (101) and rutile (110) XRD peaks.

High resolution transmission electron microscopy images were acquired with a JEOL 3010-UHR instrument operating at 300 kV, equipped with a 2 k × 2 k pixels Gatan US1000 CCD camera.

N_2 adsorption-desorption experiments were carried out at 77 K (Micromeritics ASAP 2020 instrument) to determine the Brunauer-Emmett-Teller (BET) surface area. The surface area of the samples was determined after outgassing at RT, overnight.

For the photodegradation test, the same quantities of S-doped TiO_2 and TiO_2 powder (used as a reference) were dispersed in aliquots of a methylene blue (MB) water solution 12.5 mg/L and kept in the dark for 1 h at RT. After exposure, for increasing times, to a solar lamp (SOL2/500S lamp, Honle UV technology, Munchen, Germany) ranging from ultraviolet to infrared radiation (295–3000 nm), the dispersions were centrifuged for 30 min at 10,000 rpm. Photocatalytic degradation of MB was investigated by means of UV-Vis spectroscopy in the transmission mode. The integrated intensity of the adsorbed MB manifestations (C) was used to obtain C/C_0 vs. time plots, where C_0 is the concentration at the initial intensity before illumination (Figure S4).

4. Conclusions

In this work, S-doped TiO_2 samples were synthesized by means of H_2S treatment at 673 K. From several ex situ investigations, including XRD, Raman spectroscopy, and HRTEM, the structure and morphology of samples were obtained. In particular, even if XRD and Raman analyses, due to detection limits, did not give sensitive information concerning the effects of H_2S dosage on TiO_2, HRTEM images showed remarkable changes of TiO_2 particle shapes as a consequence of H_2S, which causes strong erosion of the faces, corners, and edges of the nanoparticles. This impressive action was also confirmed by FTIR spectra, where remarkable differences in the relative intensity of all peaks in S-doped TiO_2, when compared to pure TiO_2, were observed. Moreover, the changes in the OH groups range can be ascribed to oxygen-sulfur exchange phenomena, explained with the dissociative adsorption of H_2S on TiO_2.

Finally, UV-Vis spectroscopy demonstrated how also the electronic structure of TiO_2 can be deeply modified by H_2S. The red shift of the S-doped TiO_2 absorption edge can be explained with additional extrinsic electronic levels introduced by the sulfur doping. This affected the optical properties of TiO_2, whose absorption edge was extended to the visible-light region.

From all these considerations, it can be safety concluded that H_2S treatment of TiO_2, to achieve an S-doped TiO_2 material, has a deep and strong effect on TiO_2, such as on the morphological, surface, electronic, and optical properties. However, the photocatalysis efficiency of TiO_2 in the degradation of methylene blue is not improved by the H_2S treatment at the adopted conditions, due to the strongly defective surface of S-doped TiO_2 which decreases the charge separation efficiency.

Supplementary Materials: The following are available online at www.mdpi.com/2073-4344/7/7/214/s1, Figure S1: EDAX spectrum of TiO_2/H_2S, Figure S2: Particle size distributions (PSDs) of TiO_2 P25 (left panel) and of S-TiO_2 (right panel), Figure S3: Time dependence of C/C_0 upon solar light exposure of S-TiO_2 (red line) as compared to the TiO_2 P25 (black dotted line), for photodegradation of methylene blue, Figure S4: FTIR spectra, recorded before CO dosage, of TiO_2 (black curve) and TiO_2/H_2S (red curve), Table S1: Mean crystal sizes of anatase and rutile nanoparticles in TiO_2 P25 and S-TiO_2 samples, as obtained from XRD peak broadening (black and red patters in Figure 1).

Acknowledgments: This work was supported by MIUR (Ministero dell'Istruzione, dell'Università e della Ricerca), INSTM Consorzio and NIS (Nanostructured Interfaces and Surfaces) Inter-Departmental Centre of University of Torino. The authors thank the vibrational Raman spectroscopy laboratory of Chemistry Department and in particular Alessandro Damin, for the precious support in Raman experiments.

Author Contributions: Sara Cravanzola, Federico Cesano and Fulvio Gaziano conceived, designed and performed the experiments and characterizations, analyzing the data; Sara Cravanzola, Federico Cesano and Domenica Scarano wrote the paper. All authors read and approved the paper.

Conflicts of Interest: The authors declare no conflict of interest.

References

1. Nakata, K.; Fujishima, A. TiO$_2$ photocatalysis: Design and applications. *J. Photochem. Photobiol. C* **2012**, *13*, 169–189. [CrossRef]

2. Schneider, J.; Matsuoka, M.; Takeuchi, M.; Zhang, J.; Horiuchi, Y.; Anpo, M.; Bahnemann, D.W. Understanding TiO$_2$ photocatalysis: Mechanisms and materials. *Chem. Rev.* **2014**, *114*, 9919–9986. [CrossRef] [PubMed]

3. Kubacka, A.; Diez, M.S.; Rojo, D.; Bargiela, R.; Ciordia, S.; Zapico, I.; Albar, J.P.; Barbas, C.; Martins dos Santos, V.A.P.; Fernández-García, M.; et al. Understanding the antimicrobial mechanism of TiO$_2$-based nanocomposite films in a pathogenic bacterium. *Sci. Rep.* **2014**, *4*, 1–9. [CrossRef] [PubMed]

4. Yin, Z.F.; Wu, L.; Yang, H.G.; Su, Y.H. Recent progress in biomedical applications of titanium dioxide. *Phys. Chem. Chem. Phys.* **2013**, *15*, 4844–4858. [CrossRef] [PubMed]

5. Song, Y.-Y.; Schmidt-Stein, F.; Baue, S.; Schmuki, P. Amphiphilic TiO$_2$ nanotube arrays: An actively controllable drug delivery system. *J. Am. Chem. Soc.* **2009**, *131*, 4230–4232. [CrossRef] [PubMed]

6. Banerjee, S.; Dionysioub, D.D.; Pillai, S.C. Self-cleaning applications of TiO$_2$ by photo-induced hydrophilicity and photocatalysis. *Appl. Catal. B* **2015**, *176–177*, 396–428. [CrossRef]

7. Ao, C.H.; Lee, S.C. Indoor air purification by photocatalyst TiO$_2$ immobilized on an activated carbon filter installed in an air cleaner. *Chem. Eng. Sci.* **2005**, *60*, 103–109. [CrossRef]

8. Cesano, F.; Pellerej, D.; Scarano, D.; Ricchiardi, G.; Zecchina, A. Radially organized pillars of TiO$_2$ nanoparticles: Synthesis, characterization and photocatalytic tests. *J. Photochem. Photob. A Chem.* **2012**, *242*, 51–58. [CrossRef]

9. Chatterjee, D.; Mahata, A. Visible light induced photo-degradation of organic pollutants on dye adsorbed TiO$_2$ surface. *J. Photochem. Photobiol. A* **2002**, *153*, 199–204. [CrossRef]

10. Cravanzola, S.; Cesano, F.; Gaziano, F.; Scarano, D. Carbon domains on MoS₂/TiO₂ system via catalytic acetylene oligomerization: Synthesis, structure and surface properties. *Front. Chem.* **2017**, submitted.
11. Cravanzola, S.; Muscuso, L.; Cesano, F.; Agostini, G.; Damin, A.; Scarano, D.; Zecchina, A. MoS₂ nanoparticles decorating titanate-nanotube surfaces: Combined microscopy, spectroscopy, and catalytic studies. *Langmuir* **2015**, *31*, 5469–5478. [CrossRef] [PubMed]
12. Uddin, M.J.; Cesano, F.; Bertarione, S.; Bonino, F.; Bordiga, S.; Scarano, D.; Zecchina, A. Tailoring the activity of Ti-based photocatalysts by playing with surface morphology and silver doping. *J. Photochem. Photob. A Chem.* **2008**, *196*, 165–173. [CrossRef]
13. Asahi, R.; Morikawa, T.; Ohwaki, T.; Aoki, K.; Taga, Y. Visible-light photocatalysis in nitrogen-doped titanium oxides. *Science* **2001**, *293*, 269–271. [CrossRef] [PubMed]
14. Hong, X.; Wang, Z.; Cai, W.; Lu, F.; Zhang, J.; Yang, Y.; Ma, N.; Liu, Y. Visible-light-activated nanoparticle photocatalyst of iodine-doped titanium dioxide. *Chem. Mater.* **2005**, *17*, 1548–1552. [CrossRef]
15. Dozzi, M.V.; D'Andrea, C.; Ohtani, B.; Valentini, G.; Selli, E. Fluorine-doped TiO₂ materials: Photocatalytic activity vs. time-resolved photoluminescence. *J. Phys. Chem. C* **2013**, *117*, 25586–25595. [CrossRef]
16. Cravanzola, S.; Jain, S.M.; Cesano, F.; Damin, A.; Scarano, D. Development of a multifunctional TiO₂/MWCNT hybrid composite grafted on a stainless steel grating. *RSC Adv.* **2015**, *5*, 103255–103264. [CrossRef]
17. Ansari, S.A.; Khan, M.M.; Ansaric, M.O.; Cho, M.H. Nitrogen-doped titanium dioxide (N-doped TiO₂) for visible light photocatalysis. *New J. Chem.* **2016**, *40*, 3000–3009. [CrossRef]
18. Cesano, F.; Bertarione, S.; Damin, A.; Agostini, G.; Usseglio, S.; Vitillo, J.G.; Lamberti, C.; Spoto, G.; Scarano, D.; Zecchina, A. Oriented TiO₂ nanostructured pillar arrays: Synthesis and characterization. *Adv. Mater.* **2008**, *20*, 3342–3348. [CrossRef]
19. Ramandi, S.; Entezari, M.H.; Ghows, N. Sono-synthesis of solar light responsive S–N–C–tri doped TiO₂ photo-catalyst under optimized conditions for degradation and mineralization of diclofenac. *Ultrason. Sonochem.* **2017**, *38*, 234–245. [CrossRef] [PubMed]
20. Brindha, A.; Sivakumar, T. Visible active N, S co-doped TiO₂/graphene photocatalysts for the degradation of hazardous dyes. *J. Photochem. Photobiol. A* **2017**, *340*, 146–156. [CrossRef]
21. Cesano, F.; Agostini, G.; Scarano, D. Nanocrystalline TiO₂ micropillar arrays grafted on conductive glass supports: Microscopic and spectroscopic studies. *Thin Solid Films* **2015**, *590*, 200–206. [CrossRef]
22. Uddin, M.J.; Daramola, D.E.; Velasquez, E.; Dickens, T.J.; Yan, J.; Hammel, E.; Cesano, F.; Okoli, O.I. A high efficiency 3D photovoltaic microwire with carbon nanotubes (CNT)-quantum dot (QD) hybrid interface. *PSS RRL* **2014**, *8*, 898–903. [CrossRef]
23. Hui, F.; Bu, C. DFT description on electronic structure and optical absorption properties of anionic S-doped anatase TiO₂. *J. Phys. Chem. B* **2006**, *110*, 17866–17871.
24. Smith, M.F.; Setwong, K.; Tongpool, R.; Onkaw, D.; Naphattalung, S.; Limpijumnong, S.; Rujirawat, S. Identification of bulk and surface sulfur impurities in TiO₂ by synchrotron X-ray absorption near edge structure. *Appl. Phys. Lett.* **2007**, *91*, 142107. [CrossRef]
25. Liu, G.; Sun, C.; Smith, S.C.; Wang, L.; Lu, G.Q.; Cheng, H.-M. Sulfur doped anatase TiO₂ single crystals with a high percentage of {0 0 1} facets. *J. Colloid Interface Sci.* **2010**, *349*, 477–483. [CrossRef] [PubMed]
26. Nam, S.-H.; Kim, T.K.; Boo, J.-H. Physical property and photo-catalytic activity of sulfur doped TiO₂ catalysts responding to visible light. *Catal. Today* **2012**, *185*, 259–262. [CrossRef]
27. Yang, G.; Yan, Z.; Xiao, T. Low-temperature solvothermal synthesis of visible-light-responsive S-doped TiO₂ nanocrystal. *Appl. Surf. Sci.* **2012**, *258*, 4016–4022. [CrossRef]
28. Dozzi, M.V.; Livraghi, S.; Giamello, E.; Selli, E. Photocatalytic activity of S- and F-doped TiO₂ in formic acid mineralization. *Photochem. Photobiol. Sci.* **2011**, *10*, 343–349. [CrossRef] [PubMed]
29. Chen, Y.; Jiang, Y.; Li, W.; Jin, R.; Tang, S.; Hu, W. Adsorption and interaction of H₂S/SO₂ on TiO₂. *Catal. Today* **1999**, *50*, 39–47.
30. Wei, F.; Ni, L.; Cui, P. Preparation and characterization of N–S-codoped TiO₂ photocatalyst and its photocatalytic activity. *J. Hazard. Mater.* **2008**, *156*, 135–140. [CrossRef] [PubMed]
31. Li, H.; Zhang, X.; Huo, Y.; Zhu, J. Supercritical preparation of a highly active S-doped TiO₂ photocatalyst for methylene blue mineralization. *Environ. Sci. Technol.* **2007**, *41*, 4410–4414. [CrossRef] [PubMed]
32. Travert, A.; Maugé, F. IR study of hydrotreating catalysts in working conditions: Comparison of the acidity present on the sulfided phase and on the alumina support. *Stud. Surf. Sci. Catal.* **1999**, *127*, 269–277.

33. Barba, D.; Cammarota, F.; Vaiano, V.; Salzano, E.; Palma, V. Experimental and numerical analysis of the oxidative decomposition of H_2S. *Fuel* **2017**, *198*, 68–75. [CrossRef]

34. Jüngst, E.; Nehb, W. Hydrogen sulfide to sulfur (Claus process). In *Handbook of Heterogeneous Catalysis*; Wiley-VCH: Weinheim, Germany, 2008; pp. 2609–2623.

35. Huang, W.-F.; Chen, H.-T.; Lin, M.C. Density functional theory study of the adsorption and reaction of H_2S on TiO_2 rutile (110) and anatase (101) surfaces. *J. Phys. Chem. C* **2009**, *113*, 20411–20420. [CrossRef]

36. Arrouvel, C.; Toulhoat, H.; Breysse, M.; Raybaud, P. Effects of P_{H2O}, P_{H2S}, P_{H2} on the surface properties of anatase-TiO_2 and g-Al_2O_3: A DFT study. *J. Catal.* **2004**, *226*, 260–272. [CrossRef]

37. Fahmi, A.; Ahdjoudj, J.; Minot, C. A theoretical study of H_2S and MeSH adsorption on TiO_2. *Surf. Sci.* **1996**, *352–354*, 529–533. [CrossRef]

38. Markovits, A.; Ahdjoudj, J.; Minot, C. Theoretical study of the TiO_2 and MgO surface acidity and the adsorption of acids and bases. *Mol. Eng.* **1997**, *7*, 245–261. [CrossRef]

39. Abbasi, A.; Sardroodi, J.J. Adsorption and dissociation of H_2S on nitrogen-doped TiO_2 anatase nanoparticles: Insights from DFT computations. *Surf. Interface* **2017**, *8*, 15–27. [CrossRef]

40. Ma, H.L.; Yang, J.Y.; Dai, Y.; Zhang, Y.B.; Lu, B.; Ma, G.H. Raman study of phase transformation of TiO_2 rutile single crystal irradiated by infrared femtosecond laser. *Appl. Surf. Sci.* **2007**, *253*, 7497–7500. [CrossRef]

41. Mino, L.; Spoto, G.; Bordiga, S.; Zecchina, A. Particles morphology and surface properties as investigated by HRTEM, FTIR, and periodic DFT calculations: From pyrogenic TiO_2 (P25) to nanoanatase. *J. Phys. Chem. C* **2012**, *116*, 17008–17018. [CrossRef]

42. Martra, G. Lewis acid and base sites at the surface of microcrystalline TiO_2 anatase: Relationships between surface morphology and chemical behaviour. *Appl. Catal. A* **2000**, *200*, 275–285. [CrossRef]

43. Spoto, G.; Morterra, C.; Marchese, L.; Orio, L.; Zecchina, A. The morphology of TiO_2 microcrystals and their adsorptive properties towards CO: A HRTEM and FTIR study. *Vacuum* **1990**, *41*, 37–39. [CrossRef]

44. Signorile, M.; Damin, A.; Budnyk, A.; Lamberti, C.; Puig-Molina, A.; Beato, P.; Bordiga, S. MoS_2 supported on P25 titania: A model system for the activation of a HDS catalyst. *J. Catal.* **2015**, *328*, 225–235. [CrossRef]

45. Cesano, F.; Bertarione, S.; Uddin, M.J.; Agostini, G.; Scarano, D.; Zecchina, A. Designing TiO_2 based nanostructures by control of surface morphology of pure and silver loaded titanate nanotubes. *J. Phys. Chem. C* **2010**, *114*, 169–178. [CrossRef]

46. Bordiga, S.; Scarano, D.; Spoto, G.; Zecchina, A.; Lamberti, C.; Otero Areán, C. Infrared study of carbon monoxide adsorption at 77 K on faujasites and ZSM-5 zeolites. *Vib. Spectrosc.* **1993**, *5*, 69–74. [CrossRef]

47. Mino, L.; Ferrari, A.M.; Lacivita, V.; Spoto, G.; Bordiga, S.; Zecchina, A. CO adsorption on anatase nanocrystals: A combined experimental and periodic DFT study. *J. Phys. Chem. C* **2011**, *115*, 7694–7700. [CrossRef]

48. Davydov, A.; Chuang, K.T.; Sanger, A.R. Mechanism of H_2S oxidation by ferric oxide and hydroxide surfaces. *J. Phys. Chem. B* **1998**, *102*, 4745–4752. [CrossRef]

49. Murphy, A.B. Band-gap determination from diffuse reflectance measurements of semiconductor films, and application to photoelectrochemical water-splitting. *Sol. Energy Mater. Sol. Cells* **2007**, *91*, 1326–1337. [CrossRef]

50. Li, N.; Zhang, X.; Zhou, W.; Liu, Z.; Xie, G.; Wang, Y.; Du, Y. High quality sulfur -doped titanium dioxide nanocatalysts with visible light photocatalytic activity from non-hydrolytic thermolysis synthesis. *Inorg. Chem. Front.* **2014**, *1*, 521–525. [CrossRef]

51. Ho, W.; Yu, J.C.; Lee, S. Low-temperature hydrothermal synthesis of S-doped TiO_2 with visible light photocatalytic activity. *J. Solid State Chem.* **2006**, *179*, 1171–1176. [CrossRef]

52. Chen, X.; Burda, C. The electronic origin of the visible-light absorption properties of C-, N- and S-doped TiO_2 nanomaterials. *J. Am. Chem. Soc.* **2008**, *130*, 5018–5019. [CrossRef] [PubMed]

53. Tang, X.; Li, D. Sulfur-doped hhighly ordered TiO_2 nanotubular arrays with visible light response. *J. Phys. Chem. C* **2008**, *112*, 5405–5409. [CrossRef]

54. Umebayashi, T.; Yamaki, T.; Itoh, H.; Asai, K. Band gap narrowing of titanium dioxide by sulfur doping. *Appl. Phys. Lett.* **2002**, *81*, 454–456. [CrossRef]

© 2017 by the authors. Licensee MDPI, Basel, Switzerland. This article is an open access article distributed under the terms and conditions of the Creative Commons Attribution (CC BY) license (http://creativecommons.org/licenses/by/4.0/).

![catalysts logo] *catalysts*

MDPI

Article

Using Density Functional Theory to Model Realistic TiO₂ Nanoparticles, Their Photoactivation and Interaction with Water

Daniele Selli, Gianluca Fazio and Cristiana Di Valentin *

Dipartimento di Scienza dei Materiali, Università di Milano-Bicocca, via R. Cozzi 55, 20125 Milano, Italy; daniele.selli@unimib.it (D.S.); g.fazio3@campus.unimib.it (G.F.)
* Correspondence: cristiana.divalentin@unimib.it

Received: 16 October 2017; Accepted: 20 November 2017; Published: 24 November 2017

Abstract: Computational modeling of titanium dioxide nanoparticles of realistic size is extremely relevant for the direct comparison with experiments but it is also a rather demanding task. We have recently worked on a multistep/scale procedure to obtain global optimized minimum structures for chemically stable spherical titania nanoparticles of increasing size, with diameter from 1.5 nm (~300 atoms) to 4.4 nm (~4000 atoms). We use first self-consistent-charge density functional tight-binding (SCC-DFTB) methodology to perform thermal annealing simulations to obtain globally optimized structures and then hybrid density functional theory (DFT) to refine them and to achieve high accuracy in the description of structural and electronic properties. This allows also to assess SCC-DFTB performance in comparison with DFT(B3LYP) results. As a further step, we investigate photoexcitation and photoemission processes involving electron/hole pair formation, separation, trapping and recombination in the nanosphere of medium size by hybrid DFT. Finally, we show how a recently defined new set of parameters for SCC-DFTB allows for a proper description of titania/water multilayers interface, which paves the way for modeling large realistic nanoparticles in aqueous environment.

Keywords: nanospheres; simulated Extended X-ray Adsorption Fine-Structure (EXAFS); excitons; trapping; titania/water interface; SCC-DFTB; B3LYP

---

1. Introduction

TiO₂ nanoparticles are fundamental building blocks in photocatalysis [1–4]. Their theoretical description is indeed relevant and requires the size of the model to be as realistic as possible, for direct comparison with experimental samples.

TiO₂ nanoparticles are most typically obtained from sol-gel synthesis. Several studies have proven that shape and size can be successfully tailored by controlling the conditions of preparation and by using ad-hoc surface chemistry [5–7]. The minimum energy shape was predicted by Barnard et al. [8] by Wulff construction, for dimensions below 10 nm, to be a decahedron in the anatase phase, exposing mainly (101) and small (001) facets. However, growth determining factors are pH and particle density. An excessive dilution may cause a partial dissolution of titania nanocrystals leading to the formation of spherical nanoparticles [9]. Those, analogously to nanotubes and nanorods, are characterized by a high curvature profile and, thus, expected to be more reactive towards molecular adsorption.

The majority of the computational first-principles studies are devoted to bulk or surface slabs of anatase TiO₂ [10]. Few works have dealt with the decahedral faceted nanoparticles [11–16] but none with spherical ones. Modelling nanoparticles of realistic size (few nanometers) by first-principles calculations is very demanding and a global optimization is hardly feasible [17]. Ours is a multistep/scale approach [18] where we propose first to apply a less expensive but still rather accurate method based on density functional theory (DFT), which is the self-consistent-charge density functional tight-binding (SCC-DFTB) [19], to perform a global structure optimization search of the nanoparticles; then to run a further DFT relaxation to determine structural and electronic properties with first-principles level accuracy. For the latter, we use hybrid functionals since they are known to better describe electronic structure details of TiO_2 materials [20–22]. SCC-DFTB has been demonstrated to be a powerful tool for the quantum mechanics study of many system involving TiO_2 [23–26]. The method retains most of the physics of standard DFT at an extremely reduced computational cost.

Furthermore, we would like to describe the interaction of such nanoparticles with light and their photoactivation producing energy carriers (excitons) and charge carriers (electrons and holes). The aim is to improve the general understanding of the processes at the basis of light energy conversion into chemical species with intrinsic redox potential that are those triggering the redox reaction at the oxide surface [14,27–30].

It is generally accepted that water, as the surrounding environment where titania nanoparticle work in photocatalytic processes, plays an active role [31–39]. It is, therefore, fundamental to describe accurately the dynamical water layers arrangement on the surface and how water molecules may enter the photoactivated reaction chain [40].

In the following, we will present a critical review of our work, relative to the topics highlighted and discussed above: in Section 2, we present the Computational methodology; in Section 3, we describe how to obtain realistic spherical nanoparticles models; in Section 4, we discuss the description of the photoexcitation processes; and, in Section 5, we analyze how the water environment can be modeled with sufficient accuracy.

2. Computational Details

To tackle the surface complexity of the TiO_2 spherical nanoparticles and maintain a high degree of accuracy, different levels of theory are necessary. Nowadays, density functional theory (DFT) is the most used method to properly describe equilibrium geometries and electronic structures. However, many interesting features of the system are accessible only at certain size and time scale. For example, to explore the potential energy surface related to the different configurations of the TiO_2 spherical nanoparticles through molecular dynamics and simulated annealing processes, an approximated method has to be used. The self-consistent-charge density functional tight-binding (SCC-DFTB) approach is a DFT-based quantum mechanical method, which retains a quantum description of the system at a considerably reduced computational cost. Thus, in addition to geometry optimizations and electronic structure calculations, SCC-DFTB also enables molecular dynamic simulations for large systems with a reasonable time length.

2.1. Electronic Structure Calculations

The choice of a specific density functional is based on the aim of the study. Standard generalized-gradient approximation (GGA) functionals may be sufficient to describe equilibrium geometries or adsorption energies, however a correct description of the electronic structure of semiconducting oxides requires the inclusion of a certain portion of exact exchange. The use of such hybrid functionals in a plane-wave code is extremely cumbersome, thus localized basis function codes are preferred. The CRYSTAL14 code [41] has been used for most of the density functional theory (DFT) calculations, employing all-electron Gaussian basis sets [O 8-411(d1) Ti 86-411 (d41) and H 511(p1)] and the B3LYP [42,43] and the HSE06 [44] hybrid functionals. For periodic systems, reference DFT calculations were carried out with the Quantum ESPRESSO [45] simulation package, using the PBE functional [46], ultrasoft Vanderbilt pseudopotentials and a plane-wave basis set with a cut off of 30 Ry (300 Ry for the charge density).

The optimized lattice parameters for bulk TiO_2 anatase are 3.789 Å and 3.766 Å for a and 9.777 Å and 9.663 Å for c, respectively, for B3LYP and HSE06, which are in good agreement with the experimental values [47].

A $6\sqrt{2} \times 6\sqrt{2} \times 1$ TiO$_2$ anatase bulk supercell with 864 atoms was employed to model the exciton and the related distortions in the bulk. The (101) anatase surface has been taken as a reference for: (i) surface energies; and (ii) interaction with water. (i) We employed the CRYSTAL14 code and a minimal cell slab of ten triatomic layers with 60 atoms, where the periodicity was set along the $[10\bar{1}]$ and $[010]$ directions and not in the direction perpendicular to the surface. (ii) A 1×2 supercell three-triatomic-layer slab (72 atoms) has been used within the Quantum ESPRESSO code, where the replicas in the direction perpendicular to the surface were separated by 20 Å in order to avoid any interaction between images. For the k-point sampling, a $1 \times 1 \times 6$, a $8 \times 8 \times 1$ and a $2 \times 2 \times 1$ Monkhorst–Pack grid was used for the bulk, the minimal slab cell and the 1×2 slab supercell, respectively.

Anatase TiO$_2$ nanospheres have been carved from a bulk supercell following the procedure already described in a previous work by some of us [17]. Nanoparticles have been considered as molecules in the vacuum with no periodic boundary conditions. Therefore, when an excess electron or hole is introduced in the system, no background of charge is necessary. In the case of open-shell systems, spin polarization is taken into account.

Trapping energies (ΔE_{trap}) for excitons, extra electrons and holes are calculated as the total energy difference between the trap optimized geometry and the delocalized solution in the neutral ground state geometry.

The total densities of states (DOS) of the nanoparticles have been simulated with the convolution of Gaussian functions ($\sigma = 0.005$ eV) peaked at the value of the Kohn-Sham energies of each orbital. Projected densities of states (PDOS) are built using the following procedure, based on the molecular orbitals coefficients in the linear combination of atomic orbitals (LCAO): summing the squares of the coefficients of all the atomic orbitals centered on a certain atom type results, after normalization, in the relative contribution of each atom type to a specific eigenstate. Then, the various projections are obtained from the convolution of Gaussian peaks with heights that are proportional to the relative contribution. The zero energy for all the DOS is set to the vacuum level, i.e., the energy of an electron at an infinite distance from the surface of the system.

2.2. SCC-DFTB Approach

The self-consistent-charge density functional tight-binding method (SCC-DFTB) is based on the approximation of the Kohn-Sham (KS) DFT formalism. Assuming a second-order expansion of the KS-DFT total energy with respect to the electron density fluctuations, the SCC-DFTB total energy is defined as:

$$E_{tot}^{SCC-DFTB} = \sum_i n_i \varepsilon_i + \frac{1}{2} \sum_{\alpha\beta} v_{rep}^{\alpha\beta}(R_{\alpha\beta}) + \frac{1}{2} \sum_{\alpha\beta} \gamma_{\alpha\beta} \Delta q_\alpha \Delta q_\beta \tag{1}$$

where the first term contains the one-electron energies ε_i from the diagonalization of an approximated Hamiltonian matrix and represents the attractive part of the energy, whereas the second term approximates the short-range repulsive energy, given by the sum of the pairwise distance-dependent potential $v_{rep}^{\alpha\beta}(R_{\alpha\beta})$ between the pair of atoms α and β, and Δq_α and Δq_β are the charges induced on the atoms α and β, which interact through a Coulombic-like potential $\gamma_{\alpha\beta}$. For more information on the details of the SCC-DFTB method, see Refs. [19,48,49].

For all the SCC-DFTB calculations, we employed the DFTB+ simulation package [50]. We initially made used of the "matsci-0-3" Slater–Koster parameters, which have been shown to be well-suited for the study of anatase TiO$_2$ in Ref. [24]. Subsequently, to better describe the titania/water/water interface, we combined the "matsci-0-3" parameters for Ti-O and Ti-Ti interactions with the parameters in the "mio-1-1" set [19] for O-O, O-H and H-H interactions in what we have named as "MATORG" set. Furthermore, we modified the $\gamma_{\alpha\beta}$ function to improve the description of H-bonding, using a hydrogen bonding damping function (HBD), in which a $\zeta = 4$ parameter has been used [51]. In this work, we refer to this HBD modified Slater–Koster parameters set as "MATORG+HBD" [40]. From now on, DFTB will be used as a shorthand for SCC-DFTB.

To perform the simulated annealing procedure, we carried out Born–Oppenheimer DFTB molecular dynamics (MD) within the canonical ensemble (NVT). The integration of the Newton's equations of motion has been done with the Velocity Verlet algorithm, using a relative small time step of 0.5 fs to ensure reversibility. A Nosé–Hoover chain thermostat with a time constant of 0.03 ps was applied to reach the desired temperature during the temperature-annealing simulations. The simulation time length of the annealing processes was made commensurate to the size of the nanosphere. Thus, we used a simulation time up to 45 ps for the 1.5 nm, 24 ps for the 2.2 nm, 14 ps for the 3.0 nm and 11 ps for the 4.4 nm nanosphere. In the case of the titania/water interface, each MD simulation has been performed for 25 ps.

For the molecular dynamics of the titania/water interface, a 1×3 supercell anatase (101) slab (108 atoms) with a monolayer (ML), a bilayer (BL) and a trilayer (TL) of water, composed of 6, 12 and 18 water molecules, respectively, was used. The desired temperature of the thermostat was set to a constant low value (160 K) to avoid the desorption of surface water molecules.

2.3. Structural Analysis

The extended X-ray adsorption fine structure (EXAFS) simulated spectra has been simulated via a Gaussian convolution of peaks ($\sigma = 0.0005$ Å) centered at the length of the distance between each Ti atom and other atoms (O or Ti) in the first, second, and third coordination shells. Projections have been computed considering only the distances centered on specific Ti atoms with a certain coordination sphere. In the text, we also report the surface-to-bulk ratio, defined as the ratio between the number of Ti and O atoms at the surface of the nanosphere and the number of Ti and O atoms in the bulk.

3. Modelling Realistic TiO$_2$ Nanoparticles

We carved TiO$_2$ spherical nanoparticles from large bulk anatase supercells. The radius of the sphere is set to a desired value and only atoms within that sphere are considered, whereas those outside the sphere are removed. Some very low coordinated Ti sites are found to be left at the surface of the model that must be removed or saturated with OH groups; analogously monocoordinated O must be removed or saturated with H atoms. Therefore, we use a number of water molecules to achieve the chemical stability of the nanoparticle. We try to keep the number of water molecules as low as possible. Since we aim at modelling nanoparticles of realistic size, we range from spheres with diameter of 1.5 nm up to 4.4 nm. These contain from 300 up to almost 4000 atoms. The exact stoichiometry of the prepared nanoparticles [(TiO$_2$)$_{101}$·6H$_2$O, (TiO$_2$)$_{223}$·10H$_2$O, (TiO$_2$)$_{399}$·12H$_2$O, and (TiO$_2$)$_{1265}$·26H$_2$O] is reported in Figure 1.

Structural relaxation by geometry optimization from the "as-carved" and chemically stabilized models is not an efficient approach because we found that it leads to local minimum structures, which are far from the global minimum one. For this reason, we have drastically changed our approach and decided to use a less computationally expensive, but still rather accurate, DFT-based method (DFTB) and to run some molecular dynamics simulations starting from the "as-carved" structures at increasing temperature (up to 700 K in some cases). This approach allows moving from the local minimum structure basin, close to the "as-carved" structure, and to further sample the configuration space. The thermally equilibrated structures obtained with this approach, and then fully relaxed, are much more stable to any surface modification (i.e., addition of a molecular adsorbate) because those are true global minima on the potential energy surface of the TiO$_2$ nanospheres. This multi-step procedure can be rather easily and reasonably applied to nanospheres of increasing size, and certainly up to the one with a 4.4 nm diameter (~4000 atoms). Once the fully relaxed nanospheres are prepared, we can investigate structural and electronic properties at the DFTB level of theory. However, to assess the accuracy of DFTB in this specific context, we must perform a benchmark study against hybrid DFT model calculations. Those were obtained by full atomic relaxation starting from the DFTB thermally annealed and optimized spheres. Note that we performed this further DFT(B3LYP) optimization for all four nanospheres. For the very large one (~4000 atoms) it was an extremely expensive procedure, but we consider it worth because a successful benchmark against DFT will assess DFTB reliability for the investigation of structural and electronic properties of spherical TiO$_2$ nanoparticles, as the basis for future developments.

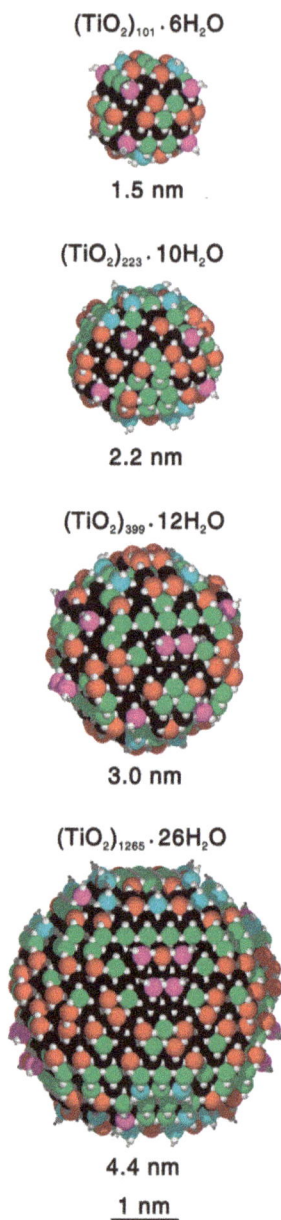

Figure 1. DFT(B3LYP) optimized structures, after simulated annealing at DFTB level, of the different nanospheres considered in this work. For each one, the stoichiometry, the approximate diameter and the position of the Ti atoms with different coordination are reported (Ti$_{4c}$ in red, Ti$_{5c}$ in green, Ti$_{6c}$ in black, Ti$_{3c}$(OH) in magenta and Ti$_{4c}$(OH) in cyan).

3.1. Structural Properties

Our study is not only meant to validate the DFTB methodology with respect to an accurate hybrid DFT method, but also to highlight how the enhanced curvature present in small TiO_2 nanospheres, in analogy to that in nanotubes and nanorods, affects both the coordination of the surface atoms and their geometrical environment. The different coordination types are illustrated in Figure 1 and their relative percentage is reported in Table 1. Flat anatase surfaces and faceted nanoparticles are dominated by Ti_{6c} and Ti_{5c}, whereas it is evident that on spherical nanoparticles different types of coordination exist that are expected to play a key role in the processes of chemical adsorption. The structural distortions induced by the nanosize and by the high curvature are investigated through the analysis of the simulated direct space X-ray absorption fine structure spectra (EXAFS) for bulk and for the nanospheres models of different size, as obtained with both DFT(B3LYP) and DFTB calculations (Figure 2). Those provide the distribution of the distances for each Ti atom with the neighboring O or next neighboring Ti atoms. Figure 2a,b show the EXAFS spectra for the bulk case. Here, no distribution is observed since crystalline bulk is characterized by two Ti-O distances (two lines on the left side), which are attributed to equatorial and axial O atoms resulting from a D_{2d} point symmetry at each Ti center [DFT(B3LYP): Ti-O_{eq} = 1.946 Å and Ti-O_{ax} = 2.000 Å; DFTB: Ti-O_{eq} = 1.955 Å and Ti-O_{ax} = 1.995 Å], and single Ti$\cdots$Ti distances at regular lattice positions [DFT(B3LYP): first shell 3.092 Å, second shell 3.789 Å; DFTB: first shell 3.090 Å, second shell 3.809 Å].

Table 1. Amount of Ti atoms with different coordination for the variously sized nanospheres, as optimized at DFTB level, in terms of their number and percentage with respect to the total number of Ti atoms, is reported. Corresponding values for the optimized nanosphere at the DFT(B3LYP) level are in parenthesis, only when different from DFTB.

DFTB [DFT(B3LYP)]	Number	%	Number	%	Number	%	Number	%
Ti Site	1.5 nm		2.2 nm		3.0 nm		4.4 nm	
Ti_{4c}	20 (19)	19.8 (18.8)	36	16.1	53	13.3	106	8.4
Ti_{5c}	20 (21)	19.8 (20.8)	43 (49)	19.2 (22.0)	69 (65)	17.3 (16.3)	159	12.6
Ti_{6c_sup}	20	19.8	28 (24)	12.6 (10.8)	72 (75)	18 (18.8)	157	12.4
Ti_{6c}	29	28.7	96 [94]	43.1 (42.1)	181 (182)	43.4 (45.6)	791	62.5
$Ti_{3c}(OH)$	8	7.9	8	3.6	16	4	20	1.6
$Ti_{4c}(OH)$	4	4	12	5.4	8	2	32	2.5

Figure 2c–j reports the EXAFS simulated spectra for the nanospheres of increasing size. With respect to bulk, we register a variety of distances due to the lattice distortion and diversity of coordination sites. For this reason and to improve the level of information provided, we present a convolution of peaks and project it on each type of coordination site. In the case of DFT(B3LYP) (left colomn), the first broad peak on the left, related to first neighbor Ti-O distances, is predominantly made up of Ti_{6c}-O bond lenghts. Low coordinated sites contribute to the shorter Ti-O distances than in the bulk (red, yellow, brown lines), whereas Ti_{5c} and Ti_{6c} species contribute to the range of bulk values and to the longer Ti-O distances up to 2.2–2.3 Å (blue, green lines). It is evident that, as the size increases, the relative portion of Ti_{6c} sites increases (blue line) with a distribution of Ti-O distances that becomes increasingly more peaked at the bulk values (see Figure 2i). It is clear that the EXAFS spectrum of the nanosphere with a diameter of 4.4 nm already quite largely resembles that of bulk anatase, since the surface-to-bulk ratio (0.43) is rather reduced with respect to the other nanospheres (1.70 > 0.94 > 0.83). The other peaks, for the second and third coordination spheres of Ti$\cdots$Ti and of

Ti···O, are also quite broad, except for the largest nanosphere, where they are rather sharp and centered at the bulk values. In the case of DFTB (Figure 2b,d,f,h,j), similar considerations hold, although a noticeable difference is that the larger Ti-O_{ax} bonds concentrate at the value of about 2.25 Å. This is a surface distortion effect, which becomes progressively lower as the size of the nanosphere increases.

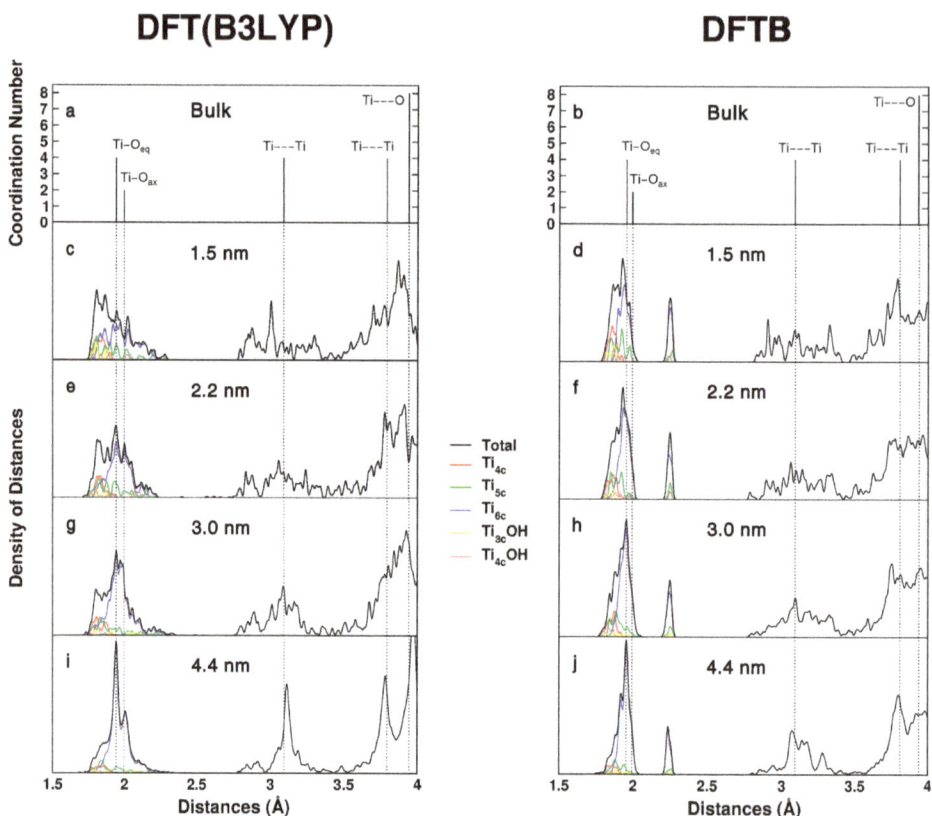

Figure 2. Distances distribution (simulated EXAFS) computed with DFT(B3LYP) and DFTB for bulk anatase (**a,b**), computed with DFT(B3LYP) for the: 1.5 nm (**c**); 2.2 nm (**e**); 3.0 nm (**g**); and 4.4 nm (**i**) nanospheres; and computed with DFTB for the: 1.5 nm (**d**); 2.2 nm (**f**); 3.0 nm (**h**); and 4.4 nm (**j**) nanospheres.

Hybrid functional B3LYP simulated EXAFS spectrum for the 2.2 nm nanoparticle was compared to the corresponding one obtained from the fully optimized 2.2 nm nanoparticle by the semilocal PBE functional in Figure S1. It is interesting to note that the two curves almost overlap, showing an excellent agreement between the two methods.

3.2. Electronic Properties

The electronic properties of a semiconducting oxide with a relatively large band gap as TiO_2 are not simply described by any quantum mechanical method. Standard DFT methods severely underestimate the band gap value, whereas hybrid DFT, as a consequence of the contribution of exact exchange in the exchange functional, provide values in closer agreement with experiments [22]. DFTB has been tested for bulk TiO_2 calculations and found to be in excellent agreement with experimental data and DFT Hubbard corrected values [23]. Therefore, DFTB is expected to perform well when investigating TiO$_2$ nanoparticles.

In the following, we present a comparison of the density of states (DOS) for the nanospheres of different size, as shown in Figure 3, that have been obtained with DFT(B3LYP) and DFTB on the corresponding fully relaxed structures. Considering that nanoparticles are finite systems, one could not really define true band states and band gaps. We have decided to distinguish between very localized states (molecular orbitals) and delocalized on several atoms of the nanoparticle (pseudo band states). These two definitions, based on a threshold value of 0.02 for the maximum squared coefficient of each eigenstate (max_c), lead to two different values of gap: the HOMO-LUMO gap and the BAND gap, which are reported in Table 2. We observe a decreasing trend with both DFT(B3LYP) and DFTB methods. The BAND gap values progressively approach the bulk Kohn–Sham value of 3.81 eV for DFT(B3LYP) and of 3.22 eV for DFTB (the experimental band gap for bulk anatase is 3.4 eV at 4 K) [52]. This is in line with experimental data [53] based on UV-Vis optical techniques the band gap of TiO$_2$ nanoparticles increases with decreasing size, due to quantum confinement effects.

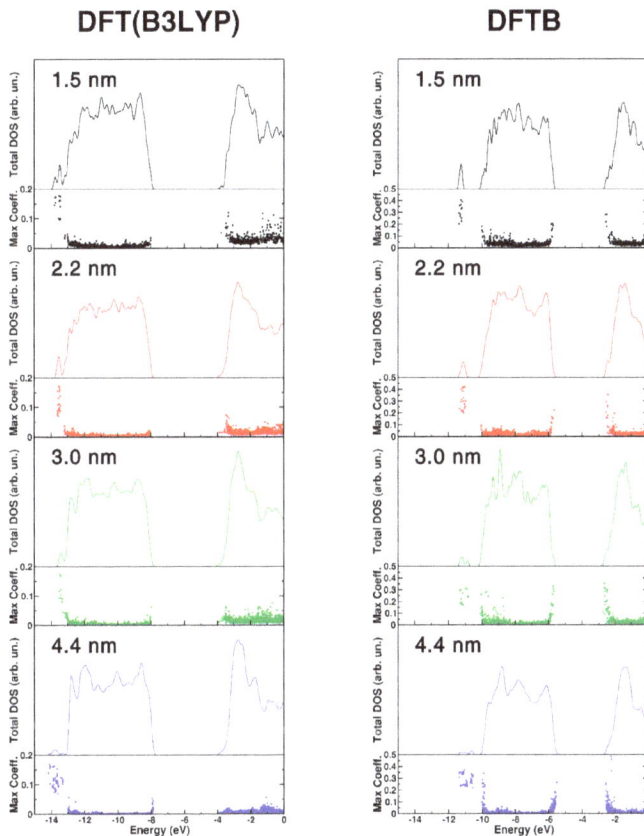

Figure 3. DFT(B3LYP) and DFTB total (DOS) density of states for different size nanosphere, 1.5 nm (black), 2.2 nm (red), 3.0 nm (green), and 4.4 nm (blue). For each nanosphere, the DOS has been normalized to the number of TiO$_2$ units to have comparable DOS intensities. The maximum atomic orbital coefficient (max_c) of each eigenstate is also reported. High values of max_c correspond to localized states, while low values correspond to delocalized states.

Table 2. HOMO-LUMO electronic gap (ΔE_{H-L}) and electronic BAND gap ΔE_{BAND} (expressed in eV) calculated for nanospheres of different size and for bulk anatase with both DFT(B3LYP) and DFTB methods.

MODEL	ΔE_{H-L}		ΔE_{BAND}	
Nanospheres	DFT(B3LYP)	DFTB	DFT(B3LYP)	DFTB
1.5 nm	4.23	3.12	4.81	3.62
2.2 nm	4.13	3.11	4.31	3.55
3.0 nm	4.00	2.95	4.13	3.42
4.4 nm	3.92	2.95	3.96	3.33
BULK	-	-	3.81	3.22

The DOS curves shown in Figure 3 for the different nanospheres further confirm a band gap opening going from the largest one (bottom panel) to the smallest one (upper panel). Additionally, we present the value of max_c for each eigenstate because we wish to highlight the degree of localization/delocalization of the states making up the DOS. The higher the value of max_c, the higher the localization. DFT(B3LYP) results show some localization at the band edges (top of the valence band and bottom of the conduction band) and in the range of OH groups in the low energy range (at about -14 eV). DFTB results are qualitatively similar, although some excess localization can be observed. In Figure S2 of the Supplementary Materials we have also reported and compared the total DOS for the anatase bulk TiO$_2$ calculated at DFT(B3LYP) and DFTB level of theory.

We may conclude this section devoted to the preparation and description of models for TiO$_2$ spherical nanoparticles with the following summarizing remarks: spherical models carved from bulk supercells must be made chemically stable by the introduction of some hydroxyl groups that saturate highly undercoordinated sites; such rough models must then undergo a simulated thermal annealing (with DFTB method) that allows to achieve global minimum stable structures; those must be then further optimized either with DFTB or, to reach even higher accuracy, with a hybrid DFT method (here B3LYP). Although some fine details are different, the general picture we obtain with DFTB is rather similar to that from DFT(B3LYP), which assesses DFTB as a reliable method to investigate nanoparticles of large realistic size (up to 4000 atoms, corresponding to a diameter of 4.4 nm). This will allow for further future development and for the study of nanoparticles' surface functionalization.

4. Modelling Photoactivation of TiO$_2$ Nanoparticles

Titanium dioxide is still considered the reference system in the research fields of photocatalysis and photovoltaics for its ability to convert light photons into chemical energy. In the very beginning of the photocatalytic process in a semiconductor, when the material is irradiated with light, an electron/hole pair or "exciton", is initially formed [54,55]. Then, if the coupling with lattice vibrations is strong enough, the exciton may become self-trapped (self-trapped exciton, STE) on few atoms of the crystal, reducing significantly its mobility. Finally, the photoexcited charge carriers may: (i) migrate towards the surface of the semiconductor as trapped electrons or holes and express their intrinsic redox activity; or (ii) recombinate radiatively and emit a photoluminescence photon.

In this regard, the quantum confinement of an exciton in a nanoparticle with a dimension of few nanometers may significantly influence its size and localization [56]. Furthermore, in a TiO$_2$ spherical nanoparticle, the close presence of highly undercoordinated surface sites may be a driving force for the process of excitons separation into electrons and holes or, on the contrary, may accelerate the radiative recombination via exciton self-trapping processes.

The study of photoexcited charge carriers in nanoparticles, although very demanding, must be performed by using a hybrid functional method (here B3LYP), since any other local or semilocal functional, and therefore also DFTB, would not properly describe the degree of electron/hole localization as a consequence of the self-interaction problem inherent in those methods [57]. DFT+U approach could be an alternative viable route; however, hole trapping was found to required very high and unphysical U values [58].

4.1. Free/Trapped Excitons and Radiative Recombination

We first recall the nature of the self-trapped exciton (STE) in bulk anatase, as shown in Figure 4 and reported in more detail in a previous work [27]. When the system is allowed to relax from the fully delocalized exciton initial condition (Figure 4b), two different self-trapped excitons can be localized, which differ for the O atom of the TiO_6 octahedron involved in the trapping: this can be either the equatorial O with respect to the electron trapping Ti (Ti^{3+}-O_{eq}^- in Figure 4c) or the axial one (Ti^{3+}-O_{ax}^- in Figure 4d).

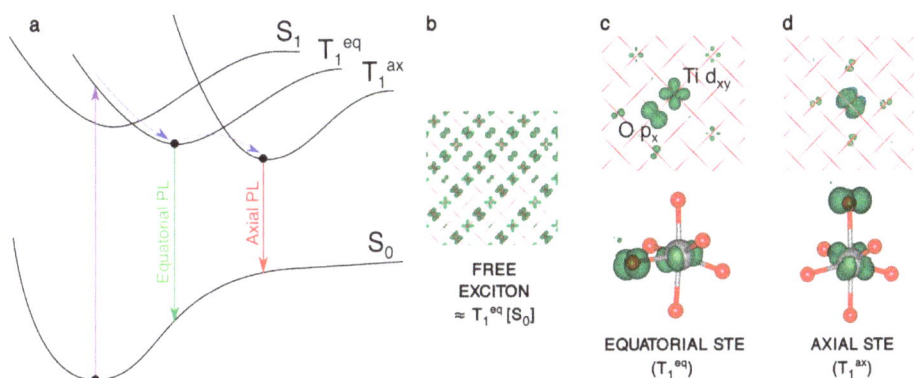

Figure 4. (a) Schematic representation of the processes involving the electron/hole couple: the vertical excitation $S_0 \rightarrow T_1$, the self-trapping relaxation in the bulk structure and the two different photoemissions $T_1^{eq} \rightarrow S_0$ and $T_1^{ax} \rightarrow S_0$. (b–d) 3D spin density plots of the anatase bulk supercell, as obtained with the B3LYP functional, for the vertical triplet state (b), trapped triplet equatorial (c) and axial (d) exciton in the bulk. The spin density isovalue is 0.01 a.u. (0.0005 a.u. for the vertical triplet).

The trapping energy (ΔE_{trap}), that is defined here and in the following as the energy difference between the fully relaxed trapped system state and the fully delocalized solution in the system ground state geometry, is more negative by -0.1 eV for the axial exciton (see Table 3). The computed photoluminescence (PL) energies for the decay of these two self-trapped excitons are given in Table 3. Noteworthy, the average PL is 2.24 eV, in very good agreement with the experimental value of 2.3 eV [59].

Table 3. Trapping Energy (ΔE_{trap}) of the exciton in the triplet state and the corresponding photoluminescence (PL) in the axial and equatorial configuration for bulk anatase and for the nanosphere. All energies are in eV.

		Bulk	Nanosphere
Ti^{3+}-O_{ax}^-	ΔE_{trap}	−0.59	−0.65
	PL	1.99	1.96
Ti^{3+}-O_{eq}^-	ΔE_{trap}	−0.49	
	PL	2.35	

In the spherical anatase nanoparticle model, the vertical excitation does not lead to a fully delocalized solution as for bulk (Figure 5a), but the resulting exciton partially localizes on a portion of the curved surface, even if the nanosphere is in its ground state optimized structure. After atomic relaxation, the exciton becomes trapped in the core of the nanosphere (Figure 5b). As for bulk, the axial solution is favored, with trapping and photoluminescence energies similar to the bulk values (Table 3). Thus, to summarize, a confinement effect in the nanosized systems is observed for the "Franck–Condon" exciton, but not for the self-trapped exciton in the core.

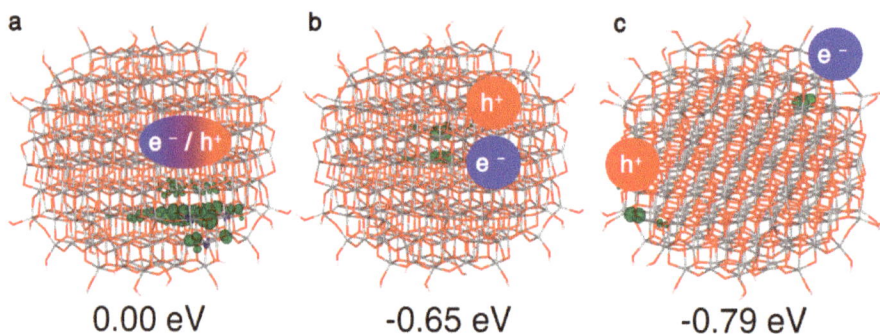

Figure 5. 3D spin density plots of the spherical TiO_2 nanoparticle, as obtained with the B3LYP functional, for the vertical triplet state (**a**), trapped triplet exciton (**b**) and state (**c**) with the hole and the electron at the best trapping sites. The spin density isovalue is 0.01 a.u. (0.002 a.u. for the vertical triplet).

As a next step, if the electron/hole couple has enough energy to separate, the charge carriers may migrate to the surface, where many trapping sites are available. Indeed, as we will discuss in Section 4.2, the most stable configuration for the electron and the hole are a subsurface Ti_{6c} site and a surface O_{2c} site, respectively. Considering these as trapping sites for the electron and the hole, the trapping energy of the separated exciton, shown in Figure 5c, amounts to −0.79 eV, significantly larger than the one of the bound exciton in the core (−0.65 eV). Therefore, there is a favorable energy gradient for the electrons and holes to separate and to move towards the surface, which is the driving force for separation and migration processes.

4.2. Separated Carriers Trapping

When the charge carriers are far apart in different regions of the nanosphere, as in Figure 5c, they behave almost like isolated charges. Thus, it is possible to study the relative stability of electron or hole trapping sites introducing a single extra electron or extra hole in the system [28].

Within an anatase spherical nanoparticle, we observed that the excess electron, when added to the system in its ground state geometry, fully delocalizes on all the Ti centers (see Figure 6a), except those in the outermost atoms on the curved surface. This is used as the reference system for the evaluation of the trapping energy (ΔE_{trap}).

After atomic relaxation in the presence of this extra electron, the spin density localizes on several Ti atoms within the central three atomic core layers (see Figure 6b), with an energy gain of −0.11 eV. In other words, a quite stable large polaron, involving few atomic layers, is formed. This partially trapped intermediate situation, also referred to as "shallow trap" [3], cannot be observed in periodic models of bulk and slabs.

We also investigated the electron localization at "deep traps", i.e., single atomic Ti sites in the nanosphere. The trapping energy and the electron localization are shown in Table 4. Unexpectedly, no trapping has been observed on undercoordinated Ti atoms, i.e., four-fold coordinated at the equator of the nanosphere ($Ti_{4c}^{equator}$), four-fold coordinated Ti with a terminal OH ($Ti_{4c}(OH)$) and five-fold coordinated (Ti_{5c}) sites, except for a very small trapping energy (−0.09 eV) in the case of the $Ti_{4c}^{equator}$ site (see Table 4).

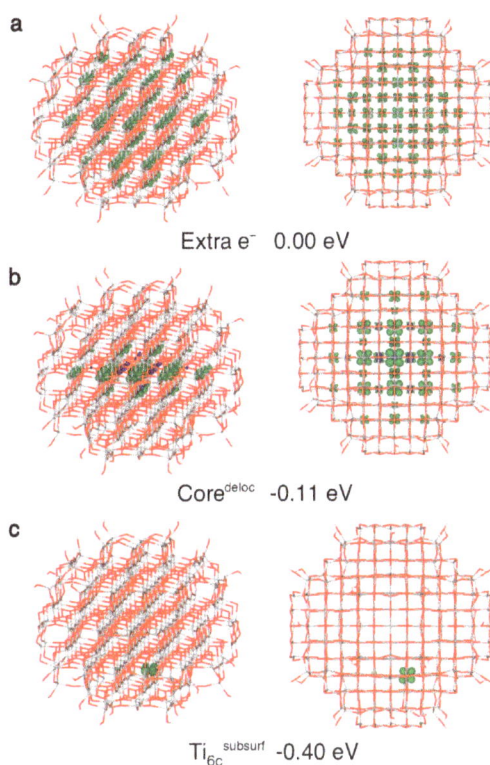

Figure 6. Front and top view of the 3D plots of spin density of: (**a**) an extra electron added to the neutral ground state structure (isovalue = 0.001 a.u.); (**b**) a trapped delocalized electron (isovalue = 0.005 a.u.); and (**c**) an electron trapped on a subsurface six-fold coordinated titanium atom (isovalue = 0.001 a.u.) in the anatase nanosphere model, as obtained with the B3LYP functional. Below each structure, the energy gain (ΔE_{trap}) relative to (**a**) is given.

Table 4. Trapping energy (ΔE_{trap}) for electrons at different sites in the spherical anatase nanoparticles, as obtained with B3LYP functional. The reference zero for ΔE_{trap} is obtained by adding one electron to the nanosphere in its neutral ground state geometry, with no atomic relaxation. The charge localization (%electron) is also given. The sites nomenclature is defined in the text.

Position	ΔE_{trap} (eV)	%electron
$Ti_{4c}^{equator}$	−0.09	88%
$Ti_{4c}(OH)$	No trapping	
Ti_{5c}	No trapping	
$Ti_{6c}^{subsurf}$	−0.40	85%
Ti_{6c}^{core}	−0.13	62%
$Core^{deloc}$	−0.11	19%

On the contrary, the best electron trap is the fully coordinated Ti_{6c} site on the subsurface ($Ti_{6c}^{subsurf}$ in Table 4 and Figure 6c) with a ΔE_{trap} of −0.40 eV. Noteworthy, the electron delocalization in the core of the nanoparticle ($Core^{deloc}$ in Table 4 and Figure 6b) has been found to be less favored than complete or full localization on a single subsurface Ti site (−0.11 vs. −0.40 eV), indicating an energy gradient, and thus a driving force, for the migration and localization of photoexcited electrons towards atoms near the surface.

Concerning the hole trapping on spherical TiO_2 nanoparticles, if one electron is removed without any atomic relaxation (vertical ionization), only some regions of the nanoparticles are involved, as shown in Figure 7a. Since this solution cannot be seen as a delocalized band-like state of a free (or untrapped) hole, its total energy cannot be used as the reference to compute trapping energies (ΔE_{trap}). Thus, in the following we will use adiabatic ionization potentials (IPs) for comparisons, since they correlate with trapping energies: the smaller the adiabatic IP, the larger the trapping energy of the considered site. In the core of the nanosphere, the hole almost completely localizes (90%) on the central three-fold coordinated O atom ($O_{3c}^{core\_ax}$ in Figure 7b and Table 5). Differently from what found for electrons and discussed above, we could not identify any delocalized "shallow trap" state for holes.

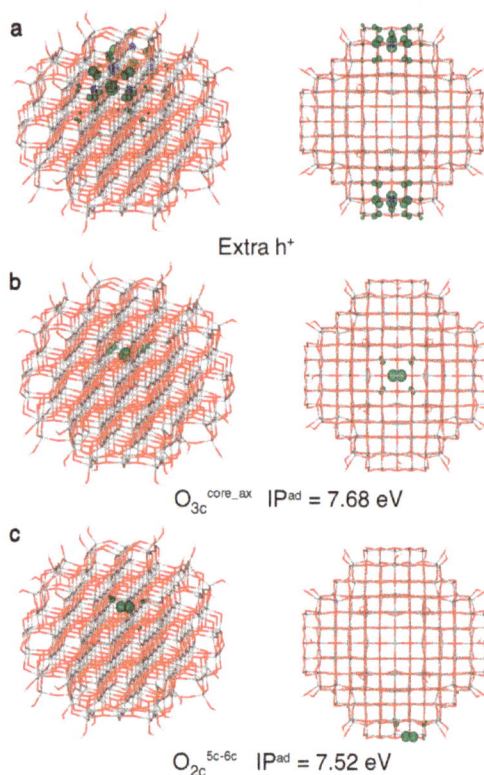

Extra h$^+$

$O_{3c}^{core\_ax}$ IP^{ad} = 7.68 eV

O_{2c}^{5c-6c} IP^{ad} = 7.52 eV

Figure 7. Front and top view of the 3D plots of spin density of: (**a**) an excess hole resulting from a vertical ionization of the model (isovalue = 0.001 a.u.); (**b**) a trapped hole in the core (isovalue = 0.01 a.u.); and (**c**) a hole trapped on a two-fold coordinated oxygen atom (isovalue = 0.01 a.u.) in the anatase nanosphere model, as obtained with the B3LYP functional. Below the structures with a trapped hole the adiabatic ionization potential (IP^{ad}) as a measure of the hole trapping ability is given.

Table 5. Adiabatic ionization potential (IP^{ad}) for holes at different sites of the spherical anatase nanoparticle, as obtained with B3LYP functional. The charge localization (%hole) is also reported. The sites nomenclature is defined in the text.

Position	IP^{ad} (eV)	%hole
O_{2c}^{4c-6c}	7.74	89%/11%
O_{2c}^{5c-6c}	7.52	92%/8%
$O_{3c}^{core\_ax}$	7.68	90%
$Ti_{3c}(OH)$	No trapping	
$Ti_{4c}(OH)$	No trapping	
$Ti_{5c}(OH)$	7.81	85%/15%

If we allow the hole to reach the curved surface of the nanosphere, several types of O sites are available for trapping. Among them, the most stable one is a two-fold O atom bridging a Ti_{5c} and a Ti_{6c} atom (O_{2c}^{5c-6c} Table 5 and Figure 7c). A second type of two-fold O on the surface, between a Ti_{4c} and a Ti_{6c} atom (see O_{2c}^{4c-6c} in Table 5), can also trap the hole but less efficiently than a O_{2c}^{5c-6c} site. It is worth underlining that the migration of holes from the nanosphere core to the surface is energetically favourable by -0.12 eV.

Finally, one should note that on the surface of a spherical nanoparticle there are several hydroxyl groups that may be stable hole trapping sites. However, as reported in Table 5, we were not able to localize the hole on the hydroxyl group of a $Ti_{3c}(OH)$ and $Ti_{4c}(OH)$ sites on the nanosphere surface. On the contrary, we could trap the hole on a $Ti_{5c}(OH)$ site, formed upon dissociation of a water molecule on a Ti_{5c}, probably because a OH bound to a fully coordinated Ti site is more electron rich than one bound to an undercoordinated Ti atom. Nonetheless, the OH trapping site is less effective by 0.29 eV than the most stable O_{2c} hole trapping site. Therefore, in vacuum, the OH groups are not good trapping sites, but the scenario may change in an aqueous medium, where water molecules may enhance the trapping properties of the hydroxyl groups, through binding as a ligand to TiOH or H-bonding to the hydrogen of the OH.

4.3. Comparison with Experiments

Experimental data on trapped charges in anatase TiO_2 are available in literature and they can be compared with calculations performed with the spherical nanoparticle models shown above. First, the calculated values of the trapping energy relative to a free electron in the conduction band are in good agreement with the experimental observations for both shallow (delocalized) [60–63] and deep (localized) [64–66] electron trapping states.

Moreover, the degree of electron localization can be probed through the hyperfine coupling constant (a_{iso}) with the next-neighboring ^{17}O in the electron paramagnetic resonance (EPR) spectrum. High values of a_{iso} are expected for localized electrons, low values for delocalized ones. Indeed, the computed a_{iso} is 6.7 MHz for an electron localized on the innermost Ti atom of the NP (see Ti_{6c}^{core} Table 4), whereas it is 3.9 MHz for an electron delocalized in the NP core ($Core^{deloc}$ in Table 4), in good agreement with experimental observations of a significant decrease of a_{iso} going from a fully localized electron on a single Ti ion (as in the $Ti^{3+}(H_2O)_6$ complex) [67] to shallow electron traps in anatase nanoparticles [68]. For the most stable hole trap on the surface, the computed EPR parameters (g = [2.004, 2.015, 2.019] G and A = [31, 30, −97] G) are in excellent quantitative agreement with the g- and A-tensor data available in the experimental literature [69,70]. Hence, we may conclude that a correct localization of both charge carriers is provided by the computational models and methods.

Finally, we employed the transition level approach [27,71] to estimate the electronic transition energies of the charged traps, since this methodology produces accurate results for excitations of defects in solids. In the case of electrons, the calculated transition from the trap level of the best electron trap ($Ti_{6c}^{subsurf}$ in Table 4) to the conduction band minimum (CBM) is 1.25 eV, in accordance with the experimental value of 1.37 eV, measured with transient absorption (TA) spectroscopy [72].

In the case of holes, we computed a transition from the valence band maximum to the trap level for the best hole trapping site ($O_{2c}{}^{5c-6c}$ in Table 5) of 2.59 eV, to be compared with a reported experimental value of 1.9 eV in the experimental TA spectrum [72]. This inconsistency between the computed and experimental results may arise because these experiments have been performed in an aqueous solution and, as mentioned in Section 4.2, the presence of water layers on the nanoparticle may influence the hole trapping ability of the system, as reported in a recent experimental work [35].

To conclude this section devoted to the study of the life path of energy carriers (excitons) and charge carriers (electron and holes) in spherical TiO$_2$ nanoparticles by hybrid DFT(B3LYP), we can summarize as follows: the photoexcited exciton self-trapping is a favorable process but electron and hole can then separate to migrate towards the surface where they can be highly stabilized. In particular, for electrons, we observed that deep trapping at subsurface fully coordinated Ti sites is favored with respect to shallow trapping in the core. In the case of holes, only deep traps were observed with a surface O$_{2c}$ (Ti$_{5c}$-O-Ti$_{6c}$) being the preferential hole trapping site. Computed electron paramagnetic resonance (EPR) parameters and optical transitions for those electron/hole traps are in good agreement with experimental data.

5. Modelling Surface Interaction with Water

Understanding the interaction of water with TiO$_2$ anatase surface [73] is essential since TiO$_2$-based technologies, including photocatalysis, normally operate in an aqueous environment. Many computational studies based on DFT methods have tackled the interaction between the most exposed anatase TiO$_2$ (101) surface and water layers [74–76] revealing how the surface complexity influences the water-titania interface.

However, the study of the dynamical behavior of real size TiO$_2$ nanoparticles (i.e., with diameter in the range 2–8 nm) [53,69,77–80] in a realistic aqueous environment and with sufficiently long simulation times, is currently not feasible with DFT methods.

As regards DFTB, from a technical point of view, its performance in the description of a certain system critically depends on the parameterization of the element-pairs interaction of the atoms involved. In the case of Ti-containing compounds, two different sets of parameters are available: "mio-1-1/tiorg-0-1" [23] and "matsci-0-3" [24]. The first set has been developed to handle the interaction of low index surfaces of both anatase and rutile with water and small organic molecule, but no assessment has been done for the anatase TiO$_2$ (101) surface. The second set has been thought to describe bulk TiO$_2$ structures and chemical reactivity of (101) anatase and (110) rutile surfaces with isolated molecule and monolayers of water. However, this set has been never tested for the description of bulk water, which is essential for a correct characterization of titania/water-multilayers interfaces.

Recently, we have shown [40] that if these two sets are properly combined in a new one, referred to as "MATORG", with some further improvement coming from the inclusion of an empirical correction [51] for a finer description of the hydrogen bonding ("MATORG+HBD"), it is possible to achieve a DFT-like description of the interaction between water-multilayers and anatase TiO$_2$ (101) surface. In the following, we will shortly present the performance of this new set of parameters for the static and dynamic description of TiO$_2$/water interface by comparison with previous DFT results. The positive assessment of DFTB methods for studying this type of solid/liquid interface is extremely important because it gives a solid basis for its application on large realistic nanoparticles in an aqueous environment.

5.1. Bulk Water and Anatase TiO$_2$ Description

The correct description of the titania/water-multilayers interface is tightly related to the method ability of properly describing the two components separately. Bulk TiO$_2$ lattice parameters have been calculated with DFTB and compared, in Table 6, with DFT and experimental values reported in literature [17,47,81,82]. The agreement is extremely good. The MATORG+HBD set gives an extremely accurate *a* value and only slightly overestimates the *c* lattice parameter and consequently the *c*/*a* ratio.

Table 6. Lattice parameters a and c and their c/a ratio for bulk TiO_2 anatase. Values computed with the DFTB and DFT methods are reported and compared with the experimental values. In parenthesis, the absolute errors referred to the experimental data are shown.

Method	Reference	a (Å)	c (Å)	c/a (Å)
DFTB-MATORG+HBD	This work	3.796 (+0.014)	9.790 (+0.288)	2.579 (+0.067)
DFT(PBE)	This work	3.789 (+0.007)	9.612 (+0.110)	2.537 (+0.025)
DFT(PBE)	Ref. [81]	3.786 (+0.004)	9.737 (+0.235)	2.572 (+0.060)
DFT(B3LYP)	Ref. [82]	3.783 (+0.001)	9.805 (+0.303)	2.592 (+0.080)
DFT(B3LYP)	Ref. [17]	3.789 (+0.007)	9.777 (+0.275)	2.580 (+0.068)
DFT(HSE06)	Ref. [17]	3.766 (−0.016)	9.663 (+0.161)	2.566 (+0.054)
Exp.	Ref. [47]	3.782	9.502	2.512

To assess the reliability of DFTB method and MATORG+HBD parameters for the description of bulk water, two different features are evaluated: the H-bond strength and the radial distribution function (RDF). We estimate the H-bond strength (ΔE_{H-bond}) by the water dimer binding energy and report it in Table 7, together with the equilibrium O-O distances (R_{O-O}) of the water dimer. DFTB results are compared with those obtained with standard and hybrid DFT [83], CCSD [84] and experiments [85,86]. The description of the H-bond with the MATORG+HBD set is extremely good, very close to the first-principles and experimental references.

Table 7. Water dimer binding energy (ΔE_{H-bond}) and oxygen–oxygen distance (R_{O-O}). Values obtained with DFTB (MATORG+HBD), higher-level methods (CCSD, PBE and B3LYP) and experiments are reported. In parenthesis, the absolute errors referred to the experimental data are shown.

Method	Reference	ΔE_{H-bond} (eV)	R_{O-O} (Å)
DFTB-MATORG+HBD	This work	0.199 (−0.037)	2.815 (−0.157)
DFT(PBE)	Ref. [83]	0.222	2.889
DFT(B3LYP)	Ref. [83]	0.198	2.926
CCSD	Ref. [84]	0.218	2.912
Exp.	Refs. [85,86]	0.236	2.972

The radial distribution functions (RDF) of oxygen–oxygen (O_w-O_w) and hydrogen–hydrogen (H_w-H_w) by DFTB with the MATORG+HBD set of parameters are compared with the experimental ones in Figure 8 [87].

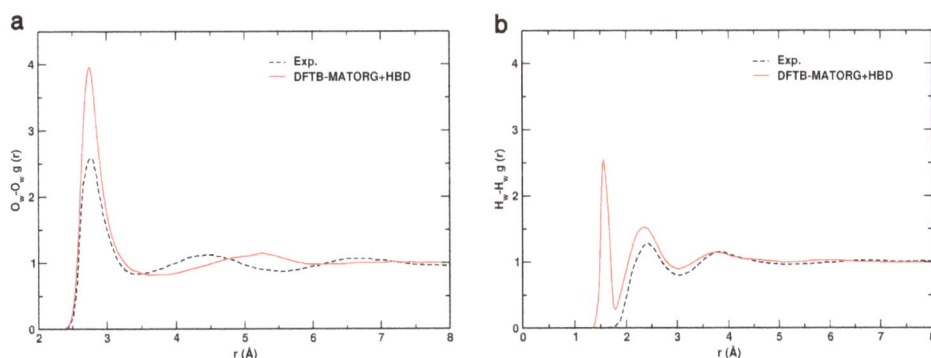

Figure 8. Comparison of the: oxygen–oxygen (O_w-O_w) (**a**); and hydrogen–hydrogen (H_w-H_w) (**b**) radial distribution functions (RDF) obtained experimentally (dashed black line) and calculated with the DFTB-MATORG+HBD method.

In the experiment, the first intermolecular peaks for $r(O_w\text{-}O_w)$ and $r(H_w\text{-}H_w)$ are found to be located at 2.77 and 2.31 Å, respectively. In excellent agreement, with the MATORG+HBD set, the first peak position of the $O_w\text{-}O_w$ RDF is at 2.75 Å (Figure 8a). The experimental/theoretical curves partially overlap for distances lower than 3.30 Å: the water density depletion zones are very similar and the second intermolecular peak is at 5.25 Å, not too far from the experimental data (4.55 Å). Regarding the $H_w\text{-}H_w$ radial distribution function (Figure 8b), the experimental curve is well reproduced by MATORG+HBD, with the first intermolecular peak located at 2.34 Å.

5.2. Static Description of TiO₂/Water Interface

The molecular (undissociated, H_2O) and dissociated (OH, H) adsorption of water on the anatase TiO_2 (101) surface has been investigated at different water coverages (low, $\theta = 0.25$ and full, $\theta = 1$). The adsorption energy per molecule (ΔE_{ads}^{mol}) has been calculated with MATORG and MATORG+HBD and compared with DFT(PBE) results and experimental measurements in Table 8.

Table 8. Values calculated with DFT and DFTB methods of the adsorption energies per molecule (ΔE_{ads}^{mol}) of water on the TiO_2 (101) anatase slab in the molecular (H_2O) and dissociated (OH, H) state. Different coverages are considered (low, $\theta = 0.25$ and full, $\theta = 1$). The experimental adsorption energy of the water monolayer on the (101) surface is also reported. The absolute errors (in parenthesis) reported for DFTB are calculated with respect to the PBE values from this work.

Method	Reference	Coverage, θ	ΔE_{ads}^{mol}, H_2O (eV)	ΔE_{ads}^{mol}, OH, H (eV)
DFTB-MATORG	This work	0.25	−1.08 (+0.41)	−0.54 (+0.22)
		1	−0.96 (+0.34)	−0.58 (+0.15)
DFTB-MATORG+HBD	This work	0.25	−0.80 (+0.13)	−0.31 (−0.01)
		1	−0.71 (+0.09)	−0.40 (−0.03)
DFT(PBE)	This work	0.25	−0.67	−0.32
		1	−0.62	−0.43
DFT(PBE)	Ref. [88]	0.25	−0.74	−0.23
		1	−0.72	−0.44
Exp.	Refs. [89,90]	1	−0.5/−0.7	

The DFTB method predicts the molecular adsorption mode of a single water molecule to be favored with respect to the dissociated one. This is in line with several experimental observations [73,89–91] and previous DFT data [10,88]. In the full coverage regime ($\theta = 1$), the MATORG set also correctly reproduces the binding energy decrease for the molecular adsorption mode and the increase for the dissociated one. However, the MATORG set tends to overestimate adsorption energies with errors up to 0.41 eV. This discrepancy is almost solved with the inclusion of the HBD correction, which reduces the error values to less than 0.13 eV.

5.3. Dynamic Description of TiO₂/Water Interface

The study of complex and realistic TiO_2 (nano)systems in aqueous environment is strongly related to the ability of the method used to describe the titania/water-multilayers dynamic behavior. To assess the performance of the MATORG+HBD set of parameters, first-principles simulations must be used as reference. We will use Car–Parrinello molecular dynamics (CPMD) DFT(PBE) simulations [75] and other DFT(PBE) structural investigations [92] of water layers on the TiO_2 (101) anatase surface, which already exist in the literature.

In Figure 9, we show the 0 K optimized geometries starting from the last snapshot of each molecular dynamics trajectory performed with the MATORG+HBD set, in the case of the fully undissociated water monolayer (ML), bilayer (BL) and trilayer (TL) of water on the (101) TiO_2 anatase surface, respectively.

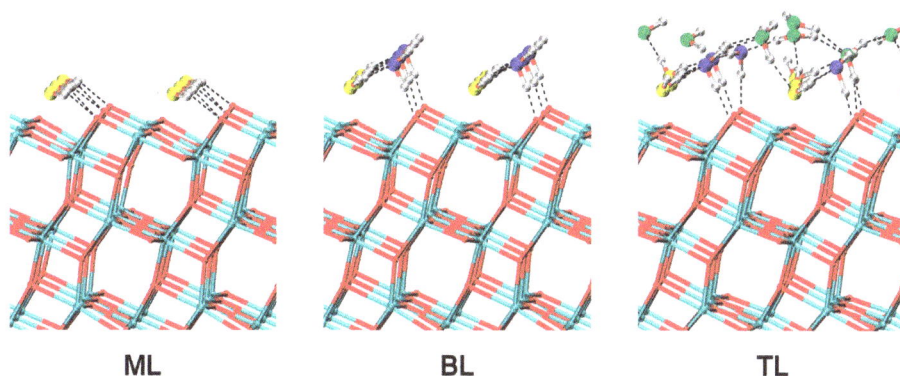

Figure 9. DFTB-MATORG+HBD structures of a monolayer (ML), a bilayer (BL) and a trilayer (TL) of water on the (101) TiO$_2$ anatase surface. Dashed lines correspond to H-bonds.

For the ML, each water molecule (in yellow in Figure 9, left panel) binds a Ti$_{5c}$ of the surface and establishes two H-bonds with the O$_{2c}$ with an adsorption energy per molecule (ΔE_{ads}^{mol} in Table 9), calculated with the MATORG+HBD method, of -0.70 eV, in very good quantitative agreement with the DFT(PBE) references [74].

Table 9. Values calculated with DFT and DFTB methods of the binding energy per molecule (ΔE_{ads}^{mol} in eV) of the water monolayer (ML), bilayer (BL) and trilayer (TL) on the (101) TiO$_2$ anatase surface after an optimization run from the last snapshot of the MD simulation. The binding energy (ΔE_{ads}^{mol}) is defined as the difference between the total energy of the titania/water interface equilibrium structure and the sum of the total energy of six isolated water molecules plus the total energy of the optimized slab with one water layer less.

Water Configuration	ΔE_{ads}^{mol} (eV)		
	DFTB-MATORG+HBD	**DFT(PBE)** [a]	**DFT(PBE)** [b]
ML	-0.70	-0.62	-0.69
BL	-0.73	-0.67	-0.65
TL	-0.53	-0.53	-0.56

[a] This work; [b] from Ref. [74].

In the BL configuration, the water molecules of the first layer are bound to Ti$_{5c}$ atoms of the surface and form two H-bonds with two molecules of the second layer. The water molecules of the second layer (in blue in Figure 9, middle panel) have only one H-bond with an O$_{2c}$ of the titania surface, with the other H atom pointing towards the vacuum. For the BL, the MATORG+HBD adsorption energy is -0.73 eV, in agreement with DFT(PBE) results (see Table 9).

The TL case is more complicated, since the third water layer (in green in Figure 9, right panel) is too mobile to allow for a unique structure definition. We added a third water layer on the BL equilibrium structure with a MATORG+HBD adsorption energy of -0.53 eV (see Table 9), again in very good agreement with the DFT(PBE) previous study.

Finally, to analyze the behavior of the titania/water-multilayers interfaces during the MD simulations, the distribution $p(z)$ of the vertical distances between the O atoms of the H$_2$O molecules and the Ti$_{5c}$ plane of the surface, together with their time evolution ($z(t)$), were extracted from the MD trajectory, as shown in Figure 10, and compared to DFT(PBE) CPMD results [75].

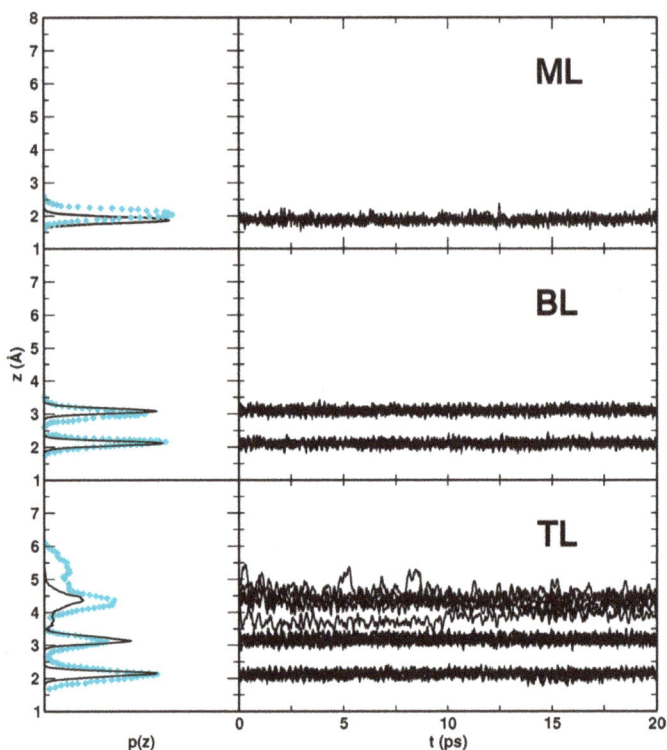

Figure 10. DFTB-MATORG+HBD distribution $p(z)$ and time evolution $z(t)$ of the distances between the water molecules (O atoms) of the monolayer (ML), bilayer (BL) and trilayer (TL) and the titania surface (Ti_{5c} atoms). In cyan diamonds, values calculated with DFT(PBE) Car–Parrinello simulations are shown.

In the case of the ML (Figure 10, top panel), the agreement with the Car–Parrinello (PBE) molecular dynamics data is satisfactory: as it can be seen from the time evolution of perpendicular distances, the molecules librate around their equilibrium site and give a total $p(z)$ distribution very similar to the reference with the peak shifted by only 0.1 Å to shorter values.

Regarding the BL molecular dynamics simulation (Figure 10, central panel), the agreement with the CPMD (PBE) is extremely good. In the CPMD DFT(PBE) case, the position of the $p(z)$ distribution peak is at 2.15 Å for the first water layer and at 2.98 Å for the second one, whereas in the MD with MATORG+HBD, those are at 2.11 Å and at 3.08 Å, respectively. The BL configuration is very stable since none of the water molecule has left its initial equilibrium position in the whole simulation time.

For the water TL (Figure 10, bottom panel) we observe again a very good agreement between the DFT(PBE) and DFTB curves: the first two water layers in the MATORG+HBD molecular dynamics are vertically ordered in their initial equilibrium. On the contrary, the third layer water molecules are very mobile interacting with the second layer through H-bond. The range of the third layer vertical distances evaluated with MATORG+HBD is $3.6 < z < 5.1$ Å, thus shorter than the one calculated with DFT(PBE) ($\sim 4 < z < 6$ Å).

To conclude this section, we have shown that the description by the parametrized DFTB method with the MATORG+HBD set of the titania/water-multilayers interface (static and dynamic calculations) is in very good agreement with DFT(PBE) results. In particular, the MATORG+HBD set correctly describes the key aspects of the multilayer water adsorption on TiO_2 surface, properly balancing the surface/water and water/water interactions. Based on this assessment and on the computational efficiency of DFTB, we conclude that this method enables the study of large and realistic TiO$_2$ nanostructured systems into an aqueous environment.

6. Concluding Remarks

In the sections above, we have presented an overview of current possibilities, as explored by our group with state-of-the-art DFT and DFT-based (DFTB) methodologies, for the description of realistic nanoparticles in water solution for photoapplications. It is evident that, when the size of the nanoparticle becomes relatively large (about 4000 atoms), to achieve a diameter size close to the smallest TiO$_2$ nanoparticles used in practical applications (4.4 nm), the feasible limit for hybrid DFT calculations is almost reached. We showed that DFTB method can be used to obtain global minimum structures and to provide a reasonable description of both structural and electronic properties of these complex systems. Additionally, DFTB method is found to also yield a satisfactory accuracy for the description of the water layers on top of TiO$_2$ surfaces, which allows introducing the water environment explicitly into the calculations. However, we have shown that the DFT level of theory is still mandatory when one wants to describe photoexcitation processes taking place in the nanoparticles or on its surface, such as exciton formation, charge carrier trapping or electron transfer to adsorbates.

Supplementary Materials: The following are available online at www.mdpi.com/2073-4344/7/12/357/s1, Figure S1: Comparison of the distances distribution (simulated EXAFS) computed with DFT(PBE), in black and DFT(B3LYP) in red, for the 2.2 nm NS produced at 300 K, Figure S2: DFT(B3LYP) and DFTB total (DOS) density of states for anatase bulk TiO$_2$. The maximum atomic orbital coefficient (max_c) of each eigenstate is also reported.

Acknowledgments: The authors are grateful to Gotthard Seifert for very fruitful discussions and to Lorenzo Ferraro for his technical help. The project has received funding from the European Research Council (ERC) under the European Union's HORIZON2020 research and innovation programme (ERC Grant Agreement No [647020]) and from CINECA supercomputing center through the computing LI05p_GRV4CUPT grant.

Author Contributions: G.F., D.S. and C.D.V. conceived the models and designed the calculations; D.S. and G.F. performed the calculations; D.S., G.F. and C.D.V. performed the data analysis; C.D.V. wrote the manuscript with the help of D.S. and G.F.

Conflicts of Interest: The authors declare no conflict of interest.

References

1. Sang, L.; Zhao, Y.; Burda, C. TiO$_2$ Nanoparticles as Functional Building Blocks. *Chem. Rev.* **2014**, *114*, 9283–9318. [CrossRef] [PubMed]

2. Ma, Y.; Wang, X.L.; Jia, Y.S.; Chen, X.B.; Han, H.X.; Li, C. Titanium Dioxide-Based Nanomaterials for Photocatalytic Fuel Generations. *Chem. Rev.* **2014**, *114*, 9987–10043. [CrossRef] [PubMed]

3. Schneider, J.; Matsuoka, M.; Takeuchi, M.; Zhang, J.; Horiuchi, Y.; Anpo, M.; Bahnemann, D.W. Understanding TiO$_2$ Photocatalysis: Mechanisms and Materials. *Chem. Rev.* **2014**, *114*, 9919–9986. [CrossRef] [PubMed]

4. Tong, H.; Ouyang, S.; Bi, Y.; Umezawa, N.; Oshikiri, M.; Ye, J. Nano-photocatalytic Materials: Possibilities and Challenges. *Adv. Mater.* **2012**, *24*, 229–251. [CrossRef] [PubMed]

5. Cargnello, M.; Gordon, T.R.; Murray, C.B. Solution-Phase Synthesis of Titanium Dioxide Nanoparticles and Nanocrystals. *Chem. Rev.* **2014**, *114*, 9319–9345. [CrossRef] [PubMed]

6. Tang, J.; Redl, F.; Zhu, Y.; Siegrist, T.; Brus, L.E.; Steigerwald, M.L. An Organometallic Synthesis of TiO$_2$ Nanoparticles. *Nano Lett.* **2005**, *3*, 543–548. [CrossRef] [PubMed]

7. Chen, C.; Hu, R.; Mai, K.; Ren, Z.; Wang, H.; Qian, G.; Wang, Z. Shape Evolution of Highly Crystalline Anatase TiO$_2$ Nanobipyramids. *Cryst. Growth Des.* **2011**, *11*, 5221–5226. [CrossRef]

8. Barnard, A.S.; Zapol, P. Effects of Particle Morphology and Surface Hydrogenation on the Phase Stability of TiO$_2$. *Phys. Rev. B* **2004**, *70*, 235403. [CrossRef]

9. Rajh, T.; Dimitrijevic, N.M.; Bissonnette, M.; Koritarov, T.; Konda, V. Titanium Dioxide in the Service of the Biomedical Revolution. *Chem. Rev.* **2014**, *114*, 10177–10216. [CrossRef] [PubMed]

10. De Angelis, F.; Di Valentin, C.; Fantacci, S.; Vittadini, A.; Selloni, A. Theoretical Studies on Anatase and Less Common TiO$_2$ Phases: Bulk, Surfaces, and Nanomaterials. *Chem. Rev.* **2014**, *114*, 9708–9753. [CrossRef] [PubMed]

11. Hummer, D.R.; Kubicki, J.D.; Kent, P.R.C.; Post, J.E.; Heaney, P.J. Origin of Nanoscale Phase Stability Reversals in Titanium Oxide Polymorphs. *J. Phys. Chem. C* **2009**, *113*, 4240–4245. [CrossRef]

12. Nunzi, F.; Mosconi, E.; Storchi, L.; Ronca, E.; Selloni, A.; Gratzel, M.; De Angelis, F. Inherent Electronic Trap States in TiO$_2$ Nanocrystals: Effect of saturation and sintering. *Energy Environ. Sci.* **2013**, *6*, 1221–1229. [CrossRef]

13. Li, Y.-F.; Liu, Z.-P. Particle Size, Shape and Activity for Photocatalysis on Titania Anatase Nanoparticles in Aqueous Surroundings. *J. Am. Chem. Soc.* **2011**, *133*, 15743–15752. [CrossRef] [PubMed]

14. Mattioli, G.; Bonapasta, A.A.; Bovi, D.; Giannozzi, P. Photocatalytic and Photovoltaic Properties of TiO$_2$ Nanoparticles Investigated by Ab Initio Simulations. *J. Phys. Chem. C* **2014**, *118*, 29928–29942. [CrossRef]

15. Nunzi, F.; Storchi, L.; Manca, M.; Giannuzzi, R.; Gigli, G.; De Angelis, F. Shape and Morphology Effects on the Electronic Structure of TiO$_2$ Nanostructures: From Nanocrystals to Nanorods. *Appl. Mater. Interfaces* **2014**, *6*, 2471–2478. [CrossRef] [PubMed]

16. Lamiel-Garcia, O.; Ko, K.C.; Lee, J.Y.; Bromley, S.T.; Illas, F. When Anatase Nanoparticles Become Bulklike: Properties of Realistic TiO$_2$ Nanoparticles in the 1–6 nm Size Range from All Electron Relativistic Density Functional Theory Based Calculations. *J. Chem. Theory Comput.* **2017**, *13*, 1785–1793. [CrossRef] [PubMed]

17. Fazio, G.; Ferrighi, L.; Di Valentin, C. Spherical versus Faceted Anatase TiO$_2$ Nanoparticles: A Model Study of Structural and Electronic Properties. *J. Phys. Chem. C* **2015**, *119*, 20735–20746. [CrossRef]

18. Selli, D.; Fazio, G.; Di Valentin, C. Modelling Realistic TiO$_2$ Nanospheres: A Benchmark Study of SCC-DFTB against Hybrid DFT. *J. Chem. Phys.* **2017**, *147*, 164701. [CrossRef] [PubMed]

19. Elstner, M.; Porezag, D.; Jungnickel, G.; Elsner, J.; Haugk, M.; Frauenheim, T.; Suhai, S.; Seifert, G. Self-Consistent-Charge Density-Functional Tight-Binding Method for Simulations of Complex Materials Properties. *Phys. Rev. B* **1998**, *58*, 7260. [CrossRef]

20. Corà, F.; Alfredsson, M.; Mallia, G.; Middlemiss, D.S.; Mackrodt, W.C.; Dovesi, R.; Orlando, R. The Performance of Hybrid Density Functionals in Solid State Chemistry. In *Principles and Applications of Density Functional Theory in Inorganic Chemistry II. Structure and Bonding*; Springer: Berlin/Heidelberg, Germany, 2004; Volume 113.

21. Muscat, J.; Wander, A.; Harrison, N.M. On the Prediction of Band Gaps from Hybrid Functional Theory. *Chem. Phys. Lett.* **2001**, *3*, 397–401. [CrossRef]

22. Labat, F.; Baranek, P.; Adamo, C. Structural and Electronic Properties of Selected Rutile and Anatase TiO$_2$ Surfaces: An ab Initio Investigation. *J. Chem. Theory. Comput.* **2008**, *4*, 341–352. [CrossRef] [PubMed]

23. Dolgonos, G.; Aradi, B.; Moreira, N.H.; Frauenheim, T. An Improved Self-Consistent-Charge Density-Functional Tight-Binding (SCC-DFTB) Set of Parameters for Simulation of Bulk and Molecular Systems Involving Titanium. *J. Chem. Theory Comput.* **2010**, *6*, 266–278. [CrossRef] [PubMed]

24. Luschtinetz, R.; Frenzel, J.; Milek, T.; Seifert, G. Adsorption of Phosphonic Acid at the TiO$_2$ Anatase (101) and Rutile (110) Surfaces. *J. Phys. Chem. C* **2009**, *113*, 5730–5740. [CrossRef]

25. Fox, H.; Newman, K.E.; Schneider, W.F.; Corcelli, S.A. Bulk and Surface Properties of Rutile TiO$_2$ from Self-Consistent-Charge Density Functional Tight Binding. *J. Chem. Theory Comput.* **2010**, *6*, 499–507. [CrossRef] [PubMed]

26. Fuertes, V.C.; Negre, C.F.A.; Oviedo, M.B.; Bonafé, F.P.; Oliva, F.Y.; Sánchez, C.G. A Theoretical Study of the Optical Properties of Nanostructured TiO$_2$. *J. Phys. Condens. Matter* **2013**, *25*, 115304. [CrossRef] [PubMed]

27. Di Valentin, C.; Selloni, A. Bulk and Surface Polarons in Photoexcited Anatase TiO$_2$. *J. Phys. Chem. Lett.* **2011**, *2*, 2223–2228. [CrossRef]

28. Fazio, G.; Ferrighi, L.; Di Valentin, C. Photoexcited Carriers Recombination and Trapping in Spherical vs. Faceted TiO$_2$ Nanoparticles. *Nano Energy* **2016**, *27*, 673–689. [CrossRef]

29. Nunzi, F.; Agrawal, S.; Selloni, A.; De Angelis, F. Structural and Electronic Properties of Photoexcited TiO$_2$ Nanoparticles from First Principles. *J. Chem. Theory Comput.* **2015**, *11*, 635–645. [CrossRef] [PubMed]

30. Nunzi, F.; De Angelis, F.; Selloni, A. Ab Initio Simulation of the Absorption Spectra of Photoexcited Carriers in TiO$_2$ Nanoparticles. *J. Phys. Chem. Lett.* **2016**, *7*, 3597–3602. [CrossRef] [PubMed]

31. Diebold, U. The Surface Science of Titanium Dioxide. *Surf. Sci. Rep.* **2003**, *48*, 53–229. [CrossRef]

32. Sclafani, A.; Herrmann, J.M. Comparison of the Photoelectronic and Photocatalytic Activities of Various Anatase and Rutile Forms of Titania in Pure Liquid Organic Phases and in Aqueous Solutions. *J. Phys. Chem.* **1996**, *100*, 13655–13661. [CrossRef]

33. Dimitrijevic, N.M.; Vijayan, B.K.; Poluektov, O.G.; Rajh, T.; Gray, K.A.; He, H.; Zapol, P. Role of Water and Carbonates in Photocatalytic Transformation of CO_2 to CH_4 on Titania. *J. Am. Chem. Soc.* **2011**, *133*, 3964–3971. [CrossRef] [PubMed]

34. Mino, L.; Zecchina, A.; Martra, G.; Rossi, A.M.; Spoto, G. A Surface Science Approach to TiO_2 P25 Photocatalysis: An In Situ FTIR Study of Phenol Photodegradation at Controlled Water Coverages from Sub-Monolayer to Multilayer. *Appl. Catal. B Environ.* **2016**, *196*, 135–141. [CrossRef]

35. Shirai, K.; Sugimoto, T.; Watanabe, K.; Haruta, M.; Kurata, H.; Matsumoto, Y. Effect of Water Adsorption on Carrier Trapping Dynamics at the Surface of Anatase TiO_2 Nanoparticles. *Nano Lett.* **2016**, *16*, 1323–1327. [CrossRef] [PubMed]

36. Panarelli, E.G.; Livraghi, S.; Maurelli, S.; Polliotto, V.; Chiesa, M.; Giamello, E. Role of Surface Water Molecules in Stabilizing Trapped Hole Centres in Titanium Dioxide (Anatase) as Monitored by Electron Paramagnetic Resonance. *J. Photochem. Photobiol. A* **2016**, *322*, 27–34. [CrossRef]

37. Addamo, M.; Augugliaro, V.; Coluccia, S.; Di Paola, A.; García-López, E.; Loddo, V.; Marcì, G.; Martra, G.; Palmisano, L. The Role of Water in the Photocatalytic Degradation of Acetonitrile and Toluene in Gas-Solid and Liquid-Solid Regimes. *Int. J. Photoenergy* **2006**, *2006*, 39182. [CrossRef]

38. Miller, K.L.; Lee, C.W.; Falconer, J.L.; Medlin, J.W. Effect of Water on Formic Acid Photocatalytic Decomposition on TiO_2 and Pt/TiO_2. *J. Catal.* **2010**, *275*, 294–299. [CrossRef]

39. Salvador, P. On the Nature of Photogenerated Radical Species Active in the Oxidative Degradation of Dissolved Pollutants with TiO_2 Aqueous Suspensions: A Revision in the Light of the Electronic Structure of Adsorbed Water. *J. Phys. Chem. C* **2007**, *111*, 17038–17043. [CrossRef]

40. Selli, D.; Fazio, G.; Seifert, G.; Di Valentin, C. Water Multilayers on TiO_2 (101) Anatase Surface: Assessment of a DFTB-Based Method. *J. Chem. Theory Comput.* **2017**, *13*, 3862–3873. [CrossRef] [PubMed]

41. Dovesi, R.; Saunders, V.R.; Roetti, C.; Orlando, R.; Zicovich-Wilson, C.M.; Pascale, F.; Civalleri, B.; Doll, K.; Harrison, N.M.; Bush, I.J.; et al. *CRYSTAL14 User's Manual*; University of Torino: Torino, Italy, 2014.

42. Becke, A.D. Density-Functional Thermochemistry. III. The Role of Exact Exchange. *J. Chem. Phys.* **1993**, *98*, 5648. [CrossRef]

43. Lee, C.; Yang, W.; Parr, R.G. Development of the Colle-Salvetti Correlation-Energy Formula into a Functional of the Electron Density. *Phys. Rev. B* **1988**, *37*, 785–789. [CrossRef]

44. Krukau, A.V.; Vydrov, O.A.; Izmaylov, A.F.; Scuseria, G.E. Influence of the Exchange Screening Parameter on the Performance of Screened Hybrid Functionals. *J. Chem. Phys.* **2006**, *125*, 224106. [CrossRef] [PubMed]

45. Giannozzi, P.; Baroni, S.; Bonini, N.; Calandra, M.; Car, R.; Cavazzoni, C.; Ceresoli, D.; Chiarotti, G.L.; Cococcioni, M.; Dabo, I.; et al. QUANTUM ESPRESSO: A Modular and Open-Source Software Project for Quantum Simulations of Materials. *J. Phys. Condens. Matter* **2009**, *21*, 395502. [CrossRef] [PubMed]

46. Perdew, J.P.; Burke, K.; Ernzerhof, M. Generalized Gradient Approximation Made Simple. *Phys. Rev. Lett.* **1996**, *77*, 3865–3868. [CrossRef] [PubMed]

47. Burdett, J.K.; Hughbanks, T.; Miller, G.J.; Richardson, J.W., Jr.; Smith, J.V. Structural-Electronic Relationships in Inorganic Solids: Powder Neutron Diffraction Studies of the Rutile and Anatase Polymorphs of Titanium Dioxide at 15 K and 295 K. *J. Am. Chem. Soc.* **1987**, *109*, 3639–3646. [CrossRef]

48. Elstner, M.; Seifert, G. Density Functional Tight Binding. *Philos. Trans. R. Soc. A* **2014**, *372*, 20120483. [CrossRef] [PubMed]

49. Seifert, G.; Joswig, J.-O. Density-Functional Tight Binding—An Approximate Density-Functional Theory Method. *WIREs Comput. Mol. Sci.* **2012**, *2*, 456–465. [CrossRef]

50. Aradi, B.; Hourahine, B.; Frauenheim, T. DFTB+, a Sparse Matrix-Based Implementation of the DFTB Method. *J. Phys. Chem. A* **2007**, *111*, 5678–5684. [CrossRef] [PubMed]

51. Hu, H.; Lu, Z.; Elstner, M.; Hermans, J.; Yang, W. Simulating Water with the Self-Consistent-Charge Density Functional Tight Binding Method: From Molecular Clusters to the Liquid State. *J. Phys. Chem. A* **2007**, *111*, 5685. [CrossRef] [PubMed]

52. Tang, H.; Levy, F.; Berger, H.; Schmid, P.E. Urbach tail of anatase TiO_2. *Phys. Rev. B* **1995**, *52*, 7771. [CrossRef]

53. Liu, Y.; Claus, R.O. Blue Light Emitting Nanosized TiO_2 Colloids. *J. Am. Chem. Soc.* **1997**, *119*, 5273–5274. [CrossRef]

54. Koch, S.W.; Kira, M.; Khitrova, G.; Gibbs, H.M. Semiconductor Excitons in New Light. *Nat. Mater.* **2006**, *5*, 523–531. [CrossRef] [PubMed]

55. Tang, H.; Berger, H.; Schmid, P.E.; Lévy, F. Photoluminescence in TiO$_2$ Anatase Single Crystals. *Solid State Commun.* **1993**, *87*, 847–850. [CrossRef]

56. El-Sayed, M.A. Small Is Different: Shape-, Size-, and Composition-Dependent Properties of Some Colloidal Semiconductor Nanocrystals. *Acc. Chem. Res.* **2004**, *37*, 326–333. [CrossRef] [PubMed]

57. Mori-Sánchez, P.; Cohen, A.J.; Yang, W. Localization and Delocalization Errors in Density Functional Theory and Implications for Band-Gap Prediction. *Phys. Rev. Lett.* **2008**, *100*, 146401. [CrossRef] [PubMed]

58. Deskins, N.A.; Dupuis, M. Intrinsic Hole Migration Rates in TiO$_2$ from Density Functional Theory. *J. Phys. Chem. C* **2009**, *113*, 346–358. [CrossRef]

59. Najafov, H.; Tokita, S.; Ohshio, S.; Kato, A.; Saitoh, H. Green and Ultraviolet Emissions from Anatase TiO$_2$ Films Fabricated by Chemical Vapor Deposition. *Jpn. J. Appl. Phys.* **2005**, *44*, 245–253. [CrossRef]

60. Panayotov, D.A.; Yates, J.T., Jr. n-Type Doping of TiO$_2$ with Atomic Hydrogen—Observation of the Production of Conduction Band Electrons by Infrared Spectroscopy. *Chem. Phys. Lett.* **2007**, *436*, 204. [CrossRef]

61. Yamakata, A.; Ishibashi, T.; Onishi, H. Time-Resolved Infrared Absorption Spectroscopy of Photogenerated Electrons in Platinized TiO$_2$ Particles. *Chem. Phys. Lett.* **2001**, *333*, 271–277. [CrossRef]

62. Durrant, J.R. Modulating Interfacial Electron Transfer Dynamics in Dye Sensitised Nanocrystalline Metal Oxide Films. *J. Photochem. Photobiol. A* **2002**, *148*, 5–10. [CrossRef]

63. Martin, S.T.; Hermann, H.; Hoffmann, M.R. Time-Resolved Microwave Conductivity. Part 1.—TiO$_2$ Photoreactivity and Size Quantization. *J. Chem. Soc. Faraday Trans.* **1994**, *90*, 3323–3330. [CrossRef]

64. Beermann, N.; Boschloo, G.; Hagfeldt, A. Trapping of electrons in nanostructured TiO$_2$ studied by photocurrent transients. *J. Photochem. Photobiol. A* **2002**, *152*, 213. [CrossRef]

65. Boschloo, G.; Fitzmaurice, D. Spectroelectrochemical Investigation of Surface States in Nanostructured TiO$_2$ Electrodes. *J. Phys. Chem. B* **1999**, *103*, 2228–2231. [CrossRef]

66. Szezepankiewicz, S.H.; Moss, J.A.; Hoffmann, M.R. Slow Surface Charge Trapping Kinetics on Irradiated TiO$_2$. *J. Phys. Chem. B* **2002**, *106*, 2922–2927. [CrossRef]

67. Maurelli, S.; Livraghi, S.; Chiesa, M.; Giamello, E.; Van Doorslaer, S.; Di Valentin, C.; Pacchioni, G. Hydration Structure of the Ti(III) Cation as Revealed by Pulse EPR and DFT Studies: New Insights into a Textbook Case. *Inorg. Chem.* **2011**, *50*, 2385–2394. [CrossRef] [PubMed]

68. Chiesa, M.; Paganini, M.C.; Livraghi, S.; Giamello, E. Charge Trapping in TiO$_2$ Polymorphs as Seen by Electron Paramagnetic Resonance Spectroscopy. *Phys. Chem. Chem. Phys.* **2013**, *15*, 9435–9447. [CrossRef] [PubMed]

69. Micic, O.I.; Zhang, Y.; Cromack, K.R.; Trifunac, A.D.; Thurnauer, M.C. Trapped Holes on Titania Colloids Studied by Electron Paramagnetic Resonance. *J. Phys. Chem.* **1993**, *97*, 7227–7283. [CrossRef]

70. Brezová, V.; Barbieriková, Z.; Zukalová, M.; Dvoranová, D.; Kavan, L. EPR Study of 17O-Enriched Titania Nanopowders under UV Irradiation. *Catal. Today* **2014**, *230*, 112–118. [CrossRef]

71. Gallino, F.; Pacchioni, G.; Di Valentin, C. Transition Levels of Defect Centers in ZnO by Hybrid Functionals and Localized Basis Set Approach. *J. Chem. Phys.* **2010**, *133*, 144512. [CrossRef] [PubMed]

72. Shkrob, I.A.; Sauer, M.C., Jr. Hole Scavenging and Photo-Stimulated Recombination of Electron–Hole Pairs in Aqueous TiO$_2$ Nanoparticles. *J. Phys. Chem. B* **2004**, *108*, 12497–12511. [CrossRef]

73. He, Y.; Tilocca, A.; Dulub, O.; Selloni, A.; Diebold, U. Local Ordering and Electronic Signatures of Submonolayer Water on Anatase TiO$_2$(101). *Nat. Mater.* **2009**, *8*, 585–589. [CrossRef] [PubMed]

74. Tilocca, A.; Selloni, A. Vertical and Lateral Order in Adsorbed Water Layers on Anatase TiO$_2$(101). *Langmuir* **2004**, *20*, 8379–8384. [CrossRef] [PubMed]

75. Tilocca, A.; Selloni, A. DFT-GGA and DFT+U Simulations of Thin Water Layers on Reduced TiO$_2$ Anatase. *J. Phys. Chem. C* **2012**, *116*, 9114–9121. [CrossRef]

76. Aschauer, U.J.; Tilocca, A.; Selloni, A. Ab Initio Simulations of the Structure of Thin Water Layers on Defective Anatase TiO$_2$ (101) Surfaces. *Int. J. Quant. Chem.* **2015**, *115*, 1250–1257. [CrossRef]

77. Zhu, Y.; Ding, C.; Ma, G.; Du, Z. Electronic State Characterization of TiO$_2$ Ultrafine Particles by Luminescence Spectroscopy. *J. Solid State Chem.* **1998**, *139*, 124–127. [CrossRef]

78. Dimitrijevic, N.M.; Saponjic, Z.V.; Rabatic, B.M.; Poluektov, O.G.; Rajh, T. Effect of Size and Shape of Nanocrystalline TiO_2 on Photogenerated Charges. An EPR Study. *J. Phys. Chem. C* **2007**, *111*, 14597–14601. [CrossRef]

79. Luca, V. Comparison of Size-Dependent Structural and Electronic Properties of Anatase and Rutile Nanoparticles. *J. Phys. Chem. C* **2009**, *113*, 6367–6380. [CrossRef]

80. Colombo, D.P., Jr.; Roussel, K.A.; Saeh, J.; Skinner, D.E.; Cavaleri, J.J.; Bowman, R.M. Femtosecond Study of the Intensity Dependence of Electron-Hole Dynamics in TiO_2 Nanoclusters. *Chem. Phys. Lett.* **1995**, *232*, 207–214. [CrossRef]

81. Lazzeri, M.; Vittadini, A.; Selloni, A. Structure and Energetics of Stoichiometric TiO_2 Anatase Surfaces. *Phys. Rev. B* **2001**, *63*, 155409. [CrossRef]

82. Labat, F.; Baranek, P.; Domain, C.; Minot, C.; Adamo, C. Density Functional Theory Analysis of the Structural and Electronic Properties of TiO_2 Rutile and Anatase Polytypes: Performances of Different Exchange-Correlation Functionals. *J. Chem. Phys.* **2007**, *126*, 154703. [CrossRef] [PubMed]

83. Xu, X.; Goddard, W.A. Bonding Properties of the Water Dimer: A Comparative Study of Density Functional Theories. *J. Phys. Chem. A* **2004**, *108*, 2305–2313. [CrossRef]

84. Tschumper, G.S.; Leininger, M.L.; Hoffman, B.C.; Waleev, E.F.; Schaefer, H.F., III; Quack, M. Anchoring the Water Dimer Potential Energy Surface with Explicitly Correlated Computations and Focal Point Analyses. *J. Chem. Phys.* **2002**, *116*, 690–701. [CrossRef]

85. Curtiss, L.A.; Frurip, D.J.; Blander, M. Studies of Molecular Association in H_2O and D_2O Vapors by Measurement of Thermal Conductivity. *J. Chem. Phys.* **1979**, *71*, 2703–2711. [CrossRef]

86. Odutola, J.A.; Dyke, T.R. Partially Deuterated Water Dimers: Microwave Spectra and Structure. *J. Chem. Phys.* **1980**, *72*, 5062–5070. [CrossRef]

87. Soper, A.K.; Benmore, C.J. Quantum Differences between Heavy and Light Water. *Phys. Rev. Lett.* **2008**, *101*, 065502. [CrossRef] [PubMed]

88. Vittadini, A.; Selloni, A.; Rotzinger, F.P.; Grätzel, M. Structure and Energetics of Water Adsorbed at TiO_2 Anatase (101) and (001) Surfaces. *Phys. Rev. Lett.* **1998**, *81*, 2954–2957. [CrossRef]

89. Egashira, M.; Kawasumi, S.; Kagawa, S.; Seiyama, T. Temperature Programmed Desorption Study of Water Adsorbed on Metal Oxides. I. Anatase and Rutile. *Bull. Chem. Soc. Jpn.* **1978**, *51*, 3144–3149. [CrossRef]

90. Beck, D.D.; White, J.M.; Ratcliffe, C.T. Catalytic Reduction of CO with Hydrogen Sulfide. 2. Adsorption of H_2O and H_2S on Anatase and Rutile. *J. Phys. Chem.* **1986**, *90*, 3123–3131. [CrossRef]

91. Herman, G.S.; Dohnàlek, Z.; Ruzycki, N.; Diebold, U. Experimental Investigation of the Interaction of Water and Methanol with Anatase-TiO_2(101). *J. Phys. Chem. B* **2003**, *107*, 2788–2795. [CrossRef]

92. Zhao, Z.; Li, Z.; Zou, Z. Structure and Properties of Water on the Anatase TiO_2(101) Surface: From Single-Molecule Adsorption to Interface Formation. *J. Phys. Chem. C* **2012**, *116*, 11054–11061. [CrossRef]

© 2017 by the authors. Licensee MDPI, Basel, Switzerland. This article is an open access article distributed under the terms and conditions of the Creative Commons Attribution (CC BY) license (http://creativecommons.org/licenses/by/4.0/).

MDPI

St. Alban-Anlage 66

4052 Basel

Switzerland

Tel. +41 61 683 77 34

Fax +41 61 302 89 18

www.mdpi.com

Catalysts Editorial Office

E-mail: catalysts@mdpi.com

www.mdpi.com/journal/catalysts

www.ingramcontent.com/pod-product-compliance
Lightning Source LLC
Chambersburg PA
CBHW051849210326
41597CB00033B/5826